Current Methods in Cellular Neurobiology

Current Methods in Cellular Neurobiology

Volume I: Anatomical Techniques

Edited by

JEFFERY L. BARKER

National Institute of Neurological
and Communicative Disorders and Stroke
National Institutes of Health
Bethesda, Maryland

JEFFREY F. McKELVY

State University of New York
Stony Brook, New York

A Wiley-Interscience Publication
JOHN WILEY & SONS
New York • Chichester • Brisbane • Toronto • Singapore

Library of Congress Cataloging in Publication Data:

Main entry under title:

Current methods in cellular neurobiology.

"A Wiley-Interscience publication."
Includes index.
Contents: v. 1. Anatomical techniques.
1. Neurobiology—Technique. I. Barker, Jeffery L., 1943- . II. McKelvy, Jeffrey F. [DNLM: 1. Neurochemistry—Methods. 2. Neurophysiology—Methods. WL 102 C976]

QP356.C87 1983 599'.0188 83-1282
ISBN 0-471-09328-9 (v. 1)

Printed in the United States of America

10 9 8 7 6 5 4 3 2 1

Contributors

Huda Akil
Department of Psychiatry and Mental Health Research Institute
University of Michigan
Ann Arbor, Michigan

Richard D. Broadwell
Division of Neuropathology
Department of Pathology
University of Maryland School of Medicine
Baltimore, Maryland

Gloria E. Hoffman
Department of Anatomy
University of Rochester School of Medicine
Rochester, New York

Gerald P. Kozlowski
Department of Physiology
Southwestern Medical School
University of Texas Health Science Center at Dallas
Dallas, Texas

Paul J. Marangos
Psychobiology Branch
National Institute of Mental Health
Bethesda, Maryland

Enrico Mugnaini
Laboratory of Neuromorphology
Department of Behavioral Sciences
University of Connecticut
Storrs, Connecticut

Gajanan Nilaver
Department of Neurology
College of Physicians & Surgeons
Columbia University
New York, New York

Wolfgang H. Oertel
Laboratory of Clinical Science
National Institute of Mental Health
Bethesda, Maryland
Present address:
Neurologische Klinik
Klinikum Rechts der Isar
Technische Universität München
München, West Germany

José M. Palacios
Sandoz Ltd.
Preclinical Research
Basel, Switzerland

Donald E. Schmechel
Laboratory of Clinical Science
National Institute of Mental Health
Bethesda, Maryland
Present address:
Division of Neurology
Department of Medicine
Duke University Medical Center
Durham, North Carolina

John R. Sladek, Jr.
Department of Anatomy
University of Rochester School of Medicine
Rochester, New York

Carolyn B. Smith
Laboratory of Cerebral Metabolism
National Institute of Mental Health
U.S. Public Health Service
Department of Health and Human Services
Bethesda, Maryland

L. W. Swanson
The Salk Institute
San Diego, California

James K. Wamsley
Department of Psychiatry
University of Utah Medical Center
Salt Lake City, Utah

Stanley J. Watson
Department of Psychiatry and Mental Health Research Institute
University of Michigan
Ann Arbor, Michigan

Series Preface

From the research in cellular neurobiology carried out over the past several years it is evident that cells comprising the vertebrate nervous system are highly organized into a vast complex of overlapping and interdigitating spatial and temporal circuits. What superficially appears as an enormous tangle likely has both design and function, which, for the moment, elude easy description. Present-day conceptualizations of the vertebrate nervous system now include the realization that many, if not all, nerve cells are quite complex and multifaceted: For example, synthesis and secretion of different transmitters by the same cell occurs; multiple binding sites for the same transmitter are present in the same membrane fraction; the same excitability mechanism can be co-regulated by different transmitters; and different excitability mechanisms can be regulated by the same transmitter in the same cell. As far as is known, these formulations are also true of the relatively simple nervous systems found in primitive *Hydra*, which does little more than eat, and in *Aplysia*, whose complex repertoire of neuronal activities has provided experimental substrate for many investigators.

Most of the current experimental strategies in cellular neurobiology are aimed at reducing the formidable complexity in such a way as to isolate a particular neuronal property or set of closely related properties for detailed study. We have begun this series by collecting in the first four volumes those methods that provide useful catechisms of logic and experimental protocol for important areas of cellular neurobiology. These volumes are not intended to be comprehensive, but rather should function as a starting point in the utilization of a particular type of strategy.

Immunological reagents have become increasingly important in many types of morphological and biochemical investigation of neuronal function, yet the immunological reactions are not always certain and their products have not been completely identified in many cases. Some of the advantages and disadvantages of using immunoreagents for analysis of specific antigens are described in various chapters in the first two volumes. Several relatively new strategies for visualizing the presence of specific receptors in the central nervous system or for determining the level of metabolic activity in defined circuits and clusters of cells in the central nervous system are also reviewed. In their own ways all of these methods have added new dimensions to an already complex wiring diagram.

Studies into the precursor structures of putative peptide transmitters using a strategy involving the production of complementary DNA libraries have led to the discovery that intricate sequences of interrelated instructions are synthesized in tandem, sometimes with multiple copies of the same sequence. Processing of these precursors into products occurs both intracellularly and extracellularly so that multiple signals can be released simultaneously or sequentially, subsequently diffusing to receptors on target cells. The application of new methods for isolation and characterization of transmitter receptors has begun to reveal some of the details regarding the molecular mechanisms involved in the transduction of chemical signals into electrical events. An array of inventive electrophysiological assays has recently been developed for investigating the elementary basis of neuronal excitability. Relatively quantitative analysis of excitable membrane properties has been carried out in vertebrate neurons recorded in carefully prepared slices and in dissociated cell culture. With these techniques it has become clear that there are numerous conductance mechanisms present in neurons, many of which appear to involve a type of ion permeation mechanism common to many other, if not all, excitable cells. Multidisciplinary analysis of nerve cells in isolation or in communication, using complementary neurobiological methods, should provide some insight into how physiological events at the level of the membrane regulate cytoplasmic functions ranging from gene transcription to the distribution of macromolecules and organelles to the far reaches of the nerve cell.

The continued development of methods in cellular neurobiology will lead quite naturally to the innovation of still newer strategies, and, when appropriate, these may form the basis of future volumes in this series.

JEFFERY L. BARKER
JEFFREY F. MCKELVY

Bethesda, Maryland
Stony Brook, New York
April 1983

Preface

Historically, study of the nervous system has been associated with a high level of concern for the morphology of individual nerve cells and the anatomical relationships between nerve cells as a basis for understanding neural integration. This continues to be true, and contemporary neurobiology is rich in morphological characterization of the nervous system based on an eclectic and rapid adoption from a variety of disciplines of techniques that can identify biochemical markers or processes in nerve cells to illuminate their anatomical relationships. This volume, we hope, reflects well the diversity of modern neuroanatomical investigation. We note, however, that we have attempted also to reflect areas of great intensity of current research activity. Accordingly, we have placed emphasis on immunocytochemistry as a powerful method for visualizing the disposition of specific molecules in the nervous system, deriving from progress in biochemistry, and for drawing inferences about functional relationships from such visualization.

Thus, Schmechel and Marangos describe how antibodies to an isozyme—expressed only in neurons and neuroendocrine cells—of the glycolytic enzyme enolase can be used to study neuronal development and the functional activity of neurons. In the contribution of Oertel, Schmechel, and Mugnaini, the purification and production of antibodies to an amino acid neurotransmitter synthetic enzyme—glutamic acid decarboxylase—is described followed by a demonstration of how such antibodies can be used to map GABA-secreting neuron systems. The prominence of neuropeptides in contemporary neurobiological research is reflected in four chapters. The chapter by Watson and Akil sets forth the strategies and attention to experimental detail necessary for the successful analysis of peptide-secreting neuron systems by light microscopic immunocytochemistry. This is followed by Kozlowski and Nilaver on the exacting technical requirements of immunocytochemistry of peptides at the electron microscopic level and the importance of such analysis to understanding the compartmentation of peptides in neurons. The study of peptidergic neurons at the ultrastructural level by immunocytochemical and enzyme cytochemical analysis is taken up by Broadwell in the context of elucidating the secretory process in this type of neuron. The simultaneous chemical identification of neurons that contain peptides—by immunocytochemistry—and those that contain biogenic amines—by histofluorescence—has been achieved by Sladek and Hoffman. This novel approach is applied by these researchers to the analysis of the morphological relationships of peptidergic and monoaminergic systems as a basis for predicting functional interactions.

It is implicit in all the volumes of this series that research in neurobiology is increasingly interdisciplinary. Hence, even within a subdiscipline such as neuroanatomy, immunocytochemistry is being combined with other marking techniques. The investigator seeking neuroanatomical relationships can, and must, consider the existence of specific biochemical processes and specific receptors as avenues for the elucidation of such relationships. This is manifest in the contributions comprising the remainder of this volume. Swanson describes the utilization of the biochemically defined processes of orthograde and retrograde axonal transport as a means of distributing a fluorescent marker for discerning the trajectories of neuronal projections, and the combination of marker dye fluorescence with peptide immunocytochemical marking. Wamsley and Palacios describe new techniques for assessing the presence and distribution of specific neurotransmitter receptors by light microscopic autoradiography, to complement knowledge of the distribution of the neurons that secrete the ligands for these receptors, and so arrive at a neuroanatomical picture that suggests functional relationships. Finally, Smith shows how a defined biochemical process in neural tissue—the preferred use of glucose for energy metabolism and the stimulation of this process during enhanced neuronal activity—can serve as a basis for resolution at the cellular level of patterns of neuronal activity related to a specific function of the nervous system.

Jeffery L. Barker
Jeffrey F. McKelvy

Bethesda, Maryland
Stony Brook, New York
June 1983

Contents

Chapter 1

Neuron-Specific Enolase (NSE): Specific Cellular and Functional Marker for Neurons and Neuroendocrine Cells

Donald E. Schmechel

Division of Neurology
Duke University Medical Center
Durham, North Carolina

Paul J. Marangos

Laboratory of Clinical Science
National Institute of Mental Health
Bethesda, Maryland

Dr. Schmechel's present address is Division of Neurology, Department of Medicine, Duke University Medical Center, Durham, North Carolina 27710.

1 INTRODUCTION

The concept that brain-specific proteins might correlate with specialized function in nervous tissue was introduced by Moore and McGregor in the 1960s. Their systematic protein maps of bovine brain yielded several soluble acidic proteins that appeared to be greatly enriched in brain compared with other tissues (1). One of the first discovered brain-specific proteins, named 14-3-2 for its chromatographic and electrophoretic properties, has amply fulfilled the expectations raised for this group. 14-3-2 is now known to be an isoenzyme of the glycolytic enzyme enolase (see

Section 2) and a specific cellular marker for neurons and neuroendocrine cells.

A number of reviews have dealt with the functional correlates of 14-3-2 protein and the identity of 14-3-2 as the neuron-specific isoenzyme (subunit designation γγ) of the dimeric enzyme enolase {2-phospho-D-glycerate hydrolase, E.C. 4.2.1.11} (2–12). Although a variety of names have been used for this protein, neuron-specific enolase (NSE) is now the accepted terminology and is used here to emphasize the cellular localization and enzymatic properties of 14-3-2.

The purpose of this chapter is to present the practical aspects of measurement and localization of NSE and encourage its use as a cellular marker. Additional material discusses the available evidence that the appearance of NSE is closely related to neuronal differentiation as well as the relation of NSE to other enolase isoenzymes encountered in brain. The following set of observations and hypotheses will serve as a framework for the chapter:

1. Neuronal differentiation is accompanied by the appearance of specific proteins.
2. The primary source of neuronal energy metabolism is glycolysis.
3. Enolase, a glycolytic enzyme, is present in neurons and neuroendocrine cells in the form of a distinct and specific isoenzyme called neuron-specific enolase (NSE).
4. During differentiation, neurons switch from non-neuronal enolase (NNE; glial and liver isoenzyme) to neuron-specific enloase (NSE), and this switch corresponds to the onset of synaptic activity.
5. Mature neurons complete the switch to NSE and do not contain significant amounts of NNE.
6. Some neuroendocrine cells and a few classes of neurons do not switch completely over to NSE and contain both neuronal (NSE) and glial (NNE) isoenzymes; such cells may contain a third enolase isoenzyme, called hybrid enolase.
7. NSE content varies significantly for different types of neurons and is related to overall levels of synaptic activity on a long-term basis; NSE content is lower for neuroendocrine cells, and therefore neuron-specific enolase is still an appropriate term for γ subunit and its homodimer γγ; NSE content may be a marker for neuronal activity complementary to such functional markers as cytochrome oxidase activity (mitochondrial enzyme) and 2-deoxyglucose uptake (glucose utilization).
8. The physiologic role of NSE may relate to its unique stability to chloride, a property not shared by other enolase isoenzymes; this

property or other structural characteristics may make it specially suited to neuronal function.

9. Impairment of NSE activity or its transport to axonal terminals might cause neuronal dysfunction and represent a possible etiology of some axonal neuropathies or neuronal aging.

2 HISTORICAL OVERVIEW

One of the first questions to arise after Moore and McGregor's pioneering discovery of the brain-specific protein 14-3-2 (NSE) was that of its cellular localization in nervous tissue. The rise in 14-3-2 levels during neurogenesis (13–15), its disappearance from peripheral nerve after Wallerian degeneration (16), and the demonstration of the axonal transport of exogenously labeled 14-3-2 (17) supported the proposal that 14-3-2 had a neuronal localization (18). "Brain-specific" 14-3-2 was therefore more precisely a neuron-specific protein, present in central and peripheral neurons.

The success of this approach led other investigators to isolate 14-3-2 homologues from the brains of several species, including rat, cat, and human (19–23). Using antisera specific to the rat 14-3-2 homologue (tentatively named nervous-system-specific protein or NSP), immunocytochemistry demonstrated the strict neuronal localization of rat 14-3-2 (24), and the term neuron-specific protein was suggested (22, 23).

At the same time, investigators studying the distribution of enolase isoenzymes in various tissues discovered a portion of brain enolase activity not recognized by antisera to the known α (liver) and γ (muscle) subunits (25–27). The brain-specific subunit and isoenzymes were characterized by biochemical analysis, and the current subunit and isoenzyme nomenclature was established for the three subunits: α (liver), β (muscle), γ (brain). The six possible enolase isoenzymes were designated αα, αγ, γγ, αβ, ββ, and βγ (27). The above studies with rat brain enolases as well as others carried out with human tissues (28, 29) suggest that separate genetic loci code each of the three subunits. Chromosomal assignment is complete for the α and β subunits (also termed ENO_1 and ENO_2) but has not yet been accomplished for the γ subunit (30).

A breakthrough uniting these independent lines of research occurred in 1975 when Bock and Dissing demonstrated that bovine 14-3-2 possesses enolase activity (31). The extension of this finding to other species was soon complete and extensive proof of the identity between 14-3-2 and the γγ subunit of enolase was furnished (12, 22, 32). The ability to biochemically characterize 14-3-2 and to place findings on its localization and tissue levels into the context of glycolysis and energy metabolism has encouraged a number of groups to study brain enolases. The large number of recent contributions are discussed in the appropriate sections below. Emphasis is on the application of specific antisera to the brain enolases (especially neuron-specific enolase) to provide reliable and easy methods of quantitation and localization in tissue. Succeeding sections present the practical aspects of these methods prior to demonstrating their usefulness in actual experiments.

3 PURIFICATION AND BIOCHEMICAL PROPERTIES OF BRAIN ENOLASES

Neuron-specific enolase (NSE) has been purified from bovine (1, 33), rat (19, 23, 34, 35), cat (23), and human brain (23, 36). The abundance of this isoenzyme in nervous tissue (probably 1–3% of total soluble protein) makes the purification relatively straightforward with good yields. NSE is present in animals and birds but not in amphibians, fish, or lower species by present criteria of immunoreactivity and chromatographic elution patterns (37).

3.1 Purification

When total brain enolase activity in mammals and birds is fractionated by ion exchange chromatography, three peaks are observed corresponding to the isoenzymes $\alpha\alpha$(non-neuronal enolase or NNE), $\alpha\gamma$(hybrid enolase), and $\gamma\gamma$(neuron-specific enolase or NSE). The most acidic peak is NSE, and published procedures for purification exploit the complete separation obtained during ion exchange chromatography and isoelectric focusing (see references cited above).

For rat brain, approximately 500 g (250 rats) of brain tissue are homogenized in batches in a Teflon-glass pestle in 30% (w/v) solution of cold 10 m*M* Tris-phosphate buffer (pH 7.5) with 2 m*M* $MgSO_4$. Magnesium is included since it is a cofactor for enolase activity and promotes enzyme stability. The homogenate is then centrifuged at 100,000*g* for 1 hr and the supernatant collected to form the "soluble fraction." Ammonium sulfate precipitation is then carried out on the soluble fraction to yield a fraction, named P60, representing protein soluble in 40–60% salt saturation. The still impure P60 fraction is then applied to a DEAE-cellulose ion exchange column, washed with starting buffer, and eluted with a 0–0.5 *M* NaCl gradient; enzyme activity is monitored to detect elution of the enolase isoenzyme peaks and protein concentration. The most acidic peak of enolase activity (NSE) is then chromatographed on Sephadex G-150 and finally purified by isoelectric focusing in sucrose with a pH gradient of 4–6. The overall yield from 500 g of rat brain is 10–20 mg of purified NSE protein. Similar methods work for the isolation of NSE from cat or human (fresh autopsy) brain (23).

The non-neuronal isoenzyme (NNE) $\alpha\alpha$ can be isolated in similar fashion by taking the least acidic enolase peak from DEAE ion exchange chromatography through molecular sieve chromatography and isoelectric focusing steps. Because of slightly different isoelectric points, rat brain NNE requires a pH 5–7 gradient and human brain NNE a pH 3–10 gradient. Hybrid enolase $\alpha\gamma$, the enolase activity peak with intermediate acidity on ion exchange chromatography, cannot be further purified by isoelectric focusing since it dissociates into two peaks of activity representing $\alpha\alpha$ and $\gamma\gamma$ (23).

Other methods have been described for enolase purification, including repeat ion exchange chromatography without isoelectric focusing and other variations (for a review, see ref. 12). Most recently, the purification of all three brain enolase isoenzymes has been accomplished with molecular sieve chromatography, Blue Sepharose CL-6B, and hydroxylapatite chromatography (35).

The homogeneity of enolase isoenzyme preparations has usually been assessed by polyacrylamide gel electrophoresis of the purified protein. Purified isoenzyme preparations must behave as a homogeneous protein during routine and denaturing (sodium dodecyl sulfate or urea) conditions, that is, they must migrate as a single band without evidence by protein detection methods of other contaminating bands (see Section 3.2). In addition, immunochemical analysis with antisera preparation to the various isoenzymes should confirm the expected cross-reactivity to the component subunits α and γ of a test protein mixture. Therefore, purified NSE is not expected to show any reaction with antisera to the non-neuronal enolase NNE nor purified NNE with antisera to NSE (see Section 3.4). Immunological assessment of purity can be carried out with Ouchterlony double immunodiffusion or by enzyme precipitation or radioimmunoassay (23; see also Section 4.2).

3.2 Biochemical Properties

Neuron-specific enolase (NSE or $\gamma\gamma$) has significantly different structural characteristics from non-neuronal enolase (NNE or $\alpha\alpha$). In accord with its behavior on ion exchange chromatography, NSE is a very acidic protein with an isoelectric point of 4.7 for human NSE and 5.0 for rat NSE (23). Amino acid analysis of purified NSE from various species shows similar profiles, with high percentages of acidic residues such as aspartate and glutamate (12, 23). Human and rat NSE are similar in molecular weight: 78,000 for the native isoenzyme γγ, which under denaturing conditions yields two subunits with identical molecular weights of 39,000.

In contrast, non-neuronal enolase (NNE or αα) is the least acidic of the brain enolase isoenzymes and is identical in properties with the single enolase isoenzyme that is found in liver. NNE in rat has an isoelectric point of 5.9, and human NNE is even more basic with an isoelectric point of 7.2 (23). The subunits of NNE are also larger than those of NSE, being of molecular weight 43,500 to yield a total molecular weight for native NNE of 87,000 (23). The third brain enolase isoenzyme, hybrid enolase or αγ, has biochemical properties intermediate to NSE and NNE, as expected from its subunit composition (38).

Although NSE and NNE have major structural differences, their enzymatic sites appear to be very similar (7, 39). The affinity of NSE and NNE for substrate (2-phosphoglycerate) and product (phosphoenol pyruvate) are essentially identical. The affinity for Mg^{2+} cofactor is slightly different for the two isoenzymes, with NSE having a threefold greater affinity than NNE (23). Enolase isoenzymes can be denatured by chloride and bromide anions in high strength, by chaotropic agents such as urea, and by heat. The presence of magnesium protects to some extent from denaturation. Heat inactivation is irreversible, but NNE shows good reversibility from urea-induced denaturation, and both NSE and NNE regain full activity after removal of halogen anion. All of the above inactivations apparently involve subunit dissociation, since all three isoenzymes can be generated from mixtures of just two isoenzymes. The most remarkable finding is that NSE is much more resistant to denaturation than NNE under all three treatments. Even without magnesium protection, NSE is resistant to heat (50°C) and salt (0.5 *M* Br) that rapidly inactivate NNE and hybrid enolase (39). Since NSE is contained in neurons (24, 40–44) and NNE in glial cells (41, 44), the above structural and functional differences may be related to a need for increased stability in the neuron (10, 39). Another apparent requirement for the neuronal isoenzyme is its association with a glycolytic complex carried in the slow phase of axonal transport (45); this is more completely discussed in Section 9.5.

3.3 Antisera Production

Antisera to neuron-specific enolase (NSE) provided the means for the identification of 14-3-2 as an enolase isoenzyme (9, 31), and they are a cornerstone for the quantitation and localization of NSE in tissue. NSE isolated from rat brain seems to be very antigenic in rabbits. Injection of 0.5–1.0 mg of purified NSE into New Zealand white rabbits produces a potent antiserum after a few months (2 to 4 total injections) (21). This schedule involves mixing the first injection with complete Freund's adjuvant and succeeding injections with incomplete Freund's adjuvant; all injections are made in multiple intradermal sites. Antisera development is monitored by serial bleedings with Ouchterlony double immunodiffusion against the antigen. In like manner, antisera to human NSE and to rat and human NNE have been prepared in rabbits. To date, NNE has not produced antisera as potent for radioimmunoassay and immunocytochemistry as NSE. Commercial antisera to NSE are currently in preparation.

3.4 Characterization of Antisera

The major need for antisera is monovalency and specificity for the particular antigen employed. Antisera to enolase isoenzymes must therefore be characterized by the usual methods of immunodiffusion, immunoelectrophoresis, and enzyme activity titration or precipitation. The early studies of brain enolase isoenzymes (24–26) showed that antisera prepared to neuron-specific enolase (γγ) did not recognize liver enolase (αα); in fact, this provided part of the evidence that the three isoenzymes are coded at separate genetic loci. Subsequent work with other antisera has shown that NSE and NNE are immunologically distinct, without any cross-reactivity by the above criteria. These findings are relevant only to polyclonal antisera; monoclonal antibody preparations might recognize some common shared site (e.g., active site) although the experience with polyclonal antisera suggests that such sites must be poorly immunogenic.

Antisera raised to NSE or NNE do cross-react with hybrid enolase, which contains both α and γ subunits. Immunotitration—performed by incubating the various enolase isoenzymes with a particular antiserum, centrifuging to remove immunoprecipitate, and measuring residual enzyme activity in the supernatant—clearly reveals this cross-reactivity (41). In this respect, anti-NSE might be termed anti-γ and anti-NNE anti-α. On the other hand, the recognition of α subunit contained in hybrid enolase (αγ) is different from that of α contained in non-neuronal enolase (αα = NNE) since the immunotitration curves do not coincide (41). This is further demonstrated by the fact that extensive absorption of anti-NNE with purified hybrid enolase can remove the cross-reactivity to αγ without significantly reducing the titer to αα (41). With the above observations in mind, the antisera will be termed anti-NSE and anti-NNE in the rest of the chapter, with a suffix indicating the source of antigen (R for rat and H for human) where this information is needed.

Immunological characterization of NSE shows that rat NSE is relatively different from bovine, cat, and human NSE (23). For example, anti-NSE-

R reacts only 25% as well with NSE from other species. In contrast, the other antiserum used by our group for immunocytochemistry is anti-NSE-H, which reacts well with cat and monkey NSE but poorly with rat NSE. Mouse NSE is not well recognized by either anti-NSE-R or anti-NSE-H. However, because of the high titer of anti-NSE-R, the relative difference in immunoreactivity across species does not prevent its effective use for immunocytochemistry in all mammals and birds.

4 MEASUREMENT AND LOCALIZATION OF BRAIN ENOLASES

4.1 Enzymatic Assay

Enzymatic analysis offers the advantage of detecting only enzymatically active molecules of enolase isoenzymes. Combined with isoenzyme separation by DEAE-cellulose chromatography, measurement of enolase activity is well suited to characterize the isoenzyme content of a single tissue specimen. This method does require more effort than radioimmunoassay techniques for many samples because a large number of fractions are generated during the isoenzyme separation, each requiring individual analysis.

Tissue samples are homogenized in a glass-Teflon apparatus after suspension in 3 to 10 volumes of 10–20 m*M* Tris-phosphate buffer (pH 7.4) with 3 m*M* $MgSO_4$. The homogenate is centrifuged at 100,000*g* for 1 hr and the supernatant is applied to a DEAE-cellulose ion exchange column (38). After washing with three column-volumes of buffer, a 0–0.5 *M* NaCl gradient is applied and enolase activity determined for the elution fractions by spectrophotometric assay (46). The DEAE column is calibrated with known isoenzyme markers before use with experimental samples.

4.2 Immunoassay

Radioimmunoassay (RIA) methods offer convenience and greater sensitivity than determination of enolase enzymatic activity but they recognize any immunoreactive protein regardless of enzymatic activity or partial denaturation or degradation. A number of procedures for radioimmunoassay of NSE are available (36, 47–51). Most involve standard double-antibody precipitation with radioactive purified antigen. The RIA procedure described below for rat and human NSE uses a double antibody method in which the second antibody is immobilized on polyacrylamide beads (Bio-Rad immunobeads), which facilitates washing, final precipitation, and radioactive counting (49). The use of iodinated antigen instead of tritiated antigen permits very sensitive assays capable of detecting even femtomole (10^{-15} m or roughly 0.1 ng) amounts of NSE present in small samples of cerebrospinal fluid or serum (36, 50, 51). Radioactive rat or human NSE can be prepared by iodination with ^{125}I-Bolton Hunter reagent (36, 50, 51) yielding specific activities of 2000 to 4000 dpm/ng protein (approximately tenfold higher specific activity than tritiation). Commercial kits containing both iodinated antigen and specific antisera are being planned.

Tissue samples are prepared as for enzymatic analysis (Section 4.1) through preparation of a high-speed soluble supernatant. Samples are diluted if necessary into the range of sensitivity of the RIA. Each assay has a final volume of 0.5 mL containing an aliquot of the sample, approximately 1 ng of labeled antigen, and a final dilution of antiserum of $1{:}10^7$ (anti-NSE-R) or $1{:}3.5 \times 10^5$ (anti-NSE-H) in 150 m*M* Tris-phosphate buffer (pH 7.4) with 2 m*M* $MgSO_4$ and 1% bovine serum albumin. The first incubation is overnight at 4°C, and then 0.3 mg of anti-rabbit IgG coupled to polyacrylamide beads is added (Bio-Rad immunobeads). After a further incubation of 1 hr at room temperature, the tubes are centrifuged at 10,000*g* for 15 min. The pellets of immunobead-bound antigen are then washed by resuspension and repeat centrifugation. Finally, the beads are suspended in 0.5 mL buffer and counted in 10 mL of Aquasol (New England Nuclear) or equivalent scintillation fluid. The resultant assays for rat and human NSE are quantitative, with a linear range of 0.1 to 2 ng (50, 51). This allows measurement of NSE levels in human cerebrospinal fluid (approximately 2ng/mL) in practical sample sizes (100–300 μL). RIA measurements show no cross-reactivity with NNE with either rat or human assay. RIA procedures using tritiated antigen are available for rat and human NNE (47, 49) and also show no cross-reactivity with NSE. In the case of monkey tissue, the high degree of cross-reactivity between human and monkey brain enolases makes it possible to apply the RIA procedures for human NNE and NSE to monkey samples (49).

RIA methods for measuring NSE and NNE are, more precisely, assays for γ (present as NSE $\gamma\gamma$ or hybrid enolase $\alpha\gamma$) and α (present as NNE $\alpha\alpha$ or hybrid enolase $\alpha\gamma$). One way of eliminating the possible contribution of hybrid enolase to NSE and NNE measurements is the partial purification of tissue samples by separating the three enolase isoenzymes before radioimmunoassay. An alternative approach has been described that uses antisera to purified NSE, NNE, and hybrid enolase from rat brain (52). This nonradioactive enzyme-linked immunoassay relies on antigen immobilization by specific $F(ab')_2$ fragments of antibody to the particular isoenzyme and then spectrophotometric assay by further addition of specific enzyme-linked Fab′ fragments. Present sensitivity of the immunoassay is 2 to 10 gigamoles (2 to 10 $\times$ 10^{-18} m or roughly 0.2 to 1 pg) of each enolase isoenzyme (53, 54). This has permitted measurement of enolase isoenzymes in individual isolated nerve cells, although questions arise about possible contamination by glial processes and recovery of enolase content (soluble cytoplasmic enzyme) during such manipulations (53). As for RIA methods, immunoassay for NNE and NSE also detects hybrid enolase. However, hybrid enolase can be specifically measured by directing the first-stage and second-stage antibody fragments to each of the brain subunits α and γ. The immunoassay for hybrid enolase therefore shows no cross-reactivity with NNE or NSE (53), and homodimer values ($\gamma\gamma$–NSE or $\alpha\alpha$–NNE) can be determined by subtracting the cross-reacted value of hybrid enolase in a given sample. For a given problem in measurement of enolase isoenzymes, the selection of RIA or enzyme-linked immunoassay will depend on the availability of the techniques and the exact requirements of the experiment. RIA methods offer enough sensitivity for most tissue measurement, including such small samples as brain micropunches and such low-level samples as cerebrospinal fluid. Where necessary, isoenzyme separation prior to the RIA can precisely delineate the amounts of NSE, NNE, and hybrid enolase. Enzyme-linked immunoassay offers greater sensitivity and the possibility of determining levels of hybrid enolase without isoenzyme separation, which may be appropriate for some applications.

4.3 Immunocytochemistry

We have seen that after the discovery of neuron-specific enolase (14-3-2), even before its recognition as an enolase isoenzyme (31), a variety of experimental models demonstrated its probable neuronal localization (Section 2). The development of immunocytochemistry during this same period furnished the necessary tools for direct visualization of NSE localization in order to test the strong but indirect evidence from biochemical analysis. In particular, the ability of the unlabeled antibody enzyme method of Sternberger (55) to reveal antigen location in permanent tissue sections enables immunocytochemistry to be an effective and powerful complement to biochemical approaches. The anatomical localization of NSE in peripheral tissues, for example, revealed that NSE is also contained in neuroendocrine cells (56) throughout the body; biochemical analysis of the same tissues is complicated in most cases by the small fraction of neuroendocrine cells in a given organ.

4.3.1 Tissue Preparation

Any bird or mammalian species can be used as a tissue source. Anti-NSE-R immunocytochemistry has been successful in the chicken (Figure

14), mouse, rat (Figure 1C), cat, human (Figure 4A), and a variety of prosimians and primates (Figure 1). Since antigen loss during tissue fixation and processing may be on the order of 30 to 50% (57) and since NSE is a soluble cytoplasmic enzyme, special care was taken in the initial investigations to maximize immunoreactivity and number of cell structures staining. In particular, immersion fixation of primary tissue cultures of mouse spinal cord neurons was used to compare different fixatives and to circumvent the variable of perfusion success in intact animals and the variable of tissue thickness in immersed specimens (58). The importance of tissue preparation is underlined by the fact that some of the initial work on NSE localization reported that not all neurons contain NSE (24, 59, 60).

The original fixation described for NSE localization was a neutral mixture of paraformaldehyde and picric acid (24, 61) with both cross-linking and coagulative properties. Subsequent comparison of this fixative with other variations in mouse spinal cord cultures (monolayer fixed by immersion) suggested that the addition of glutaraldehyde and a slight lowering of pH to the range of 5–6 increased immunoreactivity, decreased nuclear staining, and resulted in reliable staining of all neurons. With the addition of higher glutaraldehyde concentrations at neutral pH, immunoreactivity decreased, apparently because of altered immunogenicity rather than lability of the antigen (58). For these reasons, most of the immunocytochemistry done by our group utilizes the following fixative: 4% paraformaldehyde and 1% glutaraldehyde in 0.1 *M* sodium acetate buffer (pH 6) with 0.2% picric acid and also 2% sucrose.

For perfusion fixation, experimental animals are deeply anesthetized and perfused through the aorta with normal saline until clearing of the return from the right heart followed by the above fixative (3–5 mL/g). Since the fixative has coagulative properties, inadequate rinsing of the vascular tree usually results in poor fixation. At the time of dissection 15 to 30 min after perfusion, well-fixed brains are extremely firm and yellow; poorly fixed brains are best rejected at this point. Post-fixation is carried out for 3 to 6 hr at room temperature or at 4°C; immunoreactivity is relatively insensitive to fixation time beyond this point. Tissue can then be stored in Tris-buffered saline (TBS, 50 m*M* Tris-HCl buffer pH 7.6 with 150 m*M* NaCl) at 0–4°C for up to a year without loss of immunoreactivity (the solution becomes yellowed by leaching of picric acid from the tissue).

Further work in our laboratories as well as reports from other groups now indicates that NSE localization can be successfully performed under a number of different fixation and tissue processing techniques, ranging from tissue prepared by normal hospital pathology protocols to various fixatives used in immunocytochemistry (Table 1). At present we are

Table 1. Fixation Protocols for NSE Localization by Immunocytochemistry[a]

Fixative[b] (Reference)	Perfusion	Immersion	Light	EM	NNE	Tissue Culture
4% PF, 0.2% PA, pH 7.4 (24, 41, 56, 65)	Yes	Yes	Yes	—[c]	Yes	56, 65
4% PF, 1% G, 0.2% PA, pH 6 (56, 66, 67)	Yes	Yes	Yes	Yes	Yes	56, 67
0.5% PF, 1.5% G, neutral pH (62)	—	Yes	Yes	Yes	No	62
8% Formalin-saturated $HgCl_2$ (62)	—	Yes	Yes	No	Yes	62
10% Formalin, 1% PA, pH 2 (Bouin's, 56)	No	Yes	Yes	—	Yes	56
4% PF, 0.2–0.5% G, pH 7.4 (56, 60, 61)	Yes	—	Yes	Yes	—	56
4% PF, 0.1% G, pH 7.4 (42–44, 56)	Yes	—	Yes	Yes	Yes	56
Routine histological fixatives (63, 64)	No	Yes	Yes	—	—	—
Freeze-drying, vapor fixation (63, 69)	No	Yes	Yes	—	—	—

[a] For each fixative, the use for perfusion, immersion, light microscopy, electron microscopy, NNE localization, and tissue culture are indicated.

[b] PF = paraformaldehyde; G = glutaraldehyde; PA = picric acid. Solution pH is given but see text or references for details of buffers.

[c] Insufficient or no information.

employing a paraformaldehyde–glutaraldehyde mixture administered by warm-cold perfusion, which was developed for glutamic acid decarboxylase immunocytochemistry (62, 63). Other workers have adapted fixations for specific situations such as aggregating cell cultures (64) or have simply used material already available from routine histological processing (65, 66). Compared with nervous tissue, there seems to be an increased resilience of NSE immunoreactivity in peripheral neuroendocrine tissues even when fixed by immersion. Poor fixation or excessive exposure to solvents during embedding often results in the loss of reliable neuronal staining.

Once fixation is accomplished, tissue may be either sectioned without embedding on the vibratome, cryostat, or freezing microtome or embedded first in paraffin or polyester wax. The advantage of sectioning tissue on a vibratome or tissue chopper is that no freezing artifacts are introduced and the material may be used for both light and electron microscopy. Tissue prepared for electron microscopy is usually processed for immunocytochemistry before embedding (55), with an effort to shorten the total time required.

4.3.2 Antigen Visualization

Neuron-specific enolase (NSE) can be visualized in tissue using specific antisera by means of the unlabeled antibody–enzyme method of Sternberger (55; 24, 41, 44, 53, 56, 67–71), immunofluorescence (44, 72, 73), peroxidase-conjugated second antibody (42), protein-A peroxidase (74, 75), or biotin-avidin methods (44). We have favored and present here the unlabeled antibody–enzyme method of Sternberger, which provides permanent sections in contrast with immunofluorescence. The only advantage of immunofluorescent (or nonperoxidase) methods is for tissues such as bone marrow. In such cases, the high levels of endogenous peroxidases obscure the actual immuno-linked peroxidase reaction and cannot be totally inactivated prior to staining. But the low concentrations of glutaraldehyde necessary for reliable enolase localization tend to produce high background in immunofluorescent methods.

The unlabeled antibody–enzyme method of Sternberger can be carried out on tissue-mounted slides, tissue culture coverslips, or free-floating tissue sections. After rinsing in buffer, tissue is processed in the following manner (55, 68):

1. 0.25% Triton-X in Tris-buffered saline (TBS; 50 m*M* Tris-HCl buffer pH 7.6 with 0.9% NaCl) at room temperature for 5 min. This step is optional for sectioned material but improves staining intensity

for monolayer cultures (presumably by ensuring access of reagents to intracellular location of NSE).

2. 3% H_2O_2–10% methanol in buffer at room temperature for 10 min. This step is optional, but seems to decrease endogenous peroxidase activity when background levels in a particular tissue are high. Other methods have been described to minimize nonspecific peroxidase activity (76) or to avoid excess binding of immunoreagents to tissue (sodium periodate–borohydride; 42, 77).
3. 10% normal sheep serum (NSS), 0.1 *M* D,L-lysine in TBS at room temperature for 1 hr or at 4°C overnight. Lysine was selected to react with any activated groups still present after fixation (57); the use of TBS probably makes its addition unnecessary. This "blocking step" is essential for tissues fixed with glutaraldehyde and can also be performed with 10% NSS alone or with ovalbumin (76).
4. Anti-NSE at an appropriate dilution in 1–3% NSS at room temperature for 1 hr or at 0–4°C for 12 to 48 hr. Antisera dilutions are made up in 1–3% NSS in TBS as discussed below.
5. Anti-rabbit IgG (prepared in sheep or goat) diluted 1:50 to 1:100, depending on the particular batch in 1–3% NSS–TBS. Linking antibody incubation is at room temperature for 30 to 60 min.
6. Rabbit peroxidase-antiperoxidase complex (PAP), usually at 1:50 to 1:100 in TBS at room temperature for 30–60 min.
7. 0.05% 3,3′-diaminobenzidine, 0.01% H_2O_2 in TBS for 2 to 15 min with agitation. The reaction is carried out under direct observation and is terminated by rinsing before staining of control tissue occurs. All steps are separated by multiple (3–5) changes of TBS at roughly 5 min intervals. Antiserum is stored in small aliquots at 1:20 dilution in 1% NSS TBS and diluted just prior to use; protein-containing solutions are either centrifuged or ultrafiltered. Waste benzidine solutions are treated with sodium hypochlorite.

The above procedure is designed for antiserum to NSE raised in rabbits. For each experiment, control sections are exposed at step 4 to preimmune or normal rabbit serum at the same dilution as NSE; these are necessary as controls for nonspecific staining of sections. Excessive background staining is most often due to poor fixation or inadequate rinsing. The possibility of immunoreagent failure is best assessed by making parallel runs and substituting buffer at that particular step.

Dilutions of anti-NSE used for immunocytochemistry depend on the tissue studied, the antiserum titer, and the time of incubation at step 4. Dilutions of 1:500 to 1:2000 are used for tissue cultures, especially for short incubation times. For longer times (12 to 24 hr), NSE can be localized in peripheral neuroendocrine tissues at dilutions of 1:2000 to 1:5000 or more. For central nervous system, NSE localization can be accomplished with dilutions up to 1:64,000.

The above dilutions are for anti-NSE-R, which is used for most immunocytochemistry. In primate and human material, anti-NSE-H can also be used at somewhat lower dilution corresponding to its lower effective titer for radioimmunoassay.

Initial immunocytochemistry of any tissue is best carried out with a variety of fixatives and with a full range of anti-NSE dilutions. Nonstaining of expected NSE-containing structures may be a result of (i) lack of NSE content, (ii) loss of immunoreactivity through overfixation, for example, the use of high glutaraldehyde concentrations at neutral pH, (iii) loss of NSE during fixation or processing (e.g., solvent extraction, paraffin embedding), or (iv) possible theoretical decrease of reactivity in regions of antibody–antigen excess with steric hindrance of subsequent immunoreagents. In addition, nonstaining with anti-NSE may be a threshold effect for a given cell or cell class at a given dilution and incubation time. When anti-NSE dilutions are carried out to a point where there is no longer excess of specific immunoglobulin compared with total NSE antigenic sites, immunocytochemistry can be used to compare qualitatively NSE content between cell classes (see Section 9).

The methods of tissue preparation and immunocytochemistry for NSE are fully applicable to non-neuronal enolase (NNE). Anti-NNE has had a lower effective titer for immunocytochemistry than NSE, which is mirrored in the NNE radioimmunoassay (see Section 4.2). Anti-NNE-R is usually diluted from 1:400 to 1:1000 for rat tissue and anti-NNE-H from 1:500 to 1:2000. At the lower ranges of dilution, there is often appreciable background staining even of the control serum and diaminobenzidine incubation times must be carefully monitored. Anti-NNE immunocytochemistry also reveals a considerable number of glial cells with intranuclear staining. This presumably reflects a greater lability of NNE during fixation, since both NSE and NNE are cytoplasmic enzymes; nuclear staining is never encountered with NSE immunocytochemistry.

Ultrastructural localization of NSE has been reported by a number of groups (42–44, 59, 64, 67). All studies to date have employed pre-embedding staining by immunocytochemistry. The major difference with light microscope immunocytochemistry has to do with careful fixation and expeditious and careful handling of the tissue; sections may be mounted under plastic coverslips after wafer embedding and used for light microscopy with desired areas removed for ultrastructural analysis (55).

4.3.3 *Combined Anatomical and Biochemical Approaches*

The ability to characterize the enzymatic activity, immunoreactive protein levels, and apparent anatomical localization for NSE is of distinct advantage in the analysis of tissue. Immunocytochemistry offers localization on the cellular or subcellular level and may avoid artifacts of tissue disruption; biochemical studies complement this information with quantitation of the amount and activity of NSE. Where possible, studies should be designed to exploit this advantage, especially now that RIA methods permit the measurement of NSE levels in microsamples. Such a combined approach was essential to the demonstration of NSE content in neuroendocrine cells (56) and the analysis of NSE appearance during development (68, 80). In addition, NSE is perhaps unique among cellular markers since another enolase isoenzyme, non-neuronal enolase (NNE), is also a cellular marker for most glial cells (41).

5 NEURON-SPECIFIC ENOLASE (NSE) LOCALIZATION IN THE NERVOUS SYSTEM

5.1 Specific Cellular Marker for Neurons

Using specific antisera to NSE, the proposed neuronal localization of this enolase isoenzyme (18) was directly demonstrated by immunocytochemistry in rat brain (24). Subsequent studies have confirmed the lack of NSE immunoreactivity in glial cells and other non-neuronal cell classes in intact nervous tissue (40–44, 53, 59, 60, 68). As expected for a soluble cytoplasmic enzyme, neurons show immunoreactivity for NSE in dendrites, axon, and cell body (Figure 1). Ultrastructural localization of NSE demonstrates association of immunoprecipitate with various internal membranes and organelles of neurons (42–44, 59, 64, 67). Such patterns have been reported for other soluble antigens (55, 79–81) and may be due in part to antigen translocation during fixation. Monolayer cultures may minimize this artifact (67).

Present methods of tissue preparation and immunocytochemistry reveal that not only is NSE a specific marker for neurons, it is also a reliable marker, present in all neurons (41–44, 68). NSE is therefore unique in that it is a general neuronal marker as opposed, for example, to the various neurotransmitter enzymes contained in neuronal subclasses. In a given cell class such as cerebellar granule cells (Figure 1E), all neurons are consistently stained to the same intensity. In any given area, NSE

Figure 1. (A and B) Adult rhesus monkey spinal cord in region of dorsal glial septum and posterior columns, showing staining of axonal profiles with NSE (A) and glial structures with NNE (B). Anti-NSE 1:2000; anti-NNE 1:400. Magnification ×160. (C) Rat cerebral cortex with several NSE(+) pyramidal neurons among numerous truncated axonal and dendritic profiles. Anti-NSE 1:2000. Magnification ×860. (D) Photomicrograph of decussation of medial lemniscus in adult rhesus monkey stained with immunocytochemistry for NSE. Anti-NSE 1:1000. Magnification ×500. (E) Rhesus monkey cerebellum in sagittal section, demonstrating NSE(+) staining of multiple neuronal types, including granule cells, stellate and basket cells, and Purkinje cells. Axons (parallel fibers cut in cross-section in molecular layer), cell bodies, dendrites (note Purkinje cells), and synaptic specializations (glomeruli scattered among granule cells) are uniformly NSE(+). Anti-NSE 1:2000. Magnification ×350.

immunoreactivity is confined to neuronal processes and is absent from glial elements (Figure 1A,B). Axonal processes are immunoreative, even when myelinated, whether they are tangential or parallel to the plane of section, as long as immunoreagents have free access during incubation (Figure 1D). Finally, the proportion of staining intensity is relatively even throughout the neuron; the cell body is always well visualized (Figure 1C,E) in comparison with the surrounding neuropil.

Large amounts of NSE are present in neurons and many neuroendocrine cells and are easily detected by immunocytochemistry. By varying the antiserum dilution and the sensitivity of the staining techniques, apparent differences in NSE content can be demonstrated for various classes of neurons; this is discussed in detail in Section 9. Immunocytochemistry for NSE has also confirmed the presence of lesser amounts of NSE in some cell classes and structures that are clearly neuronal or neuroendocrine in nature; see Sections 6 and 7. Sympathetic ganglia neurons have not been successfully stained from material prepared by perfusion or immersion fixation (82), but they are NSE immunoreactive in cocultures of sympathetic ganglia and pineal (70). The low NSE content of sympathetic ganglia (see Section 5.3) and the staining of cultured sympathetic neurons suggest that the lack of immunoreactivity of *in vivo* neurons from sympathetic ganglia represents a technical failure due to low NSE content. Other examples of nonstaining for NSE in neural structures are the special sensory epithelial cells for hearing and olfaction (83); both are primitive cell classes that may be devoid of NSE content, although definitive biochemical analysis is not available.

5.2 Distinct and Separate Localization of Non-Neuronal Enolase (NNE)

Non-neuronal enolase, which is identical to liver enolase, is a useful cellular marker for glial cells in nervous tissue (42, 44, 68). The localization of NNE provides a negative image of unstained neurons and neuronal processes surrounded by positively stained glial cells (Figure 2). All types of glial cells are NNE immunoreactive: choroid plexus cells, ependymal cells and tanycytes, Schwann cells, Bergmann glial cells of cerebellum and Müller cells of retina, radial glial cells, both fibrillary and protoplasmic astrocytes, and oligodendroglia of both interfascicular and perineuronal types. Whether microglia contain NNE is not yet clear. Vascular endothelial cells are unstained by anti-NNE (R and H). Oligodendroglia are NNE immunoreactive, but not as uniformly or strongly as cells of astrocytic lineage (10, 11, 41, 84, 85). Immunoreactivity of interfascicular oligodendroglia (Figure 1D; ref. 85), Schwann cells (Figure 2E; ref. 85), and perineuronal satellite oligodendroglia (Figure 3B; ref. 10) supports the presence of NNE isoenzyme content in these cells. Bulk-enriched oligodendroglia preparations (86) contain NNE and no NSE when analyzed biochemically. The NNE immunoreactivity of interfascicular oligodendroglia in particular is very weak and erratic compared with astrocytic

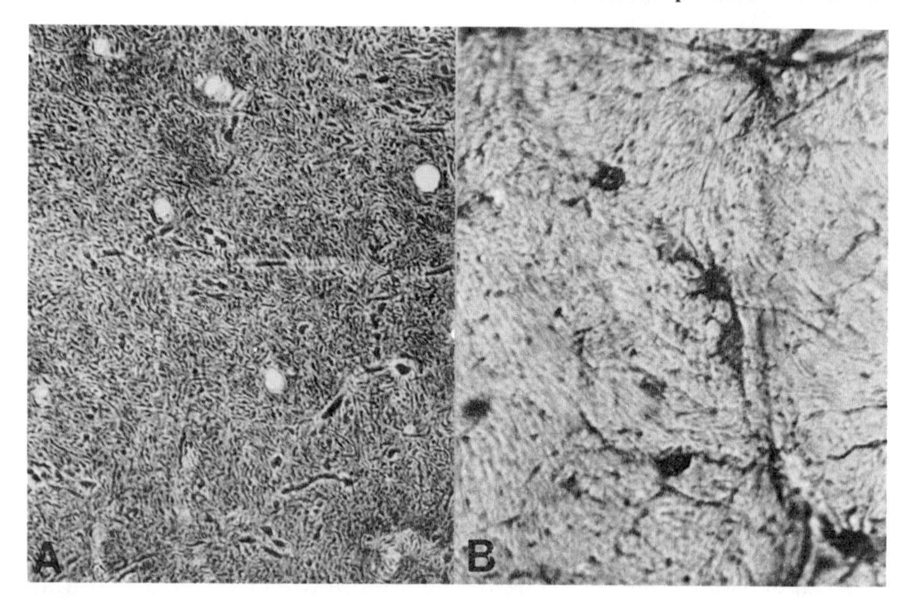

Figure 2. Adult rhesus monkey optic nerve cut in cross-section. Immunocytochemistry for NSE reveals axonal profiles (A) while NNE antisera detect glial processes and cells (B) (most may be astrocytic, see text). Anti-NSE 1:2000; anti-NNE 1:1000. Magnification ×500.

cell types; other groups of workers have reported that oligodendroglia are not immunoreactive for NNE (44). The variable staining results may reflect low NNE content in oligodendroglia and may be resolved by higher titer antisera to NNE or monoclonal antibodies. A low content of NNE in oligodendroglia would be in accord with the presence of the pentose monophosphate shunt for direct oxidative metabolism of glucose which is important to lipid and fatty acid metabolism (44).

NNE therefore has a complementary distribution to NSE (Figures 1, 2) in the nervous system, being present in glial cells of all types (low threshold levels in oligodendroglia). The original proposal that brain enolases (NSE and NNE) are specific markers of neuronal and glial cells (41) has, not surprisingly, required slight modification. Besides its presence in liver and other peripheral tissues, NNE is now known to be present in proliferating cell populations in developing nervous tissue (68). NNE cannot therefore be considered a specific glial marker in the same sense as the glial fibrillary acidic protein (GFAP) is (87). The usefulness of NNE for the study of glial cells remains to be explored. NSE is a specific marker for neurons and neuroendocrine cells, but further overlap with NNE distribution occurs for some neuroendocrine cells (see Section 6), tissue culture or tumor samples of transformed cells (Section 7), and young postmitotic neurons prior to differentiation (Section 8). Given the

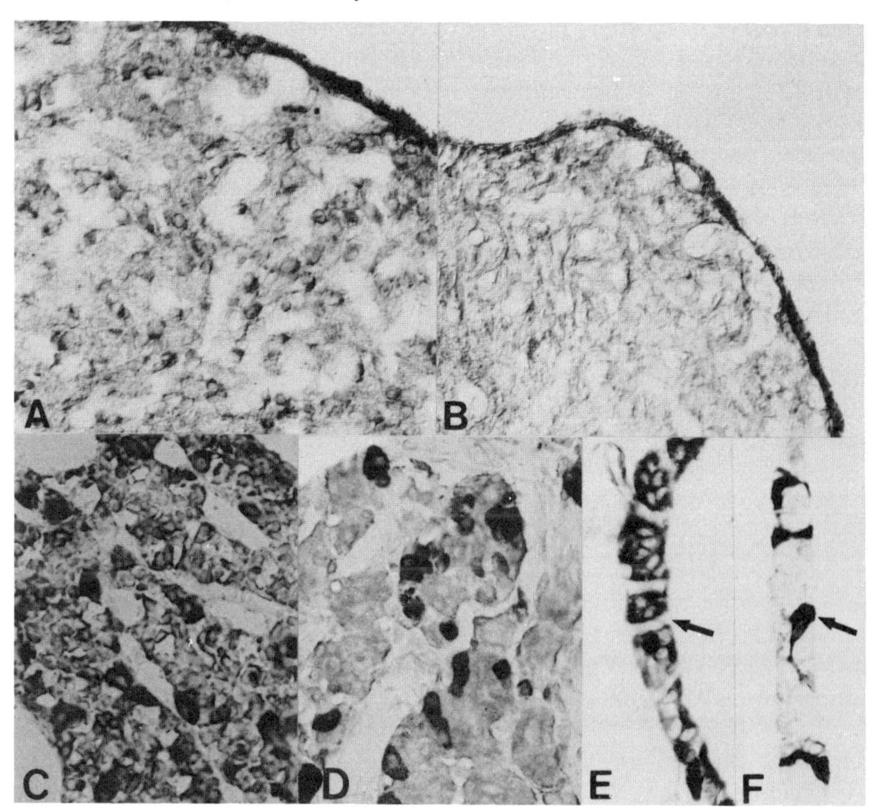

Figure 3. (A and B) Area postrema of adult rhesus monkey, stained for NSE content (A) and NNE content (B). Multiple NSE(+) cells seen in (A) are not revealed in NNE material (B) and presumably represent population of neuroendocrine cells. Anti-NSE 1:2000; Anti-NNE 1:400. Magnification ×220. (Darkened edge of tissue represents nonspecific deposition of DAB reaction product along 20 μm thick edge of section). (C) Rat pineal gland stained for NSE content (polyester wax section), revealing numerous NSE(+) pinealocytes. Anti-NSE 1:1000. Magnification ×220. (D) Rat thyroid, showing NSE immunoreactivity in neuroendocrine cells surrounded by nonstaining follicular cells; these are presumably calcitonin cells. Anti-NSE 1:1000; polyester wax section. Magnification ×220. (E and F) Cryostat sections of rhesus monkey thyroid tissue stained for NNE content (E) and NSE content (F). Follicular wall has been teased free of surrounding tissue and reveals numerous NNE(+) follicular cells with intercalated nonstaining cells (E, indicated by arrow) that may be identical to NSE(+) neuroendocrine cells (F, arrow), which are seen with anti-NSE amidst nonstaining follicular cells. Anti-NSE 1:1000; Anti-NNE 1:400. Magnification ×330.

switch during neuronal differentiation from NNE to NSE (68, 78), the occurrence of NNE in stellate and basket cells of the cerebellum, in Cajal-Retzius cells, and conceivably in other limited classes of mature neurons (68) is not surprising. In contrast to the report that many neurons contain

NNE (53), we and others have found to date that all or the vast majority of neurons make a complete switch from NNE to NSE and do not demonstrate appreciable immunoreactivity for NNE (see, for example, 41, 44, 64, 68). Further definition of the NNE content of neurons might be best resolved by using independent techniques, such as examining for the presence of labeled NNE after radioactively tagged axonal transport.

We have emphasized that NSE is a specific marker for differentiated neurons (41, 68) and that NNE is predominantly localized in glial cells. No specific method exists for localization of hybrid enolase, the other enolase isoenzyme detected in biochemical analysis of disrupted nervous tissue. One would expect cells containing hybrid enolase to be immunoreactive for NSE *and* NNE, although even this finding would be necessary but not sufficient to support the actual presence of hybrid enolase *in vivo*. All glial cells contain NNE ($\alpha\alpha$) without NSE ($\gamma\gamma$); only a few limited classes of neurons apparently contain a significant amount of NNE ($\alpha\alpha$) and therefore possibly hybrid enolase ($\alpha\gamma$) in mature nervous tissue (see the preceding paragraph and Section 9.5). Quantitation of NSE and NNE levels by RIA in mature nervous tissue is therefore essentially a specific quantitation of neuronal and glial enolase. Since hybrid enolase can be generated during the reassociation of mixtures of NSE and NNE (39), the high proportion of hybrid enolase seen in brain homogenates may be largely an artifact of tissue disruption. Hybrid enolase is likely to be present in some neuroendocrine cells, transformed cell lines, and other cell classes known to contain both NSE and NNE (see Sections 7 and 8). Using enzyme-linked immunoassays for NSE, NNE, and hybrid enolase, Kato and coworkers have reported that microdissected Purkinje cells contain nearly 40% of their enolase content as NNE or hybrid enolase (53). Although this is an attractive approach to profiling the enolase isoenzyme content of neurons, the possible effects of retained glial processes, incomplete recovery of enolase or an admixture of isoenzymes during disruption might result in apparent content of α subunit. At present, the extent and amount of α subunit (as NNE or hybrid enolase) present in mature neurons is controversial.

5.3 NSE and NNE Levels in Nervous Tissue

The availability of specific RIA methods for both NSE and NNE of rat and human (also applicable to monkey) allows a precise quantitation of enolase isoenzymes in rat, monkey, and human tissues (Tables 2 and 3). For the adult central and peripheral nervous systems, the strict immunocytochemical localization of NSE in neurons and NNE in glial cells (except stellate-basket cells of cerebellum) means that NSE is a measure

Table 2. Tissue Levels of NSE and NNE in Rat (ng/mg Protein)[a]

Tissue	NSE	NNE	NNE/NSE	NNE + NSE
Thalamus	17,500	18,200	1.04	35,700
Cerebral cortex	14,700	14,700	1.00	29,400
Midbrain	12,100	14,600	1.20	26,700
Cerebellum	11,800	9,800	0.83	21,600
Striatum	10,000	10,700	1.07	20,700
Hippocampus	9,800	11,700	1.20	21,500
Optic nerve	9,100	11,800	1.50	19,600
Cervical spinal cord	6,900	9,600	1.40	16,500
Medulla	6,800	12,300	1.80	19,100
Olfactory bulb	6,300	16,000	2.54	19,300
Pyramidal tract	5,600	12,600	2.25	18,200
Posterior pituitary	3,400	4,630	1.30	8,030
Pineal gland	2,650	18,400	6.90	21,050
Trigeminal nerve	1,150	2,400	2.10	3,550
Anterior pituitary	900	4,450	4.90	5,350
Superior cervical ganglion	650	2,700	4.15	3,350
Adrenal gland	240	8,200	34.20	8,440
Thyroid gland	150	3,330	22.20	3,480
Pancreas	18	—[b]	—	—
Skeletal muscle	9	3,560	396.	3,569
Liver	4	6,800	1,700.	6,804

[a] Levels are given in ng enolase/mg protein and represent the mean of 2 or 3 determinations on samples pooled from 4 to 6 animals. Variance approximately 10%.
[b] No information.

of neuronal enolase content and NNE a measure of glial (predominantly astrocytic) enolase content. The highest NSE content in rat brain is the thalamus, containing an amount of NSE approximating 1.75% of total soluble protein; NNE content in thalamus is likewise high, representing 1.82% of total soluble protein (Table 2). Another feature seen in the analysis of rat brain is the low NSE content of the hippocampus and olfactory bulb; NNE content for these regions is not similarly low, yielding a high NNE/NSE ratio. This probably reflects the presence of proliferating cell zones postnatally in the hippocampus and the olfactory bulb and delayed neuronal differentiation (see Section 8). In contrast to the central nervous system, the overall enolase content of peripheral nervous system is low (Table 2), especially for superior cervical ganglia. Low NSE content for a given region may reflect low NSE content of individual neurons (apparently true for sympathetic and dorsal root ganglia neurons; see

Sections 5.1 and 7.1) or the presence of non-neuronal tissue or both. Low levels of NNE may reflect differences in glial NNE content (for example, the large numbers of oligodendroglia in myelinated tracts) or possibly non-nervous tissue (for example, connective tissue in peripheral nerves). High NNE/NSE ratio is certainly typical of regions of white matter (Tables 2, 3), nervous tissue during development (see Section 8), and peripheral ganglia. In superior cervical ganglia, the low total NSE content, high NNE/NSE ratio, and the difficulty in staining neurons with anti-NSE suggests low levels of NSE in individual neurons. Levels of NSE in superior cervical ganglia are still, however, two orders of magnitude higher than non-nervous, non-endocrine tissue such as liver or muscle.

Table 3. Tissue Levels of NSE and NNE in Monkey Nervous System (ng/mg Protein)[a]

Tissue	NSE	NNE	NNE/NSE	NNE + NSE
A. Gray matter, CNS[b]				
Cerebellar cortex	12,300	14,800	1.10	28,100
Deep cerebellar nuclei	11,700	30,800	2.60	42,500
Visual cortex (area 17p)[c]	7,800	11,600	1.50	19,400
Visual cortex (area 17m)[d]	14,400	11,700	0.80	26,100
Visual cortex (area 18)	20,500	22,700	1.10	43,200
Sensorimotor cortex	13,200	19,600	1.48	32,800
Temporal cortex	11,400	21,900	1.90	33,300
Frontal cortex	12,800	19,600	1.48	32,400
Parietal cortex	15,500	15,400	0.99	30,900
Dentate gyrus (hippocampus)	9,200	20,900	2.27	30,100
Subiculum (hippocampus)	16,000	20,000	1.25	36,000
Caudate-putamen	19,700	23,100	1.20	42,800
Globus pallidus	14,600	21,200	1.45	35,800
Thalamus	11,700	20,500	1.75	32,200
Hypothalamus	10,300	23,800	2.30	34,100
Substantia nigra	9,800	20,500	1.75	32,200
Superior colliculus	13,000	21,400	1.65	34,400
Inferior colliculus	13,800	25,100	1.80	38,900
Olfactory bulb	12,900	20,200	1.57	33,100
B. White matter, CNS				
Anterior commissure	10,600	18,000	1.70	28,600
Corpus callosum	12,100	29,100	2.40	41,200
Pyramidal tract (pons)	10,600	25,300	2.40	35,900
Pyramidal tract (obex)	10,200	36,000	3.53	46,200
Optic nerve	4,200	13,800	3.26	18,000

Table 3. *(continued)*

Tissue	NSE	NNE	NNE/NSE	NNE + NSE
C. Mixed regions, CNS				
Pons	13,000	30,700	2.36	43,700
Medulla	9,500	35,000	3.68	44,500
Cervical spinal cord	4,900	20,000	4.08	24,900
Thoracic spinal cord	6,040	20,800	3.44	26,840
Sacral spinal cord	5,670	22,200	3.90	27,870
D. Peripheral nervous system				
Sensory roots	2,370	6,060	2.56	8,430
Motor roots	2,080	5,000	2.40	7,080
Dorsal root ganglia	1,150	5,100	4.45	6,250
Superior cervical ganglia	810	16,350	20.74	17,160
Sciatic nerve	770	5,770	7.45	6,540
Phrenic nerve	230	2,700	11.74	2,930
Vagus nerve	60	1,230	20.80	1,290

[a] Levels are given in ng enolase/mg protein; both NSE and NNE radioimmunoassays would potentially recognize hybrid enolase (see Section 4.2).
[b] CNS = central nervous system.
[c] Area 17p = peripheral field portion of visual cortex area 17.
[d] Area 17m = macular field portion of visual cortex area 17.

Table 3 presents quantitative levels of NSE and NNE in selected regions of monkey brain and peripheral nervous system. In several regions, total enolase content is in the range of 4 to 5% of total soluble protein. The following observations are possible.

1. Like rat brain, some gray matter regions are low in NSE content, such as dentate gyrus (hippocampus), 1° visual cortex (peripheral field), and substantia nigra.
2. The highest NSE levels are seen in association with cortex (parieto-occipital lobe) and caudate-putamen; the highest total enolase levels (NNE + NSE) are seen in these same two regions as well as in deep cerebellar nuclei, in corpus callosum, and in pyramidal tract.
3. There are marked differences among white matter tracts and roots: corpus callosum and pyramidal tract are higher in NSE and total enolase content than are optic nerve, anterior commissure, and especially sensorimotor roots of spinal cord.
4. Particularly low levels of NSE are seen in dorsal root ganglia and superior cervical ganglia.

5. High NNE/NSE ratios (greater than 2) are typical of white matter and mixed white-gray matter regions of the central nervous system as well as all regions of the peripheral nervous system.
6. Moderate variation of NSE content is seen in cerebral cortex; for visual cortex, there is an increase in both NSE and total enolase content from primary visual cortex of the peripheral field to primary visual cortex of the macular field to visual association cortex (area 18) of greater than twofold.

The measurement of enolase levels in nervous tissue has revealed patterns suggestive of varying amounts of NSE among different regions and neuronal cell classes. The local NSE content of neocortical slabs (cortical layers without white matter) varies by a factor of 2.5 (item 6 above). The NSE content of central neurons may be 5 to 10 times greater than that of dorsal root ganglion or superior cervical ganglion neurons; this difference in NSE content is supported by immunocytochemistry under conditions where these two neuronal cell classes are uniformly processed (58). NSE or total enolase content is not clearly related to local glucose utilization as measured by the deoxyglucose method (88). Some similarities exist; for example, superior cervical ganglia have a rate of glucose utilization five times slower than the most active brain regions (average values), which corresponds to the low NSE content of superior cervical ganglia compared with brain. On the other hand, white matter has very low glucose utilization rates (comparable to superior cervical ganglia), whereas NSE and total enolase levels in some white matter regions can exceed gray matter regions (Table 3). It is perhaps not surprising that only rough correlations exist between levels of an enzyme such as NSE and transient local glucose utilization for nervous tissue. Glucose utilization apparently fluctuates with functional activity on a moment-to-moment basis and is localized in certain portions of the neuron (e.g., axon terminals) where energy requirements are greatest (88). In contrast, NSE is distributed throughout the neuron, although levels may be increased by long-term changes in synaptic or functional activity (see Section 9). The methods used to localize and quantitate NSE in neurons reveal static levels of enzyme protein and therefore may complement kinetic measurements such as glucose utilization rates in the evaluation of neuronal functional activity.

5.4 NSE Localization in Neuronally Derived Tumors

The discussion of the localization and quantitation of NSE is not complete without mention of the role of NSE in the study of nervous system tumors. The possibility of using NSE as a marker for tumors of neuronal derivation has been recognized since the discovery of NSE (89, 90). A recent paper has demonstrated the presence of NSE immunoreactive giant cells in the subependymal tumors seen in tuberous sclerosis (66). The presence of NSE in human neuroblastomas (92) and the ability to demonstrate NSE (albeit at reduced levels) in virally fused multinucleate neurons (75) show that NSE is an effective marker for cell lineage in tumors. The potential of NSE for use in the study of nervous system tumors remains to be fully defined and exploited.

6 NSE LOCALIZATION IN THE NEUROENDOCRINE SYSTEM

The immunocytochemistry and quantitation by radioimmunoassay of NSE in a variety of tissues quickly revealed that NSE is not restricted to neurons. Table 2 (rat) and Table 4 (tissue comparison) list a number of sites other than nervous tissue where NSE levels are significantly higher than muscle or liver. Specific immunocytochemistry shows that NSE is in fact contained in central and peripheral neuroendocrine cells throughout the body (56).

6.1 Central Neuroendocrine System

The presence of NSE in all central neuroendocrine cells is underlined by the finding that the NSE level in the rat pineal (Table 2) is higher than in superior cervical ganglia. In the monkey and human (Table 4), NSE levels in the pineal approach that of brain itself. Specific immunocytochemistry with anti-NSE and anti-NNE demonstrates that pinealocytes are strongly immunoreactive for NSE (Figure 3C) whereas NNE staining is apparently confined to supporting cells. NSE staining is also seen in anterior pituitary cells, pars intermedia cells, and the endings of the hypothalamo-neurohypophysial system in the posterior lobe (pars nervosa).

Many of the central neuroendocrine cells are contained in the circumventricular organs—area postrema, subcommissural organ, subfornicial organ, organ vasculosum of the lamina terminalis, supraoptic nucleus, and paraventricular nucleus—as well as in the pituitary and pineal glands. Cells in the area postrema are NSE immunoreactive but apparently do not contain NNE (Figure 3A,B); all of the other circumventricular organs contain NSE immunoreactive cells, but only in the case of pineal gland, area postrema, supraoptic nucleus and paraventricular nucleus does it appear that neuroendocrine cells do not contain NNE. Analysis in other

Table 4. NSE Levels in Neuroendocrine Tissues (ng/mg Protein)[a]

A. NSE-rich tissues (dense population of neuroendocrine cells)	
Cerebral cortex	7800–20,500 (R, M) for comparison
Pineal gland	2700 (R); 8000–9000 (M, H)
Posterior pituitary	2000–3400 (R, M, H)
Anterior pituitary	900–1,400 (R, M)
Adrenal medulla	700–900 (R, M, H)
Platelets	500 (H)
B. NSE-poor tissues (diffuse population of neuroendocrine cells)	
Thyroid	150–250 (R, H)
Thymus	100 (R)
Pancreas	20–50 (R, M)
Lung	20 (H)
C. NSE-poor tissues (no NSE-containing cells by immunocytochemistry*)[b]	
Red blood cells	20 (H)
Liver	3–15 (R, M, H)
Muscle	3–10 (R, M, H)
D. Neuroendocrine tumors (APUDomas)	
Oat-cell carcinoma	3900–5600 (H)
Glucagonoma	380–1300 (H)
Insulinoma	360 (H)
Rectal carcinoma	15 (H) for comparison
Pheochromocytoma	650 (R) cultured cells

[a] Levels are given in ng NSE/mg protein and, in most cases, represent the mean value of several determinations. R = rat; M = rhesus monkey; H = human autopsy or biopsy material.

[b] NSE immunocytochemistry in liver and muscle reveal peripheral nerves but no local populations of NSE(+) cells.

regions is incomplete owing to greater difficulties with background staining in periventricular structures.

Another example of NSE immunoreactivity is cells intercalated between ependymal cells of the central canal of spinal cord and also in the region of the third ventricle. Such cells are present in both rat and monkey; in a single 25-μm section, two to three such cells are usually seen in the ependyma of the central canal. They are easily detected by virtue of the nonstaining ependymal cells on either side and they apparently extend across the full width of the ependyma. CSF-contacting cells are also seen in the lateral ventricles during development (see Section 9 and Figure 9), where they are profuse and again starkly delineated against nonstaining cells around them. Such cells in adult animals have been considered central neuroendocrine derivatives (92).

NSE levels have only been determined for pineal gland, anterior pituitary, and posterior pituitary, although they are feasible for other circumventricular organs by using microdissection or punch techniques with radioimmunoassay. As for brain regions, there is considerable variation, since the NSE level for monkey and human pineal gland is equivalent to brain NSE levels while levels in the pituitary are much lower. The NSE levels for central neuroendocrine tissues are higher than for peripheral neuroendocrine tissues, which contain a more diffuse distribution of NSE(+) cells.

6.2 Peripheral Neuroendocrine System

Both peripheral and central neuroendocrine cells are characterized by their silver-staining properties, their content of one of more neuropeptides, the occasional presence of biogenic amines, and their property of amine precursor uptake and decarboxylation (APUD); the list of such cells is extensive and coincides in large part with known derivatives of the neural crest (93). The diffuse distribution of peripheral neuroendocrine cells in glandular tissue, gut, and other organs makes radioimmunoassay for NSE less striking than the actual visualization of NSE content by immunocytochemistry. The low levels of NSE in muscle and liver represent the contribution of NSE contained in peripheral nerves. Significantly higher levels of NSE are seen in a number of tissues (Table 4) but are still approximately 0.01% of the total soluble protein in adrenal and thyroid glands. Analysis of adrenal NSE content reveals that medulla has 10 times higher levels than adrenal cortex (56). Immunocytochemistry for NSE easily demonstrates that NSE is localized to adrenal medullary chromaffin cells, being present only in nerve fibers traversing the adrenal cortex. The initial survey confirmed that neuroendocrine cells in pancreas, thyroid and adrenal gland contain NSE and proposed that NSE would be a useful generalized marker for cental and peripheral neuroendocrine cells and their associated tumors called APUDomas (56).

Subsequent work has extended these findings to include all known APUD cells of the gastrointestinal tract (94) and lung (95). Using specific antisera to the various peptides found in the diffuse neuroendocrine system of gut, Tapia and coworkers have positively characterized these cell types and confirmed their NSE immunoreactivity (65). NSE immunocytochemistry provides a powerful method for demonstrating neuroendocrine cells, more reliable and specific than silver stains (65, 96) and easier than the methods involved in APUD criteria (93). Adrenal gland and thyroid gland show the typical delineation of NSE(+) cells against the nonstaining surrounding tissue (Figure 3D,F).

In some cases, peripheral neuroendocrine cells may contain both NNE and NSE. Adrenal medulla contains much more NNE than NSE (Table 4), and immunocytochemistry for NNE demonstrates that adrenal medullary chromaffin cells are immunoreactive. In thyroid gland of monkey, there is some indication that NSE-containing calcitonin (C) cells are not NNE immunoreactive (Figure 3E). There are probably other examples of peripheral neuroendocrine cells that contain predominantly NSE and little or no NNE; a precise definition of NNE content will require cell fractionation with RIA determinations on purified neuroendocrine cells and better, higher titer antisera to NNE.

Neuroendocrine cells can be reliably demonstrated in all tissues with NSE immunocytochemistry. Reticular epithelial cells of thymus are NSE-reactive in immersion-fixed human thymus specimens (Figure 4A). The staining appearance of NSE immunoreactivity in rat, monkey, and human material corresponds to the reticular appearance of these cells in the medullary region of thymus (Figure 4). Reticular epithelial cells are neuroendocrine cells that secrete thymopoietin peptides and interact with T-lymphocytes. NSE immunoreactivity resembles staining with monoclonal antisera to a complex ganglioside (GQ), which also marks neuroendocrine cells (97).

6.3 Neuroendocrine Tumors

The neuronal characteristics of neuroendocrine cells include shared antigens with neurons (e.g., complex ganglioside GQ), the presence of neuron-specific enolase isoenzyme, shared neuropeptides and neurosecretory behavior (98), and electrical properties (99). The relation of NSE content to the differentiated state of neuroendocrine cells is as yet undefined, but other evidence suggests that NSE content in neural crest tumors is closely related to the degree of differentiation (100, 101).

NSE immunocytochemistry has demonstrated that NSE is a reliable marker for neuroendocrine tumors or APUDomas and is more sensitive than staining for neuropeptides (65), which are sometimes depleted by high secretory rates. NSE levels by RIA in a variety of tumors—insulinomas, oat-cell carcinomas, gut carcinoid, and other APUDomas—show significant levels of NSE, ranging up to 5000 ng/mg protein (Table 4; ref. 65). Prelimary studies also demonstrate the presence of NSE in serum from such patients, in contrast to the near-absent levels of normals (102). The correlation of NSE levels with the secretory behavior of APUDomas is not yet finished; the study of NSE and neurosecretion in such tumors may well help explain NSE regulation in nervous tissue.

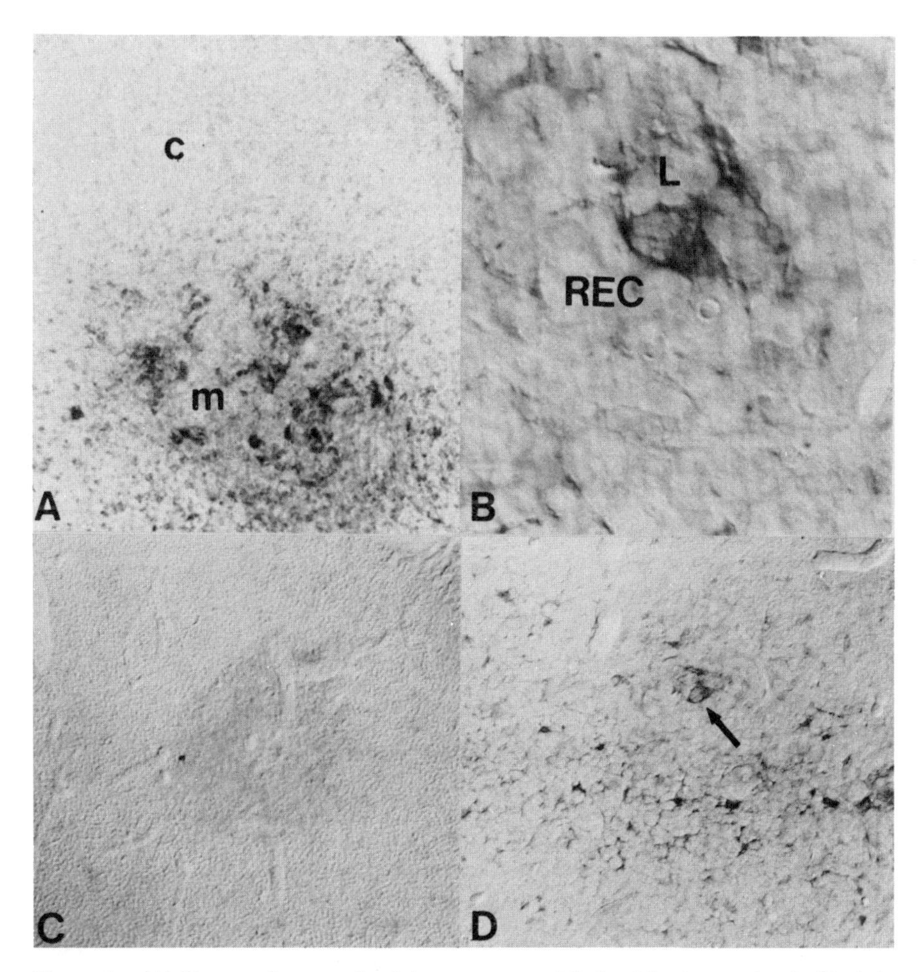

Figure 4. (A) Human thymus gland (autopsy material) fixed by immersion in Bouin's fluid, illustrating NSE(+) medullary region (*m*) compared with nonstaining cortical region (*c*); NSE immunoreactivity is contained in thymus reticular epithelial cells. Anti-NSE 1:1000. Magnification ×130. (B) Rhesus monkey thymus gland fixed by perfusion, showing reticular epithelial cell (REC) surrounded by reticulated NSE(+) processes and nonstaining lymphocytes; letter *L* indicates several lymphocytes embraced by NSE(+) processes of reticular epithelial cell origin. Anti-NSE 1:1000. Magnification ×880. (C) Section of rhesus monkey thymus gland incubated with normal serum to demonstrate low background typical of immunocytochemistry using unlabeled antibody enzyme method of Sternberger; photomicrograph contains medullary center surrounded by cortex. Normal rabbit serum 1:1000. Nomarski interference optics, magnification ×220. (D) Sections incubated with anti-NSE at same dilution show immunoreactive processes and occasional cell bodies of reticular epithelial cells; medullary center is larger than control section (C), and arrow indicates cell shown at higher magnification in (B). Anti-NSE 1:1000. Nomarski interference optics, magnification ×220.

7 NSE LOCALIZATION IN OTHER TISSUES

7.1 Primary Tissue Culture

NSE is an effective marker for neurons growing in primary cultures of nervous tissue. With the fixation and immunocytochemical methods presented in Section 4, anti-NSE is a reliable marker for neurons in culture (58). NSE-immunoreactive cells can be demonstrated by electrophysiological techniques to have membrane properties of neurons: excitability and in many cases spontaneous synaptic activity. In cocultures of mouse spinal cord neurons and dorsal root ganglia cells, NSE immunoreactivity varies consistently with strong staining of large spinal neurons and fainter staining of dorsal root ganglia cells (Figure 5). NSE content seems to correlate with levels of synaptic activity, since dorsal root ganglia cells are known to have no spontaneous electrical activity in culture (103, 104).

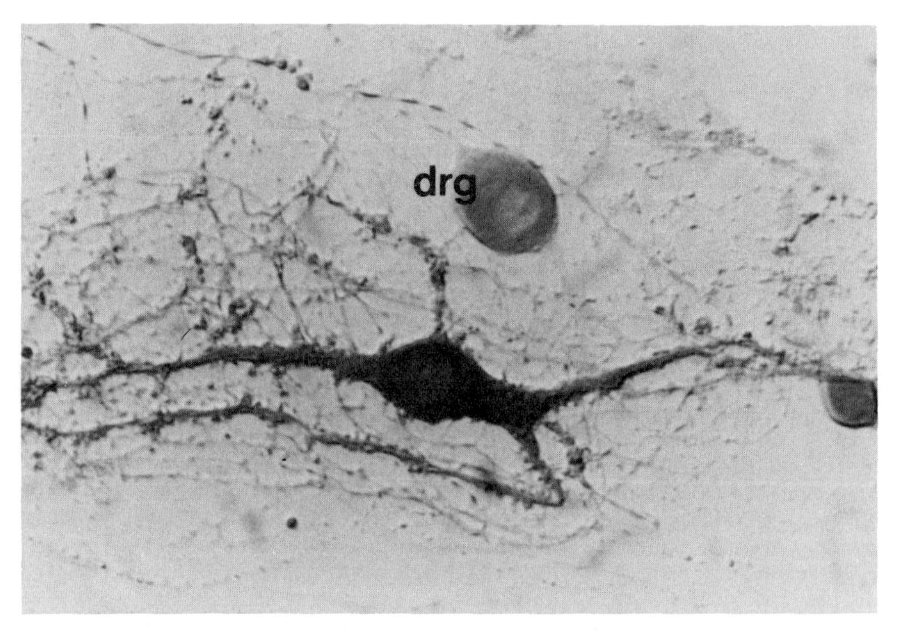

Figure 5. NSE immunoreactivity in primary tissue culture of mouse spinal cord neurons. Large multipolar spinal cord neuron is strongly NSE(+) compared with neighboring dorsal root ganglia cell (*drg*); numerous background cells (fibroblasts and occasional glial cells) are unstained. See text for correlation of NSE immunoreactivity with synaptic activity. Anti-NSE 1:1000. Magnification ×880.

NSE levels are an accepted marker for neuronal characteristics of cell separations from nervous tissue (105, 106). Immunocytochemistry with anti-NSE of primary cultures from mouse and rat brain (58, 64, 67, 69, 70) is a useful means of cell identification. The levels of NSE in such cultures is much lower than in intact nervous tissue (69, 107). In cultures of rat cerebral cortex, initial NSE levels resemble those of the source fetal tissue (17-day-old fetal cortex) but do not rise appreciably afterwards even up to 5 weeks in culture (69). In aggregating cell cultures derived from fetal rat brain, NSE levels increased 12-fold during 9 weeks in culture (64) suggesting that these culture conditions may favor neuronal differentiation. Study of cell number and staining intensity supports the contention that initial increases in NSE represent increased total number of NSE immunoreactive cells whereas a further threefold increase in NSE content represents increased levels in individual cells (64).

NSE is not only a useful marker for neurons, it also offers possible assessment of the functional activity and level of differentiation of nerve cells in culture (58, 64, 67, 69). The gradual increase of NSE levels in individual neurons during tissue culture (64) suggests continued differentiation of neurons over a period of weeks. For some nerve cells, maturation and survival may depend on synaptic activity, since a blockade of spontaneous electrical activity during the first month of culture with tetrodotoxin (TTX) results in a marked loss of normally active neurons (108). Dorsal root ganglia neurons, which are synaptically inactive and low in NSE content, are not vulnerable to TTX. After longer periods of culture (15 weeks), the susceptibility of spontaneously active neurons to TTX disappears (108). This model would afford an excellent test of the hypothesis that NSE levels are related to long-term functional activity; NSE content might be expected to decrease toward the level of dorsal root ganglia neurons for such TTX-silenced neurons.

NNE levels in tissue cultures are easily determined by RIA (64, 69). Immunocytochemistry with anti-NNE-R is possible (64) although hampered by the lower effective titer of anti-NNE compared with anti-NSE. Such studies confirm the strict localization of NSE and NNE to neuronal and glial cells respectively; no examples have been detected to date of neurons containing both NSE and NNE in culture conditions. Fibroblasts and background epithelial cells stain for neither NSE nor NNE. It is interesting that even synaptically silent neurons that have low NSE levels, such as dorsal root ganglia cells, contain no detectable NNE content (this is also true *in vivo*; see Figure 2 and ref. 85).

NSE will also have obvious utility in the analysis of primary tissue culture of other NSE-containing cell classes such as neuroendocrine cells. Pinealocytes in culture exhibit strong NSE immunoreactivity, which helps distinguish them from other background cells (70). As for neuronal cultures, NSE levels could provide quantitative assessment of differentiation for comparison with the intact animal or with different culture conditions. Culture of adrenal medullary chromaffin cells, which contain both NNE and NSE, might provide a good model for NSE regulation and expression.

7.2 Tissue Culture of Transformed Cells

Although NSE levels are correlated with synaptic activity and differentiation, NSE is nevertheless a stable marker even for transformed cells of neuronal or neuroendocrine lineage. The levels of NSE seen in some APUDomas are almost equivalent to brain tissue levels (65) and are possibly related to active neurosecretion. Lower NSE levels are present in cell lines of neuroblastoma, pheochromocytoma, and other neuronal/neuroendocrine tumor lines, including hybrids (100, 101, 109–112). Neuroblastoma cells contain both NSE and NNE with greater than 90% of enolase enzyme activity present as the non-neuronal form and the remainder as NSE (100). Treatment of these cultures with butryl cAMP produces increased cell body size and neurite outgrowth; enolase activity increases, but NSE is now present in greater quantities (twice as much) and NNE in lesser quantities (100). NSE therefore seems to correlate with increased morphological differentiation. Others have reported that basal levels of NSE in neuroblastoma cells seem to be related to cell cycle (NSE increasing with arrest of cell division) and that further elevation is related to morphological differentiation (111). NSE levels in neuroblastoma lines are lower than in pheochromocytoma (PC 12), which has NSE levels similar to those of adrenal medulla, its tissue of origin (see Table 4 and ref. 101). NSE levels in cultures of pheochromocytoma cells increase with cell density, but this increase is accelerated and enhanced by the addition of nerve growth factor (101). NSE levels are therefore low in transformed neurons in culture but are still useful as a specific marker.

Recently, very high levels of NSE have been shown in cultured transformed human small-cell lung tumors (113). NSE quantitation and immunocytochemistry promise to be useful tools in the evaluation of such cultures, particularly their response to pharmacologic manipulation. At the same time, the nature of cell transformation and dedifferentiation makes it likely that some examples will be found of cell cultures derived from glial or other non-neuronal cells which nevertheless express NSE after transformation.

7.3 Platelets and Non-Neuronal Tissues

Small quantitites of NSE can be measured in human serum and cerebrospinal fluid by radioimmunoassay for human NSE (50, 51). Increased serum levels of NSE are present in patients with APUDomas (102) and may serve as a useful diagnostic or monitoring test. In both serum and CSF, NSE is present outside of cells and presumably represents protein released from damaged or dying cells; normal background levels might also be the result of escape during neurosecretion.

In contrast with the presumed origin of NSE in serum and CSF from neurons and neuroendocrine cells, the presence of the γ subunit in platelets and megakaryocytes (114) is potentially troublesome because megakaryocytes are not considered neuroendocrine cells or thought to be derivatives of the neural crest. Platelet "NSE" levels are significantly higher than liver or muscle and are similar to adrenal medulla. Both platelets and megakaryocytes appear to stain with anti-NSE (114). Another interesting feature is that all γ subunit in platelets is present as hybrid enolase without any NSE peak, underlining the fact that cells containing both α and γ subunits may contain different proportions of enolase isoenzymes. Using enzyme-linked immunoassay, Kato and coworkers have confirmed the presence of significant amounts of γ subunit in isolated megakaryocytes (54). In contrast with the results in platelets, most of the γ subunit appeared in the form of NSE and not as hybrid enolase. The presence of NSE in platelets is not totally frustrating for the concept of NSE as a specific cellular marker for neurons and neuroendocrine cells. Platelets do have specialized properties such as amine precursor uptake, storage, and release that resemble those of APUD neuroendocrine cells; these properties have led to their use as models for monoaminergic neurons (115).

Another possible non-neuronal, non-neuroendocrine NSE (or γ subunit)-containing structure is the glycogen body in developing avian embryos (83). NSE immunoreactivity appears ontogenetically in glycogen body cells before it does in other regions such as dorsal root ganglia, and spinal cord. As for platelets, the significance of NSE content may relate to neurosecretory characteristics, or may indicate common lineage with nervous tissue, or both. Another alternative is that the presence of NSE in cells is related to the need for a particularly stable isoenzyme that is resistant to chloride ion (see Section 3.2). Since neural crest generates some NSE(−) derivatives such as Schwann cells, we feel that NSE content in neurons, neuroendocrine cells, platelets, and glycogen body cells is more likely related to common functional characteristics and is not strictly determined by cell lineage. Lest some think that NSE might stand for "non-specific enolase," it is well to remember that the vast majority of NSE is contained in neurons.

8 NNE-TO-NSE SWITCH DURING NEURONAL DIFFERENTIATION

8.1 Developmental Profile of NNE and NSE

Studies of enolase isoenzymes in developing nervous tissue have shown that NSE activity is very low until after the onset of neurogenesis in a given region and then steadily increases to adult levels (13–15, 27). In contrast, NNE activity is high from the beginning of development and even shows a depression just at the onset of neurogenesis and NSE appearance (Figure 6). Radioimmunoassay analysis of NSE and NNE content in developing nervous system provides a detailed profile of the NNE and NSE changes suggested by enzyme activity analysis (78). The

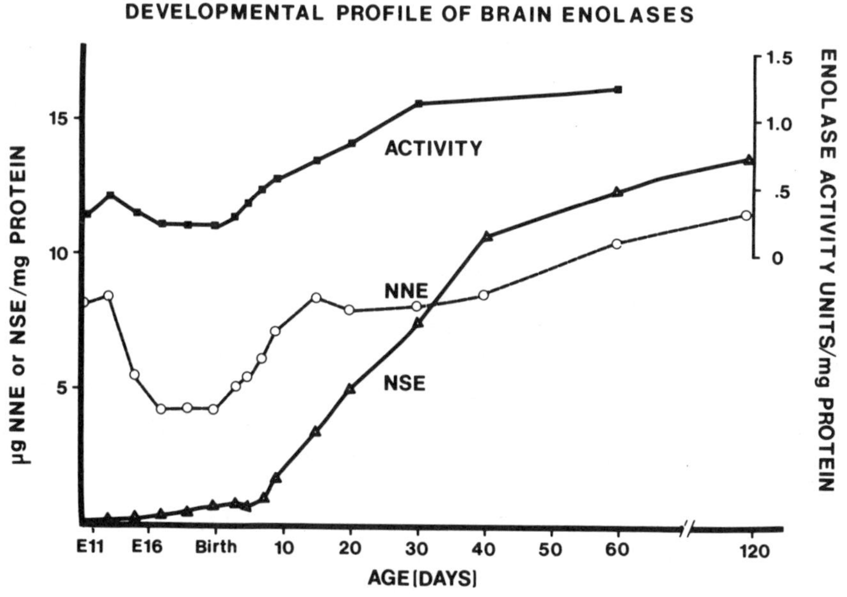

Figure 6. Developmental profile of enolase isoenzymes in rat brain. The transition from NNE to NSE and NNE is evident; a transient decrease occurs in NNE levels just at the upslope of NSE. ■, enoclase activity; ▲, NSE level by RIA; ○, NNE level by RIA in ng/mg protein.

depression in NNE levels and the subsequent appearance and increase in NSE levels are observed for whole rat brain as well as various subregions (78). Most of the NSE increase occurs postnatally and, except for brainstem, NSE levels at 25 days after birth are still only 30% of the eventual levels seen in older animals. In developing monkey visual cortex, NSE levels of cortical wall towards the end of neurogenesis at 100 days after conception (116) are extremely low. Comparison of various areas of neocortex shows that NNE/NSE ratio is correlated with gradients of neurogenesis; for example, medial cortical wall is lower than lateral wall and frontal cortex lower than occipital cortex (78). Analysis of subventricular zone (dividing precursor cells for glia and neurons) compared with cortical plate (final destination of postmitotic neurons) does not show much difference (78). This is not unexpected since most neurons in cortical plate contain small amounts of NSE and there are numerous neuronal processes adjacent to and within the subventricular zone from mature neurons in thalamus and deep white matter (see Section 9).

Analysis of the developing nervous system shows that NSE levels continue to increase long after neurogenesis is finished and that the initial rise in NSE occurs at definite time points related to the onset of neurogenesis in a given region. Biochemical analysis alone supports the assumption that a switch from NNE to NSE must occur for neurons or their precursor cells and that this NNE to NSE switch is timed differently for each region of nervous system (78). Immunocytochemistry with antisera to NNE and NSE can directly demonstrate, on a cellular basis, when the NNE to NSE switch occurs for neurons, and it is therefore a perfect complement to biochemical analysis (68, 78).

8.2 Cerebellum

Granule cell neurons in cerebellum are generated from cells of the transient external granule cell layer and migrate after final cell division across the molecular layer to their final location in the internal granule cell layer (117). The proliferating cells of the external granule cell layer give rise exclusively to granule cell neurons and stellate-basket cell neurons and therefore can be considered as a homogeneous neuronal precursor population (118). Immunocytochemistry for NNE and NSE in developing cerebellum in rat and monkey (68) shows that the external granule cell layer cells contain NNE, the non-neuronal enolase isoenzyme and no detectable NSE (Figure 7). After final cell division, prospective neurons from this layer begin to form their parallel fiber axons at the outermost boundary of the molecular layer, and their cell body migrates inward along Bergmann glial fiber guides to the internal granule cell layer (117,

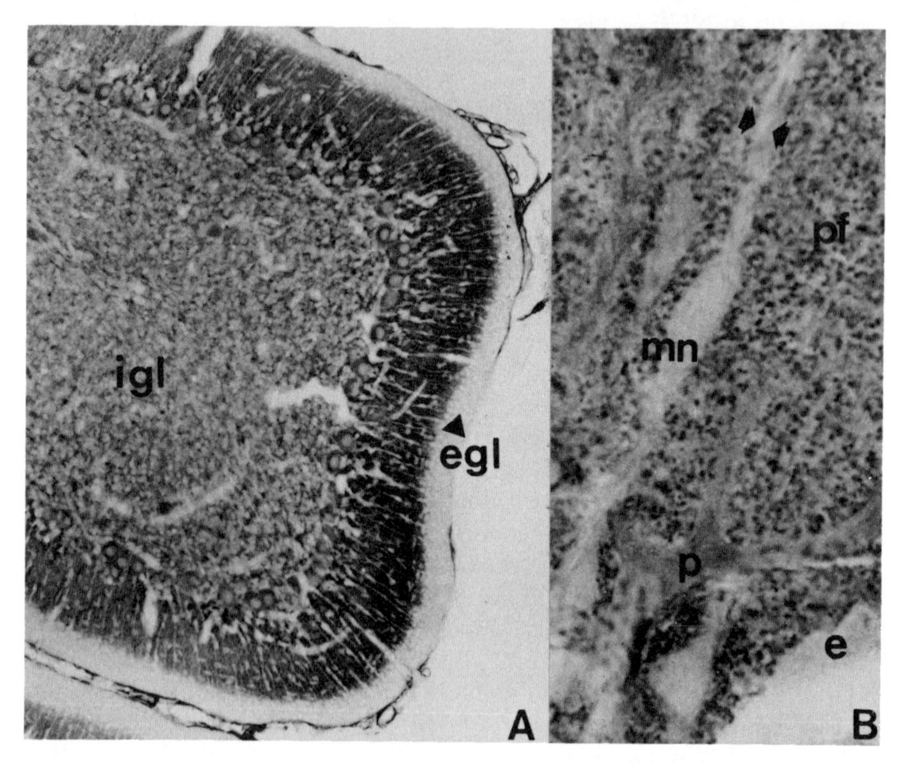

Figure 7. (A) Early postnatal rat cerebellum (2 weeks after birth) showing the NSE(−) external granule cell layer (*egl*, arrowhead), which gives rise to granule cell neurons that must migrate inwards to the internal granule cell layer (*igl*). The molecular layer is relatively dark with parallel fiber axon profiles and is traversed by many unstained Bergmann glial fibers. Anti-NSE 1:1000. Magnification ×130. (B) Migrating granule cell neurons (*mn*) are unstained by anti-NSE during their migration inwards along Bergmann glial fibers (between thick arrows). Parallel fiber axons (*pf*) are cut in cross-section in this sagittal section and are NSE(+). Purkinje cell dendrites (*p*) are also NSE immunoreactive. Endothelial cells (*e*) are unstained with either NNE or NSE antisera. Anti-NSE 1:1000. Magnification ×1080.

119). During this migration, granule cell neurons are still unstained by anti-NSE and appear to be less immunoreactive for NNE (Figure 7). Granule cell neurons located in the internal granule cell layer are uniformly NSE(+) and their parallel fiber axons in the molecular layer are NSE(+), although with decreasing intensity towards the outermost margin. The switch from NNE to NSE in granule cell neurons occurs relatively late in differentiation, after their arrival in their final location. NSE does not appear to be correlated with the eventual lineage of dividing cells, the last cell division, or even the elaboration of axonal processes and migration of young neurons, but rather with the final steps of neuronal differentiation.

8.3 Cerebral Cortex

Cerebellum offers the advantage of demonstrating the switch from NNE to NSE in a homogeneous proliferating zone (external granule cell layer) giving rise to a defined set of neurons (granule cell and stellate-basket cell neurons). Developing monkey visual cortex is better, however, for examining the question of enolase content of migrating neurons and the timing of the NNE-NSE switchover. The late stages of neurogenesis are well defined in rhesus monkey visual cortex and involve the migration of post-mitotic neurons generated from the subventricular zone across large expanses of cerebral wall to their final destination in the superficial cortical layers (116, 118, 119). In contrast to cerebellum, the migration distance is longer and results in the arrival of the post-mitotic neuron into a much more immature neural structure (118).

Immunocytochemistry for NNE and NSE in developing visual cortex graphically displays the switch from NNE to NSE (Figure 8). The cerebral wall in rhesus monkey at the 100th embryonic day (E100; gestation is 165 days) consists of proliferating zones (ventricular and subventricular), mature thalamocortical fiber tracts (optic radiations), prospective white

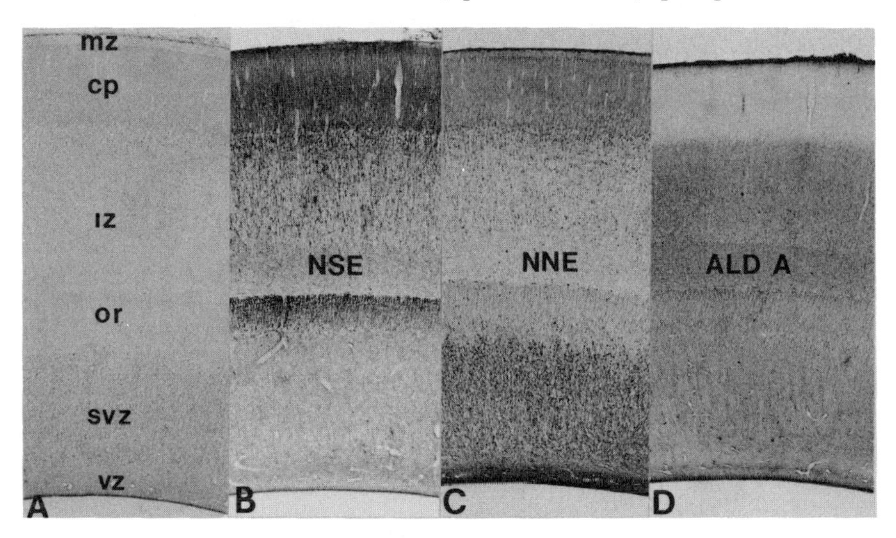

Figure 8. Coronal sections of cerebral wall from visual cortex in a rhesus monkey fetus (70 days after conception). The various layers of the cerebral wall are indicated on a control section (A): molecular zone (*mz*), cortical plate (*cp*), intermediate zone (*iz*), optic radiations (*or*), subventricular zone (*svz*) and ventricular zone (*vz*). NSE staining layers include the optic radiations, scattered cells in the intermediate zone, and a lightly stained cortical plate (B). In contrast, NNE immunoreactivity is most prominent in the two innermost zones of cell proliferation (C), and is different from aldolase subunit A (D). Control 1:400; anti-NSE 1:1000; anti-NNE 1:400; anti-aldolase A 1:400. Magnification ×36.

matter and deep cortex (intermediate zone), and developing cortical plate and overlying molecular zone (Figure 8A). Even at low power, NSE immunoreactivity is most evident in the thalamocortical afferents and outer cerebral wall (Figure 8B) in contrast with the marked NNE immunoreactivity in the proliferating cell populations of the inner cerebral wall (Figure 8C). This pattern is markedly different from the localization of other glycolytic enzymes such as aldolase (Figure 8D; ref. 120).

Unlike cerebellum, the subventricular zone in cortex generates both neurons and glial cells (118) and is clearly heterogeneous since a subpopulation of dividing cells expresses the astrocytic marker glial fibrillary acidic protein (GFAP) (87). In terms of enolase content, all dividing cells are presumed to be NNE(+) and devoid of NSE content (68). Those that are NNE(+) and GFAP(+) are presumably astrocytic precursors such as radial glial cells and destined to transform into astrocytes (87, 121). Neurons destined for superficial layers of visual cortex must therefore derive from a subclass of the population of NNE(+), GFAP(−) subventricular zone cells (68, 87). Some NNE(+), GFAP(−) cells may give rise to oligodendroglia although the exact lineage is as yet unresolved. Another class of cells is NSE(+) and intercalates through the proliferating cell zones to contact the ventricular surface (see Section 8.5). GFAP therefore defines the two cell populations GFAP(+) and GFAP(−) and NSE/NNE, and also the two cell populations NSE(−),NNE(+) and NSE(+),NNE(−) (68, 87). Since neurons are unlikely to derive from GFAP(+) dividing cells, there are a minimum of two dividing cell populations (astrocytic precursors, neuronal precursors, and possibly oliogodendroglial precursors) and two or more relatively static populations [latent astrocytes and radial glial cells and the NSE(+) ventricle-contacting neuronal population] (68, 87, 121, 122).

Farther out in the cerebral wall, the distribution of enolase isoenzymes resembles adult tissue. Neurons are NSE(+),NNE(−) although only faintly NSE(+) in the superficial, newly formed layers of the cortical plate (Figure 9A). Glial cells are uniformly NSE(−),NNE(+) (Figure 9C). The important finding, however, is that NNE immunoreactive profiles in the intermediate zone include not only astrocytes and glial cells but also migrating neurons (Figure 9B). These young post-mitotic neurons are NSE(−),NNE(+) and are clearly recognizable as a distinct class (68, 119, 123). Their proportion in the cerebral wall at various developmental stages is correlated with the final migration phase of neurogenesis (116, 117). Since the most superficially located (youngest) neurons of the cortical plate are faintly NSE(+) and NNE(−), this suggests that the NNE-to-NSE switch takes place *coincident* with the arrival of the migrating neurons to their final location. This is crucial because migrating granule cells in

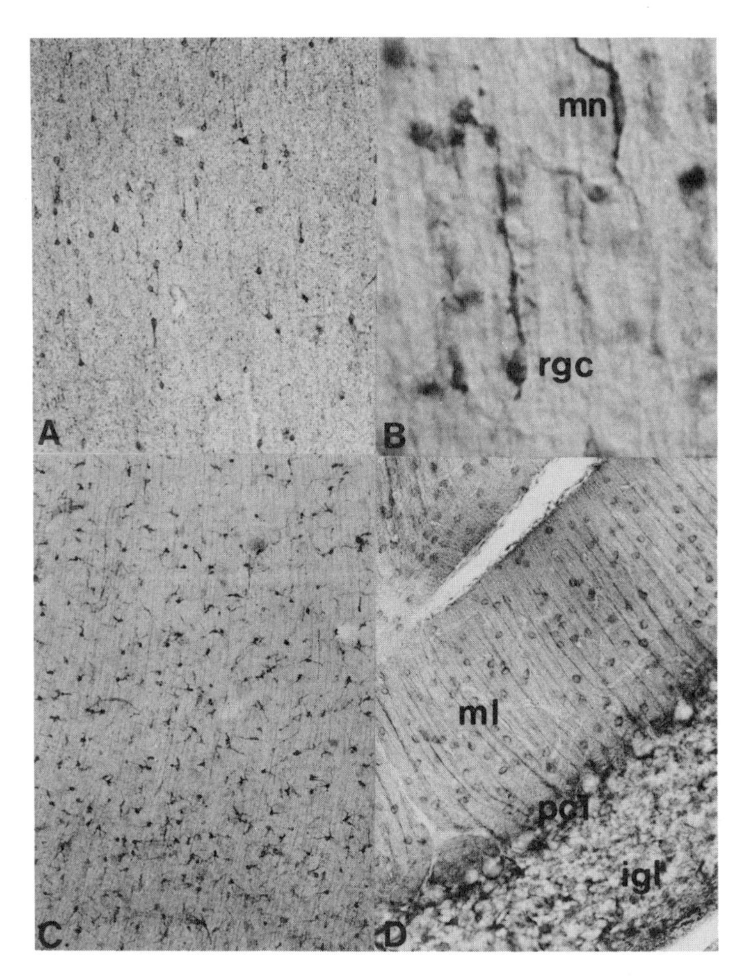

Figure 9. (A–C) Intermediate zone of cerebral wall from rhesus monkey fetus 100 days after conception. NSE(+) cells are inverted and normal pyramidal cells (A), whereas NNE immunoreactivity includes numerous astrocytes, displaced radial glial cells, and radial glial fibers (C). Radially oriented cells in the intermediate cell include numerous examples of radial glial cells (*rgc*) transforming into astrocytes and the distinctive profiles of migrating postmitotic neurons (*mn*); both are NNE immunoreactive (B). Anti-NSE 1:1000; anti-NNE 1:400. (A) and (C) magnification ×130; (B), ×800. (D) Stellate-basket cells of cerebellum do not complete the switch from NNE to NSE. Numerous NNE(+) stellate-basket cells are seen in molecular layer (*ml*) of adult rat cerebellum, whereas Purkinje cells in the Purkinje cell layer (*pcl*) and granule cells in the internal granule layer (*igl*) are unstained [NNE(−)]. Other NNE(+) elements include Bergmann glial fibers and fragments of astrocytic processes. Anti-NNE 1:1000. Magnification ×220.

cerebellum are often nonreactive to both NSE and NNE antisera. Since onset of NSE immunoreactivity only places an upper limit on the actual time when gene expression of α ceases and γ is expressed, the observation of NNE(+) migrating neurons far out in the cerebral wall defines this time-point more precisely. In cortex, therefore, NNE content persists throughout migration, and evidence of the NNE-to-NSE switch is present only on arrival at the cortical plate.

8.4 Neurons That Fail to Complete the NNE-to-NSE Switch

In adult tissue, the only neurons that appear to contain both NNE and NSE are stellate-basket cells of cerebellum (Figure 8D). This finding has held up after numerous attempts at "correction" for both rat and monkey tissues (including the use of multiple experimental animals and different bleeds of anti-NNE-R and anti-NNE-H). No other such examples have been detected elsewhere in the nervous system, specifically in the hippocampus, which has a similar developmental history (118); olfactory bulb has not been systematically examined.

This limited class of local circuit neurons therefore contains both α and γ subunits even in the adult and may contain both NNE and NSE and potentially hybrid enolase (αγ). Stellate-basket cells do have an unusual latent period during development (124), but the reasons for incomplete NNE-to-NSE switchover remain unclear. The only other examples of cells known to contain both α and γ subunits are some Cajal-Retzius cells (see Section 9), some neuroendocrine cells such as chromaffin cells of adrenal medulla, and platelets (see Sections 6 and 7.3).

Stellate-basket cells of cerebellum are the exception to the strict localization of NSE in neurons and NNE in glial cells. Other neurons lose detectable NNE content before arriving at their final location. In hippocampus, granule cell precursors along the inner aspect of the dentate gyrus are NNE(−) after final cell division, like their analogues in cerebellum. At present, it appears that (i) dividing populations giving rise to neurons are NSE(−), NNE(+); (ii) the NNE-to-NSE switch occurs late in differentiation (see Sections 8.3 and 9); (iii) the only dividing cells that may contain NSE are neuroendocrine tumors and certain transformed neuronal or neuroendocrine cell lines; (iv) the NNE-to-NSE switch is completed for most neurons. In addition, the analysis of NSE content in developing and mature brain suggest that the NNE-to-NSE switch may be dissociated in time from an increase in NSE content for selected neuronal classes (see Section 9).

8.5 Relation to Other Markers of Neuronal Differentiation

The switch from non-neuronal to neuron-specific enolase during neuronal differentiation is a late event and may prove useful in defining the level of differentiation of neurons both in intact animals and in tissue culture. The unique characteristics of NSE as a marker for the development of neurons are the ability to detect and measure both glial and neuronal isoenzymes and to correlate these levels to neuronal energy metabolism. Other markers that occur late in neuronal differentiation include GM1 ganglioside (125), which has a profile in developing cerebellum that is similar to NSE. The current explosion in the characterization of neuronal antigens (126) will undoubtedly provide more markers, including, perhaps, some of the other brain-specific isoenzymes of the glycolytic pathway.

9 NSE LEVELS AS A POTENTIAL MARKER FOR FUNCTIONAL ACTIVITY

The following section shows that (i) NSE levels are correlated with synaptic activity; (ii) NSE levels in brain continue to increase well after neurogenesis; (iii) NSE levels vary markedly among different classes of neurons in both mature and developing brain; and (iv) NSE levels decrease with dedifferentiation. In addition, the proposal that NSE and other "soluble" glycolytic enzymes are transported in a complex by slow axonal transport (45) further suggests a close correlation of NSE levels with neuronal activity.

9.1 Close Relation of NSE Levels to Synaptic Activity

The initial expression of NSE in developing neurons of cerebellum and cortex is after their migration from zones of cell proliferation to their final destination and the establishment of synaptic connections. The studies of NSE onset (68, 78) did not define further the actual onset of NSE immunoreactivity with synaptogenesis or actual synaptic activity. In the chick vestibular and auditory system, study of the cochlear ganglion and brainstem has permitted a closer correlation of NSE and functional activity. Cochlear ganglion cells are NSE(−) during innervation and axonal outgrowth, become NSE(+) first in the region of the cell body and peripheral process during ingrowth and early synaptogenesis of the central process, and finally demonstrate an NSE(+) central process during the later stages of synaptogenesis (82). Examination of other areas in the brainstem, including the vestibular nuclei, shows marked variation in NSE; vestibular axons become NSE(+) at a point corresponding to early synaptogenesis of their endings in the tangential nucleus (82). NSE immunoreactivity has also been studied in chick spinal cord and ganglia at the level of the hindlimb during early embryogenesis (83). The earliest NSE(+) structure is the glycogen body (see Section 7.3). For both spinal cord motor neurons and dorsal root ganglia cells, NSE immunoreactivity occurred late in differentiation, 1–2 days after onset of functional activity (e.g., spontaneous motor activity and intact reflex arc). The onset of NSE immunoreactivity did not coincide with first synapse formation and did not correlate with cell birthday (83). The appearance of NSE(+) neurons in cultured avian neural crest occurred at roughly the same time as synaptic activity and neurotransmitter synthesis (83). This suggests that NSE is correlated with functional synaptic activity.

After the expression of γ enolase subunit, the further increase of NSE content in neurons is dependent on synaptic activity. When superior cervical ganglia are decentralized, their NSE content does not change further when compared with the normal unoperated side (127). Changes in NSE levels must be correlated with actual localization by immunocytochemistry because many alterations of synaptic activity or functional input lead to cell death and altered growth during early development. Enucleation in chicken produces marked atrophy of the contralateral optic tectum with selective changes in cell layers and number (Figure 10). In rat and monkey, interrupting the hypoglossal nerve leads to a retrograde reaction in the hypoglossal motor neurons and a gradual recovery with reinnervation. NSE levels in hypoglossal motor neurons fall with the retrograde reaction and gradually recover with reinnervation (128). If reinnervation is prevented by implantation of the severed nerve in another nondenervated muscle, hypoglossal neurons do not regain their previous NSE content (128). This type of experiment suggests that NSE levels in specific neurons can decrease without actual cell death and that recovery is dependent on the reestablishment of functioning synaptic connections.

9.2 Unique Patterns of NSE Staining in Developing Cortex

At present, two reasonable hypotheses about NSE in developing nervous tissue are (i) NSE is present only in post-mitotic neurons (neuroendocrine cells), and (ii) the appearance of NSE is correlated to the onset of functional synaptic activity (secretory activity) (see Sections 8.3 and 9.1). In addition, the marked increase in NSE levels after neurogenesis suggests that not only the appearance of NSE (γ expression) but also the levels of NSE

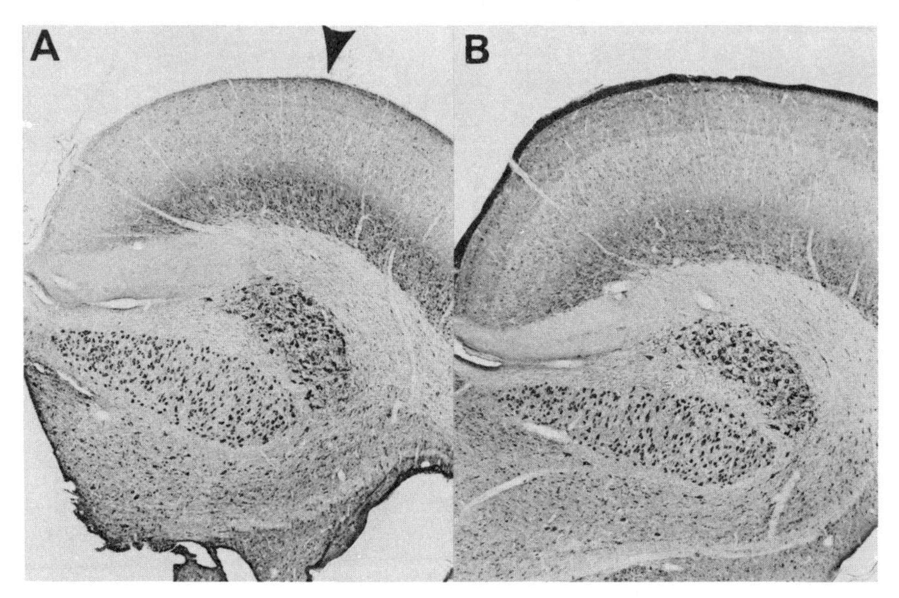

Figure 10. NSE immunoreactivity in chick optic tectum at 3 weeks of age. In chicks with one eye removed at 3 days of age, the contralateral optic lobe shows reduced size, absence of NSE(+) retinotectal fibers (arrowhead) and alteration of NSE(+) cells in tectal layers (A) compared with the normally innervated side (B). Anti-NSE 1:1000. Magnification ×36. ×36.

may be related to functional activity of given neuronal classes. It is not surprising, then, that developing visual cortex shows unique patterns of NSE staining.

At 100 days of embryonic development, developing visual cortex in rhesus monkey shows prominent bands of NSE staining in the thalamocortical afferents (optic radiations) and inner aspect of the molecular zone (Figure 11A). The cortical plate is only faintly NSE-immunoreactive, with the darkest cells being in the deeper (older) portion of the cortical plate. In the intermediate zone, from optic radiations outward to the cortical plate, numerous strongly NSE-immunoreactive pyramidal neurons are seen (Figure 11B). Many, but not all, are inverted pyramids or Martinotti cells (Figures 11B and 12).

Other strongly NSE-immunoreactive cells include Cajal-Retzius cells seen in the molecular zone (Figure 13A) and, of course, neurons in the thalamic relay nuclei (Figure 13D). Some examples of apparent Cajal-Retzius cells showing NNE immunoreactivity are detected as well (Figure 13B); it is not possible to say whether individual Cajal-Retzius cells contain both NNE and NSE, but this is likely. Along with the stellate-basket cells of the cerebellum, Cajal-Retzius cells represent a small class

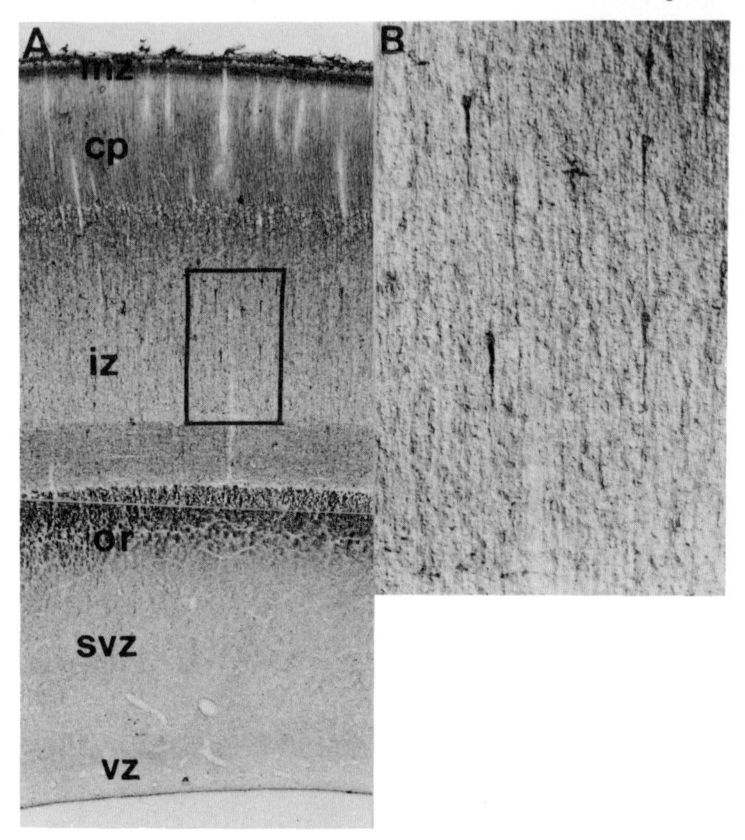

Figure 11. NSE immunoreactivity in developing visual cortex in rhesus monkey fetus 70 days after conception shows prominent staining in the inner aspect of the molecular zone (*mz*), the optic radiations (*or*), and scattered cells in the intermediate zone (*iz*). The proliferative zones (*svz* and *vz*, see Figure 8) are relatively faint, as is the cortical plate (*cp*), especially in its outermost aspect (A). Anti-NSE 1:1000. Magnification ×72. Examination of the intermediate zone shows that NSE(+) cells are in fact pyramidal neurons, many of which are inverted pyramids or Martinotti cells. (B) shows four such cells from the area of the rectangle in the intermediate zone. Anti-NSE 1:1000. Magnification ×200.

of neurons, with differentiated morphology and function, that are strongly immunoreactive for both NNE and NSE.

One of the most striking features of the cerebral wall in developing visual cortex is a multitude of strongly NSE immunoreactive fibers contacting the ventricular surface (Figure 14). They intercalate between ventricular zone cells and from end-feet at the ventricular surface. Their parent cell bodies are not seen in the vicinity of the ventricular surface, in contrast with the diffuse and numerous population of NNE(+) cells surrounding them (Figure 14D). At present, it is not possible to be definitive about the location of parent cell bodies, the cell class (proliferating or post-mitotic), or the ultimate disposition of these NSE(+) fibers. One possibility is that they are the apical processes of the numerous inverted pyramids in the cerebral wall; the intensity of their NSE staining and the number of inverted pyramids throughout the cerebral wall would be in apparent accord. As for proliferative status, there are no examples in nervous tissue of NSE immunoreactivity in a proliferating cell population. The eventual fate of these cells is unknown, but it should be noted that NSE(+) ventricle-contacting cells are present in adult spinal cord and third ventricle but are much rarer in the lateral ventricles.

The above pattern of NSE staining in developing cortex correlates with the location of early afferents and synaptic development (118, 129–131). The location and nature of intensely NSE-immunoreactive neurons (inverted pyramids and Cajal-Retzius cells) also corresponds with a proposed ontogenetic organization of early cortex in a bilayer system (129). NSE immunocytochemistry in developing nervous tissue therefore provides a potential marker for functional activity, revealing patterns of organization and cell classes (e.g., CSF-contacting cells) not visualized by ordinary methods. Other markers, such as aldolase isoenzymes, may differ markedly in some respects from NSE (Figure 8D) but support other findings (Figure 14A) (120).

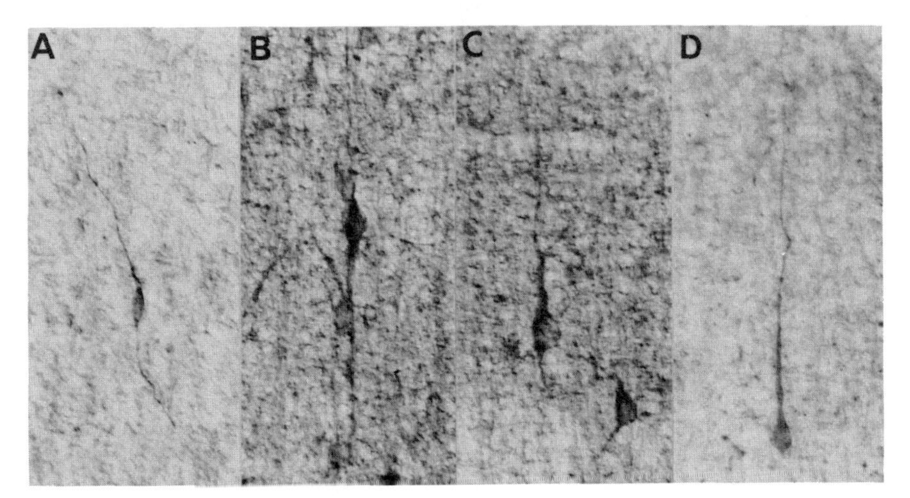

Figure 12. Neurons with strong NSE immunoreactivity in monkey visual cortex 70 days after conception include examples of bipolar cells (A), a numerous population of inverted pyramids (B), regular pyramidal neurons (C), and pyramidal neurons (D), which are immunoreactive for glutamic acid decarboxylase in preliminary experiments. Anti-NSE 1:1000. Magnification ×800.

9.3 Variation of NSE Content in Mature Nervous System

From a historical standpoint, the effort to discover whether NSE was contained in all neurons and whether NSE levels in peripheral tissues had a specific cellular localization led to methods that result in heavy and uniform staining of all neurons and neuroendocrine cells (41, 56,

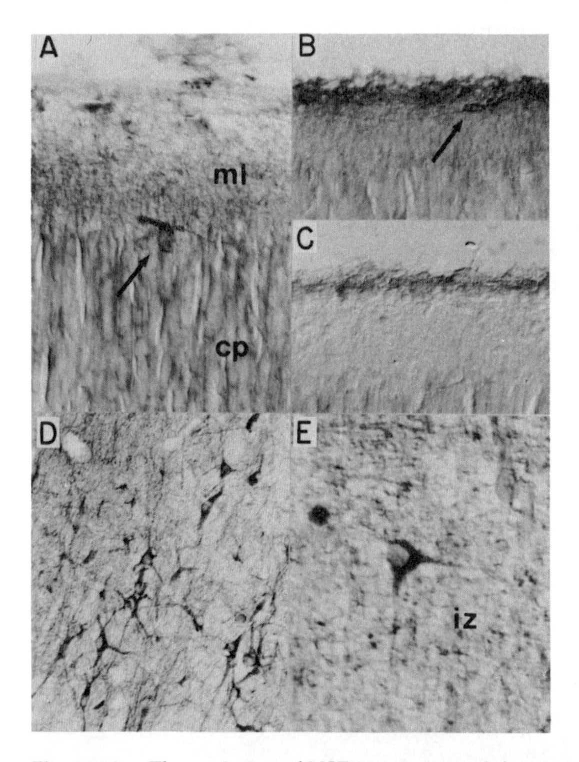

Figure 13. The variation of NSE immunoreactivity occurs in very stereotyped fashion in these sample micrographs from fetal monkey brain at 70 days after conception. Newly arrived neurons of the cortical plate (*cp*) are faintly NSE immunoreactive, having recently completed the switch from NNE to NSE; in contrast, horizontal neurons of the molecular layer (*ml*), which are consistent in morphology with Cajal-Retzius cells (arrow), are strongly NSE(+). Fibers in the inner part of the molecular layer are more reactive than superficial fibers. Like stellate and basket cells of cerebellum, Cajal-Retzius cells are strongly NSE(+) but are also immunoreactive for NNE. In (B), a horizontal Cajal-Retzius cell is NNE(+) (arrow), whereas sections incubated with control sera (C) show no staining cells; staining along the pial border is presumed nonspecific. In areas of the nervous system where neurons are further differentiated, such as thalamic relay nuclei, neurons are strongly NSE(+) at the same age (D). They compare in staining intensity to the Cajal-Retzius cells of the molecular layer and the large neurons of the intermediate zone (*iz*) of the cerebral wall (E). Anti-NSE 1:1000; anti-NNE 1:400. A, B, C, and E magnification ×500; D, ×220. Nomarski interference optics.

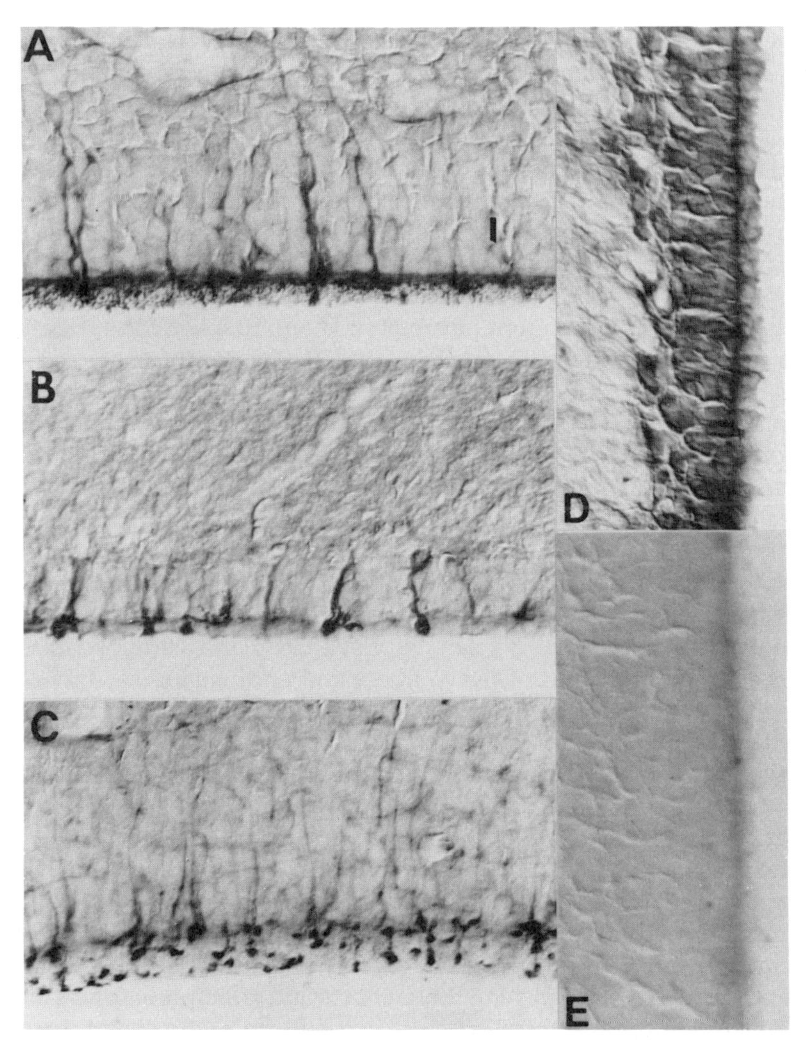

Figure 14. Closer inspection of the inner proliferative zones in developing cerebral wall shows that NNE and NSE distinguish different populations. Most ventricular zone cells are NNE immunoreactive (D). In sections stained for NSE content, numerous immunoreactive processes traverse the proliferative zone to contact the ventricular surface (B and C) with end-feet; they extend outwards in the cerebral wall and may belong to neurons of the intermediate zone or another presumably differentiated neuron/neuroendocrine cell class (see text). This cell class is also distinguished by its content of aldolase isoenzyme subunit C (A). Control sections (E) show no staining. Anti-NSE 1:1000; anti-NNE 1:400; anti-aldolase C 1:400; control sera 1:400. Magnification ×500. Nomarski interference optics.

58). The marked variation in NSE levels in adult nervous tissue (see Tables 2 and 3) suggested, however, that neurons might well vary in NSE content. When NSE immunocytochemistry is carried out by serial dilutions of antisera from 1:4000 to 1:64,000, qualitative differences are easily detected between various classes of neurons. Much like immunotitration for enzyme activity, higher dilutions eventually result in very faint or no staining for some neuronal classes while others persist in strong immunoreactivity. In adult visual cortex, the most strongly NSE immunoreactive neurons include deep pyramidal cells, particularly Meynert cells, as well as prominent neurons in layers III and IVA. The later cells can be seen to alternate in cycles of 400–500 μm in visual cortex (Figure 15) and correspond to the location of cytochrome oxidase-rich periodic staining (132, 133). NSE-rich neurons apparently lie in the upper layers of visual cortex in patches that correspond to mitochondria-rich nerve endings (133), GABAergic endings (133), patchy distribution of geniculate afferents (134), and physiologically distinctive columns (non-orientation-specific) (135). NSE-rich neurons can be conveniently identified at the light microscopic level, but comparison of nerve terminal NSE content has been more difficult. In developing cerebellum, a gradient of NSE content in parallel fibers can be visualized, and certainly gross differences in NSE content of nerve fibers and endings are seen in developing cortex (Figure 11A) and deafferented optic tectum (Figure 10). When functional deprivation of visual cortex (unilateral total ocular patching) is attempted in rhesus monkey, only scattered patchiness of geniculocortical afferents is visible in layer IV after one month (unpublished observations).

At the present time, NSE has potential usefulness as a cellular marker for functional activity in the adult animal. Since other factors may influence NSE immunoreactivity (fixation of various neuronal classes, amount of structural protein, cell size, local circuit vs. projection neurons), work is currently directed at examining these effects and attempting functional alterations in neuronal activity. The distinct pattern of the visual cortex (Figure 15) and the correspondence of NSE content during development (see Sections 9.1 and 9.2) and the retrograde cell reaction in hypoglossal nucleus (128) to synaptic activity are encouraging.

9.4 NSE and Differentiation

Since the actual biochemical control of γ expression important to the initial appearance of NSE and the maintenance of subsequent levels is unknown, we are limited at present to saying that NSE is strongly correlated with the differentiated state of neurons and related cells. Some

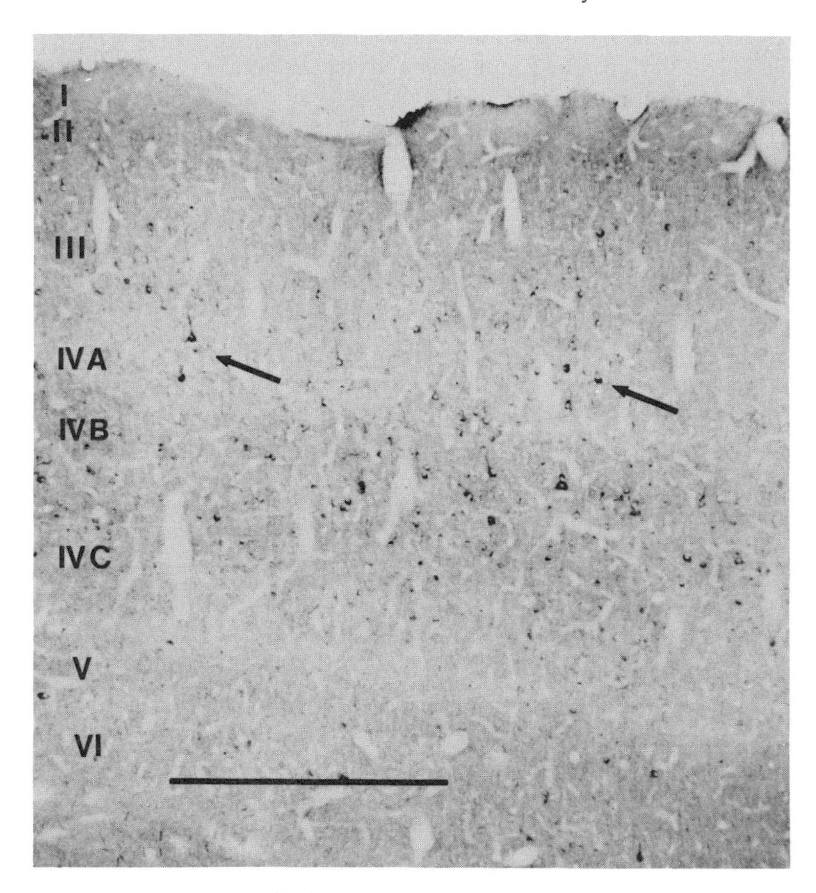

Figure 15. Adult visual cortex in rhesus monkey shows variation in NSE content when antisera dilutions are titrated over a wide range. Clusters of NSE(+) neurons are seen above layer IVB (arrows) at a spacing of approximately 400–500 μm (bar is 0.5 mm); many neurons are unstained at this dilution of antiserum. Anti-NSE 1:32,000. Magnification ×125.

of the most direct evidence has come from cell cultures (see Sections 7.1 and 7.2). NSE levels increase with pharmacological induction of morphological differentiation (100, 101) and neuronal maturation in aggregating cultures (64) and decrease with viral transformation of neurons in culture (75) and possibly neuronal aging (64). High NSE levels are associated with the expression of synaptic activity (58) and the cessation of cell division (transformed cells) (112). Low NSE levels are associated with transformed cells (Table 4, see Section 7.2) and a lack of synaptic activity (58). All of these experiments are in accord with NSE content in intact animals (see Sections 9.1 to 9.3); NSE is therefore a practical marker for neuronal differentiation and functional activity. Several groups are now involved with cell-free synthesis of the neuron-specific γ subunit of enolase (136–138), and such efforts may eventually supply the missing information on how NSE levels are regulated. Tissue culture may well be the easiest testing ground for such methods.

9.5 Relationship of NSE to Slow Axonal Transport

NSE is present in large amounts in nervous tissue (1–3% total soluble protein). If all enzymes of the glycolytic pathway were present in similar amounts, the total pathway would represent 1/3 of total brain-soluble protein. The abundance of NSE makes it readily detectable by high-resolution two-dimensional polyacrylamide gel electrophoresis of soluble brain extracts (139). Two of eight brain-specific proteins visualized in such profiles are, in fact, glycolytic enzymes: NSE and aldolase isoenzyme C_4 (139). The ease of analyzing brain homogenates for NSE has made it possible to examine the axonal transport of NSE (45). Brady and Lasek report that NSE is carried in a complex of proteins known as slow-component *b* (SCb) at a rate of 2 mm/day (45). They suggest that NSE is part of a larger enzyme complex of intermediary metabolism which also includes structural proteins such as clathrin and actin, and they review the evidence for the association of "soluble" glycolytic enzymes with structural proteins (45).

One concept of NSE is that this enzyme, compared with other enolase isoenzymes, is particularly suited to neuronal cytoplasm (39) in its increased resistance to chloride and thermal dissociation. NSE may be adapted to high intracellular chloride found in nerve cells during depolarization. The only classes of vertebrates to contain appreciable amounts of NSE are birds and mammals (37), both of which maintain high body temperatures. The association of NSE with slow axonal transport raises the possibility that isoenzyme specialization in neurons may be related to specific structural properties of NSE that permit interaction with cytoskeletal proteins or other transported enzymes. Similar considerations may account for the distribution of other glycolytic enzymes (120).

Energy metabolism in the distal axon has been proposed as a common site of attack in polyneuropathies (140). Such symmetrical central-peripheral distal axonopathies might reflect the effect of a variety of toxins and metabolic derangements on glycolytic metabolism in the axon (140). Whether by direct inhibition of enzyme activity or interference with transport, such agents would prevent the delivery of sufficient energy metabolism capacity to the distal axon even though the cell body was functioning normally. In a particular instance, acrylamide neuropathy has been proposed to involve specific inhibition of NSE (141, 142). NSE may not only represent an isoenzyme adapted for axonal transport but it may also be vulnerable to specific influences that lead to central and peripheral axonal dysfunction.

10 PRESENT STATUS OF NSE AS A SPECIFIC MARKER

Neuron-specific enolase (NSE, $\gamma\gamma$, 14-3-2) is a specific cellular marker for neurons and neuroendocrine cells (Table 5). This final section places NSE in the context of other neuronal markers and current methods in neurobiology.

10.1 What Is the Meaning of NSE Content for Traditional Concepts of the Neuron?

Neurons have distinct morphological characteristics revealed by silver staining, cell stains such as Nissl appearance, and ultrastructural appearance. In addition, electrophysiological recording offers a definitive way of distinguishing neurons from other cells as well as separating subclasses of neurons. The more recent use of immunocytochemistry for specific substances and neurotransmitters has resulted in an explosion of information relating to various subclasses of neurons. In contrast, NSE content is a basic property of neurons which serves to unify them as a cell class and provide a reliable criterion for "neuronalness" even in cell cultures, tumors, or other alterations of normal tissue. The relationship between neuroendocrine cells and neurons typified by the presence of peripheral neuropeptides in central nervous tissue is underlined by the fact that NSE is a marker for both neurons and neuroendocrine cells. NSE content is a useful, generalized marker for both cell classes, therefore, and it also suggests that the energy metabolism of neurons and of neurendocrine cells have common characteristics distinct from other cell types. The presence of NSE in cells may well reflect a differentiated function that is independent of cell lineage. For example, megakaryocytes and platelets derive from non-neural-tube cell populations but contain NSE and show some neurosecretory properties.

10.2 NSE Compared with Other Neuronal Markers

A number of other biochemical markers for neurons exist, and the use of monoclonal antibody techniques will undoubtedly add others (126).

Table 5. NSE-Containing Cells (Levels in ng NSE/ng Protein)[a]

A. Neurons (CNS: 4000–20,000; PNS: 500–1000)

Projection neurons
Local circuit neurons
Stellate-basket cells of cerebellum[b]
CSF-contacting cells (third ventricle, central canal)
Dorsal root ganglia cells
Autonomic neurons

B. Central neuroendocrine cells (pituitary, pineal: 3000–10,000)

Pineal cells
Pituitary gland cells (anterior and intermediate lobes)
Area postrema cells
Subcommissural organ cells
Subfornicial organ cells
Organ vasculosum lamina terminalis cells
Supraoptic nucleus cells
Suprachiasmatic nucleus cells

C. Peripheral neuroendocrine cells (adrenal medulla: 500)

Thyroid calcitonin cells
Thymic reticular epithelial cells
Pancreatic islet cells
Adrenal medullary chromaffin cells[b]
Gut neuroendocrine cells
Lung neuroendocrine cells (K-cells)
Merkel cells (specialized skin receptor cells)

D. Transitional cell populations

Cajal-Retzius cells (developing cortex, layer 1)[b]
CSF-contacting cells (lateral ventricles)

E. Other cells

Megakaryocytes (platelets)
Glycogen body cells (chick embryo)

F. Transformed cells (500–5000)

Neuroblastoma
Pheochromocytoma
Neuroendocrine tumors in human (APUDomas)
Visna-infected neurons (culture)
Giant cells, subependymomas (human brain; tuberous sclerosis)

[a] Levels in parentheses are ng NSE/mg protein of tissue or cells representative of that group (see Tables 2–4).
[b] Cells are immunoreactive for both NSE and NNE.

These include tetanus toxin binding (143), surface antigens specific to peripheral (144) and central neurons (145), other glycolytic enzymes such as aldolase (120, 146), neurofilament proteins (147), GM1 ganglioside (125), and GQ ganglioside (97). Other markers include specific neurotransmitter substances and their associated enzymes, which define subclasses of neurons and neuroendocrine cells. Calcium-binding proteins also seem to define limited subclasses of neurons (148).

Like tetanus toxin binding and GQ ganglioside, NSE serves as a generalized marker for neurons and neuroendocrine cells. Unlike these two markers, NSE is an intracellular label, making it better for tissue sections but not suitable for cell sorting methods. NSE appears late in neuronal differentiation, like GM1 ganglioside (125) but unlike neurofilament proteins, which are present in dividing cell populations during normal development (147). A major difference with all the above markers is that NSE represents a specific neuronally located isoenzyme of enolase and can therefore be compared with the other brain isoenzymes NNE and hybrid enolase. The developmental switch from NNE to NSE can thus be explicitly studied as a case of gene expression and may yield valuable insights into neuronal differentiation.

10.3 NSE vs. Hybrid Enolase

Throughout this chapter, NSE has been used to denote both NSE ($\gamma\gamma$) and the neuron-specific enolase isoenzyme subunit γ. It is our present belief that hybrid enolase ($\alpha\gamma$) does *not* represent a significant fraction of total enolase isoenzymes in *intact adult* nervous tissue. The occurrence of hybrid enolase in disrupted brain tissue may reflect interconversion of isoenzymes during processing (39). Immunocytochemistry of enolase isoenzymes therefore plays a vital role in determining cellular localization. At present, the only neurons that might possibly contain hybrid enolase are stellate-basket cells of cerebellum. With antisera to both rat and human NNE used in rat and monkey tissue, no other neurons are seen with detectable NNE (α subunit) immunoreactivity, even in cells with low NSE content such as dorsal root ganglia neurons. Given the NNE-to-NSE switch during neuronal differentiation and the lone example of stellate-basket cells having both NSE and NNE, it is possible that neurons contain small amounts of NNE (α subunit). The lack of its detection by immunocytochemistry (see Section 5.2) suggests that α subunit does not represent 25% of total enolase content in neurons, as recently proposed (53). Similar questions will also complicate the localization of other isoenzymes, such as aldolase (tetrameric with two subunits in brain and five isoenzymes). Independent evidence of intraneuronal localization such as axonal transport may well be needed to supplement quantitation and immunocytochemical localization.

10.4 Potentially Unique Role of NSE as Marker of Functional Activity

NSE obviously serves as a useful specific cellular marker for neurons and neuroendocrine cells. In addition, many studies have sought to correlate NSE content with functional activity of neurons given the participation of this enzyme in glycolysis. The use of 2-deoxyglucose for anatomic studies of energy metabolism represents a successful model (88) for combining anatomical studies with function. Another similar approach involves the use of cytochrome oxidase, a mitochondrial enzyme, as a marker for functional activity (132). Variation in patterns, found even in normal tissue (133), correlate with other organizational principles (134, 135). Like cytochrome oxidase, the potential uses of NSE content would be relevant to long-term alterations in functional activity of neurons, not the short-term paradigms usually employed in 2-deoxyglucose studies. Since NSE is a neuron-specific marker, unlike cytochrome oxidase, and is easily visualized in cell bodies, unlike 2-deoxyglucose, the use of NSE for functional studies may offer distinct advantages over these other two methods.

10.5 Necessity of Anatomic Approach Combined with Biochemistry

One of the features of NSE is the opportunity it offers to conduct studies by both anatomical methods with immunocytochemistry and biochemical methods as with radioimmunoassay. The combined use of both methods in experimental design has been fruitful. The most practical discovery concerning NSE, that it occurs in cells of the diffuse neurendocrine system, is an example of the complementary strength of using both biochemical measurement and immunocytochemical localization. The most intriguing discovery concerning NSE, the late switchover from NNE to NSE, also depended on immunocytochemistry demonstrating the cellular basis for the biochemical evidence for a switch in isoenzymes.

10.6 Future Directions

Future directions in the use of neuron-specific enolase (NSE) will most likely be concerned with the actual mechanisms controlling NSE expression in the neuron and neuroendocrine cell. The close correlation of NSE content with functional activity in many experimental models holds out the possibility that NSE may be a unique cellular marker for the long-term activity of neurons. Of the nine observations and hypotheses presented in the introduction to this chapter, the last three concerning the variation of NSE in neuronal populations, the possibility that NSE expression is related to its association with axonal transport, and the possible involvement of NSE in axonopathies and neuronal dysfunction, are now under active investigation.

The use of NSE in cellular neurobiology at present extends from its practical use as a specific marker for neurons and neuroendocrine cells to current experiments examining the relationship of NSE to synaptic activity, development and organization of nervous tissue, and axonal transport. The results to date range from the use of NSE content for diagnosis of neurendocrine tumors to more specialized concerns such as the dramatic variations in NSE content seen during development. NSE may well serve as a model for investigation of axonal transport of glycolytic enzymes as well as for a study of more complicated isoenzymes sets such as aldolase. The practical uses of NSE and the theoretical interest in its relation to neuronal function are far from exhausted; they bear tribute to the exciting discovery of NSE as "14-3-2," one of the first brain-specific proteins.

REFERENCES

1. B. W. Moore and D. McGregor, *J. Biol. Chem.* **240**, 1647 (1965).
2. B. W. Moore and V. J. Perez, "Specific Acidic Proteins of the Nervous System," in F. D. Carlson, Ed., *Physiological and Biochemical Aspects of Nervous Integration*, Prentice-Hall, Englewood Cliffs, N.J., 1966, pp. 343–359.
3. B. W. Moore, "Acidic Proteins," in A. Lajtha, Ed., *Handbook of Neurochemistry*, Vol. 1, Plenum Press, New York, 1969, pp. 93–99.
4. B. W. Moore, Chemistry and Biology of Two Proteins, S-100 and 14-3-2, Specific to the Nervous System," in *International Review of Neurobiology*, Vol. 15, Academic Press, New York, 1972, pp. 215–225.
5. B. W. Moore, "Brain-Specific Proteins," in D. J. Schneider, Ed., *Proteins of the Nervous System*, Raven Press, New York, 1973, pp. 1–12.
6. B. W. Moore, "Brain-Specific Proteins: S-100 Proteins, 14-3-2 Protein, and Glial Fibrillary Protein," in B. W. Agranoff and M. H. Aprison, Eds., *Advances in Neurochemistry*, Plenum Press, New York, 1975, pp. 137–155.
7. C. Zomzely-Neurath and A. Keller, "The Different Forms of Brain Enolase: Isolation, Characterization, Cell Specificity and Physiological Significance," in S. Roberts, A. Lajtha, and W. H. Gispen, Eds., *Mechanisms, Regulations and Special Functions of Protein Synthesis in the Brain*, Vol. 2, Elsevier, Amsterdam, 1977, pp. 279–298.
8. C. Zomzely-Neurath and A. Keller, *Neurochem. Res.* **2**, 353 (1977).
9. E. Bock, *J. Neurochem.* **30**, 7 (1978).

10. P. J. Marangos, D. Schmechel, A. P. Zis, and F. K. Goodwin, *Biol. Psych.* **14**, 563 (1979).

11. P. J. Marangos and D. E. Schmechel, "The Neurobiology of the Brain Enolases," in M. B. H. Youdim, D. F. Sharman, W. Lovenberg and J. R. Lagnado, Eds., *Essays in Neurochemistry and Neuropharmacology*, vol. 4, John Wiley and Sons, New York, 1980, pp. 211–247.

12. C. E. Zomzely-Neurath and W. A. Walker, "Nervous System-Specific Proteins: 14-3-2 Protein, Neuron-Specific Enolase, and S-100 Protein," in R. A. Bradshaw and V. M. Schneider, Eds., *Proteins of the Nervous System*, Raven Press, New York, 1980, pp. 1–57.

13. T. J. Cicero, W. M. Cowan, and B. W. Moore, *Brain Res.* **24**, 1 (1970).

14. T. J. Cicero, J. A. Ferrendelli, V. Suntzeff, and B. W. Moore, *J. Neurochem.* **19**, 2119 (1972).

15. T. J. Cicero and R. R. Provine, *Brain Res.* **44**, 294 (1972).

16. V. J. Perez, J. W. Olney, T. J. Cicero, B. W. Moore, and B. A. Bahn, *J. Neurochem. 17*, 511 (1970).

17. P. Marangos, C. Zomzely-Neurath, C. York, and S. C. Bondy, *Biochem. Biophys. Acta (Amsterdam)* **392**, 75 (1975).

18. T. J. Cicero, W. M. Cowan, B. W. Moore and V. Suntzeff, *Brain Res.* **28**, 25 (1970).

19. G. S. Bennett and G. M. Edelman, *J. Biol. Chem.* **243**, 6234 (1968).

20. G. S. Bennett, *Brain Res.* **68**, 365 (1974).

21. P. J. Marangos, C. Zomzely-Neurath, and C. York, *Arch. Biochem. Biophys.* **170**, 289 (1975).

22. P. J. Marangos, C. Zomzely-Neurath, D. C. M. Luk, and C. York, *J. Biol. Chem.* **250**, 1884 (1975).

23. P. J. Marangos, C. Zomzely-Neurath, and F. K. Goodwin, *J. Neurochem.* **28**, 1097 (1977).

24. V. M. Pickel, D. J. Reis, P. J. Marangos, and C. Zomzely-Neurath, *Brain Res.* **105**, 184 (1976).

25. C. C. Rider and C. B. Taylor, *Biochem. Biophys. Acta (Amsterdam)* **365**, 285 (1974).

26. C. C. Rider and C. B. Taylor, *Biochem. Biophys. Res. Commun.* **66**, 814 (1975).

27. L. Fletcher, C. C. Rider, and C. B. Taylor, *Biochim. Biophys. Acta* **452**, 242 (1976).

28. J. M. Pearce, Y. H. Edwards, and H. Harris, *Am. J. Hum. Genet.* **39**, 263 (1976).

29. Shi-Han Chen and E. R. Giblett, *Am. J. Hum. Genet.* **39**, 277 (1976).

30. V. A. McKusick, *Am. J. Med.* **69**, 267 (1980).

31. E. Bock and J. Dissing, *Scand. J. Immunol.* **4** (Suppl. 2), 31 (1975).

32. E. Bock, L. Fletcher, C. C. Rider, and C. B. Taylor, *Neurochemistry* **30**, 181 (1978).

33. A. Grasso, G. Roda, R. A. Hogue-Angeletti, B. W. Moore, and V. J. Perez, *Brain Res.* **124**, 497 (1977).

34. L. Persson, L. Rönnback, A. Grasso, K. G. Haglid, H. A. Hansson, L. Dolonius, S. O. Molin and H. Nygren, *Neurol. Sci.* **35**, 381 (1978).

35. F. Suzuki, Y. Umeda, and K. Kato, *J. Biochem.* **87**, 1587 (1980).

36., D. A. Hillin, K. Brown, P. A. M. Kynoch, C. Smith, and R. J. Thompson, *Biochim. Biophys. Acta* **628**, 98 (1980).

37. R. L. Clark-Rosenberg and P. J. Marangos, *J. Neurochem.* **35**, 756 (1980).

38. P. J. Marangos, A. P. Zis, R. L. Clark, and F. K. Goodwin, *Brain Res.* **150**, 117 (1978).
39. P. J. Marangos, A. M. Parma, and F. K. Goodwin, *J. Neurochem.* **31**, 727 (1978).
40. A. Grasso, K. G. Haglid, H. A. Hansson, L. Persson, and L. Rönnback, *Brain Res.* **122**, 582 (1977).
41. D. E. Schmechel, P. J. Marangos, A. P. Zis, M. Brightman, and F. K. Goodwin, *Science* **199**, 313 (1978).
42. O. R. Langley, M. S. Ghandour, G. Vincendon, and G. Gombos, *J. Neurocytol.* **9**, 783 (1980).
43. M. S. Ghandour, O. R. Langley, G. Gombos, G. Vincendon, "Cellular Localization of 14-3-2 Protein and Carbonic Anhydrase in Rat Cerebellum in the Optical and Electron Microscope," in R. Balazs, C. DiBenedetta, G. Gombos, and R. Porcellati, Eds., *Multidisciplinary Approach to Brain Development*, Elsevier, Amsterdam, 1980, pp. 35–36.
44. M. S. Ghandour, O. R. Langley, and A. Keller, *Exp. Brain Res.* **41**, 271 (1981).
45. S. T. Brady, and R. J. Lasek, *Cell* **23**, 515 (1981).
46. P. J. Marangos, and C. Zomzely-Neurath, *Biochem. Biophys. Res. Comm.* **68**, 1309 (1976).
47. P. J. Marangos, C. Zomzely-Neurath, and C. York, *Arch. Biochem.* **170**, 289 (1975).
48. R. Revoltella, L. Bertolini, L. Diamond, E. Vigneti, and A. Grasso, *J. Neurochem.* **26**, 831 (1976).
49. P. J. Marangos, D. Schmechel, A. M. Parma, R. L. Clark, and F. K. Goodwin, *J. Neurochem.* **33**, 319 (1979).
50. R. W. Brown, A. Ky, and R. J. Thompson, *Chem. Acta* **101**, 257 (1980).
51. A. M. Parma, P. J. Marangos, and F. K. Goodwin, *J. Neurochem.* **36**, 1093 (1981).
52. K. Kato, F. Suzuki, and Y. Umeda, *J. Neurochem.* **36**, 793 (1981).
53. K. Kato, F. Suzuki, and R. Semba, *J. Neurochem.* **37**, 998 (1981).
54. K. Kato, Y. Ishiguro, F. Suzuki, A. Ito, and R. Semba, *Brain Res.* **237**, 441 (1982)
55. L. A. Sternberger, *Immunocytochemistry*, Prentice-Hall, Englewood Cliffs, N.J., 1979.
56. D. E. Schmechel, P. J. Marangos, and M. W. Brightman, *Nature* (*London*) **276**, 834 (1979).
57. P. J. Stoward, *Fixation in Histochemistry*, Chapman and Hall, London, 1973.
58. D. E. Schmechel, M. W. Brightman, and J. L. Barker, *Brain Res.* **181**, 391 (1980).
59. B. K. Hartmann, M. Cimino, B. W. Moore, and H. C. Agrawal, *Trans. Am. Soc. Neurochem.* **8**, 66 (1977).
60. L. F. Eng and J. W. Bigsbee, *Adv. Neurochem.* **3**, 43 (1978).
61. M. Stefanini, C. De Martino and L. Zamboni, *Nature* (*London*) **216**, 173 (1967).
62. W. H. Oertel, E. Mugnaini, D. E. Schmechel, M. L. Tappaz, and I. J. Kopin, "The Immunocytochemical Demonstration of Gabaergic Neurons; Method and Application," in V. Chan-Palay and S. L. Palay, Eds., *Cytochemical Methods in Neuroanatomy*, Alan R. Liss, New York, 1982.
63. W. H. Oertel, D. E. Schmechel and E. Mugnaini, "Glutamic Acid Decarboxylase (GAD): Purification, Antiserum Production, Immunocytochemistry," in J. Barker and J. McKelvy, Eds., *Current Methods in Cellular Neurobiology*, vol. 1, John Wiley, New York, 1983.
64. B. D. Trapp, P. J. Marangos, and H. DeF. Webster, *Brain Res.* **220**, 121 (1981).

65. F. J. Tapia, J. M. Polak, A. J. A. Barbosa, S. R. Bloom, P. J. Marangos, C. Dermody, and A. G. E. Pearse, *Lancet* **1**, 808 (1981).
66. K. Stefansson and R. Wollman, *Acta Neuropathol.* **53**, 113 (1981).
67. J. D. Secchi, D. Lacaque, M. A. Cousin, D. Lando, L. Legault-Demare, and J. P. Raynaud, *Brain Res.* **184**, 455 (1980).
68. D. E. Schmechel, M. W. Brightman, and P. J. Marangos, *Brain Res.* **190**, 195 (1980).
69. C. L. Shengrund and P. J. Marangos, *J. Neurosci. Res.* **5**, 305 (1980).
70. A. G. Parfitt, J. E. Freschi, W. G. Shain, Jr., D. E. Schmechel, and D. A. Auerbach, "Pineal Cells: Culture and Co-Culture with Neurons from the Superior Cervical Ganglion," in D. C. Klein, Ed., *The Developmental Neurobiology of the Melatonin Rhythm Generating System*, Karger Press, Basel, pp. 62–83 (1983).
71. J. Gu, J. M. Polak, F. J. Tapia, P. J. Marangos, and A. G. E. Pearse, *Amer. J. Pathol.*, **104**, 63 (1981).
72. M. Hyden and L. Rönnback, *J. Neurol. Sci.* **39**, 157 (1978).
73. L. Rönnback, L. Persson, H. A. Hansson, K. G. Haglid, and A. Grasso, *Experientia* **33**, 1094 (1977).
74. M. Dubois-Dalcq, H. McFarland, and D. McFarlin, *J. Histochem. Cytochem.* **25**, 1201 (1977).
75. E. Hooghe-Peters, M. Dubois-Dalcq, and D. Schmechel, *Lab. Invest.* **41**, 247 (1979).
76. D. R. Zehr, *J. Histochem. Cytochem.* **26**, 415 (1981).
77. M. Schachner, E. T. Hedley-Whyte, D. W. Hsu, G. Schoonmaker, and A. Bignami, *J. Cell Biol.* **75**, 67 (1977).
78. P. J. Marangos, D. E. Schmechel, A. M. Parma, and F. K. Goodwin, *Brain Res.* **190**, 185 (1980).
79. V. M. Pickel, T. H. Joh, and D. J. Reis, *Proc. Natl. Acad. Sci. U.S.A.* **72**, 659 (1975).
80. V. M. Pickel, T. H. Joh, and D. J. Reis, *J. Histochem. Cytochem.* **24**, 792 (1976).
81. V. Chan-Palay, J. Y. Wu, and S. L. Palay, *Proc. Natl. Acad. Sci. U.S.A.* **76**, 2067 (1979).
82. G. D. Maxwell, M. C. Whitehead, S. M. Connolly, and P. J. Marangos, *Dev. Brain Res.*, **3**, 401 (1982).
83. M. C. Whitehead, P. J. Marangos, S. M. Connolly, and D. K. Morest, *Dev. Neurosci.*, **5**, 298 (1982).
84. M. W. Brightman, J. J. Anders, D. Schmechel, and J. M. Rosenstein, "The Lability of the Shape and Content of Glial Cells," in E. Schoffeniels, G. Frank, D. B. Tower, and L. Hertz, Eds., *Dynamic Properties of Glial Cells*, Pergamon Press, Oxford, 1978, pp. 21–44.
85. P. J. Marangos, D. E. Schmechel, and W. H. Oertel, "Neuron-Specific Enolase (NSE): A Specific Marker for the Diffuse Endocrine System," in S. R. Bloom and G. M. Polak, Eds., *Gut Hormones*, pp. 101–106.
86. D. S. Snyder, C. S. Raine, M. Farooq, and W. T. Norton, *J. Neurochem.* **34**, 1614 (1980).
87. P. Levitt, M. Cooper, and P. Rakic, *J. Neurosci.* **1**, 27 (1981).
88. L. Sokoloff, *The Relationship Between Function and Energy Metabolism: Its Use in the Localization of Functional Activity in the Nervous System*, Neurosciences Research Program Bulletin, Vol. 8, p. 19 (1980).
89. K. Haglid, C. A. Carlsson, and D. Stavrou, *Acta Neuropathol.* (*Berlin*) **24**, 187 (1973).

90. L. Fletcher, C. C. Rider, C. B. Taylor, E. O. Adamson, B. M. Luke, and C. F. Graham, *Dev. Biol.* **65**, 462 (1978).

91. H. R. Herschman and M. P. Lerner, *Nature New Biol.* **241**, 242 (1973).

92. O. C. McKenna and J. Rosenbluth, *J. Comp. Neurol.* **154**, 133 (1974).

93. A. G. E. Pearse, *J. Histochem. Cytochem.* **17**, 303 (1969).

94. P. Facer, J. M. Polak, P. J. Marangos, and A. G. E. Pearse, *Proc. R. Microsc. Soc.* **15**, 111 (1980).

95. G. A. Cole, J. M. Polak, J. Wharton, P. Marangos, and A. G. E. Pearse, *J. Pathol.* **132**, 351 (1980).

96. L. Grimelius and F. Wilander, *Invest. Cell Pathol.* **3**, 3 (1980).

97. B. F. Haynes, R. W. Warren, R. M. Buckley, J. E. McClure, A. L. Goldstein, and G. C. Eisenbarth, *Clin. Res. Abstr.*, in press (1982).

98. A. G. E. Pearse, *Chromatfin, Enterochromatfin and Related Cells*, Elsevier, Amsterdam, 1976.

99. A. S. Tischler, M. A. Dichter, B. Biales, and L. A. Greene, *New Engl. J. Med.* **296**, 919 (1977).

100. P. J. Marangos, F. K. Goodwin, A. M. Parman, C. Lauter, and E. Trams, *Brain Res.* **145**, 49 (1978).

101. S. A. Vinores, P. J. Marangos, A. M. Parma, and G. Guroff, *J. Neurochem.* **37**, 597 (1981).

102. D. N. Carney, P. J. Marangos, D. C.Ihde, P. A. Bunn, Jr., M. H. Cohen, J. D. Minna, and A. F. Gazdar, *Lancet* **1**, 583 (1982).

103. G. D. Fischbach and M. A. Dichter, *Dev. Biol.* **37**, 100 (1974).

104. B. R. Ransom, C. Christian, P. N. Bullock, and P. G. Nelson, *J. Neurophysiol.* **40**, 1151 (1977).

105. E. Bock, O. S. Jorgenson, L. Dittmann, and L. F. Eng, *J. Neurochem.* **25**, 867 (1975).

106. E. Bock and A. Hamberger, *Brain Res.* **112**, 329 (1976).

107. E. Bock, A. Yavin, O. S. Jorgensen, and E. Yavin, *J. Neurochem.* **35**, 1297 (1980).

108. G. K. Bergey, S. C. Fitzgerald, B. K. Schrier, and P. G. Nelson, *Brain Res.* **207**, 49 (1981).

109. G. Augusti-Tocco, L. Casola, and A. Grasso, *Cell Differ.* **2**, 157 (1973).

110. F. A. McMorris, A. R. Kolberf, B. W. Moore, and A. S. Perumal, *J. Cell Physiol.* **84**, 473 (1974).

111. M. Braun, A. Grasso, and W. Wechsler, *Exp. Brain Res.* **25**, 93 (1976).

112. L. Legault-Demare, Y. Zeitoun, D. Lando, N. Lamande, A. Grasso, and F. Gros, *Exp. Cell Res.* **125**, 233 (1980).

113. P. J. Marangos, A. F. Gazdar, and D. N. Carney, *Cancer Lett.* **15**, 67 (1982).

114. P. J. Marangos, I. C. Campbell, D. E. Schmechel, D. L. Murphy, and F. K. Goodwin, *J. Neurochem.* **34**, 1254 (1980).

115. A. Pletscher, "Platelets As Models for Monoaminergic Neurons," in M. B. H. Youdim, W. Lovenberg, D. F. Sharman, and J. R. Lagnado, Eds., *Essays in Neurochemistry*, vol. 3, John Wiley and Sons, New York, 1978, pp. 49–101.

116. P. Rakic, "Timing of Major Ontogenetic Events in the Visual Cortex of the Rhesus Monkey," in N. A. Buchwald and M. Brazier, Eds., *Brain Mechanisms in Mental Retardation*, Academic Press, New York, 1975, pp. 3–40.

117. P. Rakic, *J. Comp. Neurol.* **141**, 283 (1971).
118. M. Jacobson, *Developmental Neurobiology*, Plenum Press, New York, 1978.
119. P. Rakic, *Science* **183**, 524 (1974).
120. D. E. Schmechel, *Neurology*, **32**, A153, 1982.
121. D. E. Schmechel, and P. Rakic, *Anat. Embryol.* **156**, 115 (1979).
122. D. E. Schmechel and P. Rakic, *Nature* (*London*) **277**, 303 (1979).
123. P. Rakic, *Neuroscience* **4**, 184 (1981).
124. P. Rakic, *J. Comp. Neurol.* **147**, 61 (1973).
125. M. Willinger and M. Schachner, *Dev. Biol.* **74**, 101 (1980).
126. R. McKay, Ed., *Monoclonal Antibodies to Neural Antigens*, Cold Spring Harbor Symposium, Cold Spring Harbor, N.Y., 1981.
127. A. Grasso, and R. Pirazzi, *Brain Res.* **90**, 324 (1975).
128. I. Kirino, W. Oertel, M. W. Brightman, and P. Marangos, *J. Neuropathol. Exp. Neurol.* **40**, 338 (1981).
129. M. Marin-Padilla, *Anat. Embryol.* **152**, 109 (1978).
130. M. E. Molliver, I. J. Kostovic, and H. van der Loos, *Brain Res.* **50**, 403 (1973).
131. M. Rickmann, B. M. Chronwall, and J. R. Wolff, *Anat. Embryol.* **151**, 285 (1977).
132. M. Wong-Riley, *Brain Res.* **171**, 11 (1979).
133. A. E. Hendrickson, S. P. Hunt, and J.-Y. Wu, *Nature* (*London*) **292**, 605 (1981).
134. D. Fitzpatrick, Thesis work, Ph.D. dissertation, Department of Psychology, Duke University, 1982.
135. D. H. Hubel and M. S. Livingstone, *Soc. Neurosci. Abstr.* **7**, 367 (1981).
136. A. Marks, M. M. Portier, Y. Zeituan, L. Legault-Demore, J. Thisault, N. Lamande, C. Jeanlet, and F. Gros, *Biochimie* **62**, 463 (1980).
137. K. Sakimura, Y. Hoshida, Y. Nabeshine, and Y. Takahaohi, *J. Neurochem.* **34**, 687 (1980).
138. W. A. Walker and C. Zomzely-Neurath, *Neurochem. Res.*, **5**, 361 (1982).
139. P. Jackson and R. J. Thompson, *J. Neurol. Sci.* **49**, 429 (1981).
140. P. S. Spencer, M. I. Sabri, H. H. Schaumburg, and C. L. Moore, *Ann. Neurol.* **5**, 501 (1979).
141. R. D. Howland, I. L., Vyas, and H. E. Lowndes, *Brain Res.* **190**, 529 (1980).
142. R. D. Howland, *Toxicol. Appl. Pharmacol.* **60**, 324 (1981).
143. R. Mirsky, L. H. B. Wendon, P. Black, C. Stolkin, and C. Bray, *Brain Res.* **148**, 251 (1978).
144. T. Vulliamy, S. Rattray, and R. Mirsky, *Nature (London)* **291**, 418 (1981).
145. J. Cohen and S. Y. Selvendran, *Nature* (*London*) **291**, 421 (1981).
146. R. J. Thompson, P. A. M. Kynoch, and V. J. C. Willson, *Brain Res.* **232**, 489 (1982).
147. S. J. Tapscott, G. S. Bennett, and H. Holtzer, *Nature* (*London*) **292**, 836 (1981).
148. M. R. Celio and C. W. Heizmann, *Nature (London)* **293**, 300 (1981).

Chapter 2

Glutamic Acid Decarboxylase (GAD): Purification, Antiserum Production, Immunocytochemistry

Wolfgang H. Oertel and Donald E. Schmechel

Laboratory of Clinical Science
National Institute of Mental Health
Bethesda, Maryland

Enrico Mugnaini

Laboratory of Neuromorphology
Department of Biobehavioral Sciences
University of Connecticut
Storrs, Connecticut

Dr. Oertel's present address is Neurologische Klinik, (c/o Prof. Dr. A. Struppler), Klinikum Rechts der Isar, Technische Universität München, D-8000 Munich 80, Federal Republic of Germany.

Dr. Schmechel's present address is Division of Neurology, Department of Medicine, Duke University Medical Center, Durham, North Carolina, 27710.

1 INTRODUCTION

Gamma-aminobutyric acid (GABA) is a major inhibitory transmitter of the vertebrate central nervous system. Immunocytochemistry utilizing specific antisera to L-glutamic acid decarboxylase (GAD; E.C. 4.1.1.15), the biosynthetic enzyme for GABA, provides an easy means of identifying GABAergic neurons at the light and electron microscopic level. Furthermore, the ability to identify GABAergic neurons in histological material is crucial to investigations of the relationship of GABAergic neurons to other classes of neurons.

This chapter describes (i) methods for the purification and characterization of GAD, (ii) experiments supporting the identity of GAD and brain cysteine sulfinate decarboxylase form II (CSD II), (iii) the production and characterization of specific antisera against GAD, and (iv) the use of these antisera for the immunocytochemical localization of GABAergic neurons in the mammalian nervous system.

2 HISTORY

The presence of GABA in brain tissue was discovered in 1950 (1–3). In one of the reports, Roberts and Frankel (2) proposed that GABA in brain is synthesized via enzymatic decarboxylation of glutamic acid by glutamic acid decarboxylase (GAD). In 1954, Hope (4) suggested GAD might also decarboxylate cysteine sulfinate (Section 4.2) and therefore be involved in hypotaurine and taurine synthesis.

The importance of GABA in nervous system inhibition (5) encouraged further study of its biosynthetic enzyme GAD. The marked enrichment of GAD in brain and specifically in synaptosomal fractions (6) suggested that nerve cells could synthesize GABA. GAD was therefore potentially a specific marker for a proposed class of GABAergic neurons. In 1973, Wu, Matsuda, and Roberts (7) reported the purification and characterization of GAD from mouse brain. The subsequent preparation of a specific antiserum to mouse brain GAD (8) allowed visualization of GAD-containing (presumed GABAergic) neurons at the cellular (9) and subcellular (10) level in rat cerebellum. No non-neuronal elements were stained by antiserum to GAD.

To date, the production of three other antisera to GAD and their use for immunocytochemical studies has been reported (11–13). GAD immunocytochemistry is now recognized as a standard method for anatomical demonstration of GABAergic neurons (14, 15).

3 PURIFICATION

3.1 Enzyme Assay

Glutamic acid decarboxylase activity is easily assayed by the radiometric $^{14}CO_2$-trapping (16, 17) method, using carboxy-labeled glutamate as substrate. A more specific method is to measure the rate of GABA formation from uniformly labeled radioactive glutamate by separation on column chromatography (18) or by radioreceptor assay (19).

For monitoring the purification of neuronal GAD, the radiometric CO_2-trapping method is reliable and rapidly performed for multiple assays. Nonspecific decarboxylation can be decreased by using detergents for brain homogenates and does not pose a problem in later stages of purification. However, in peripheral tissue or brain areas with very low GAD levels, the rate of CO_2 release from nonspecific decarboxylation may represent a significant portion of the total CO_2 release. Specific GAD activity is then best detected by measuring GABA formation by means of column chromatographic techniques (for discussion, see refs. 20 and 21) or radioreceptor assay (19). Radioimmunoassay methods are also described for quantitating actual levels of the neuron-specific GAD protein (11) but do not directly measure enzymatic activity.

3.2 Non-Neuronal GAD

The presence of GAD enzyme activity in non-neuronal tissue (22, 23), as measured by the $^{14}CO_2$ method (16, 17), now appears due to impurities in the labeled substrate (24). Using more reliable techniques, including mass spectrometry, low GAD activity is found, for example, in astrocytes in primary cultures (25), human skin fibroblasts (19), and arachnoid vessels. In the latter tissue immunocytochemical techniques so far fail to demonstrate GAD immunoreactive structures (26).

Wu (27) has reported on the purification of non-neuronal GAD from bovine heart with biochemical and immunological properties different from those of neuronal GAD. However, purification to homogeneity, production of a specific antiserum, and immunocytochemistry will be necessary to prove the existence of a distinct non-neuronal GAD.

If such a peripheral, non-neuronal GAD exists, its level in brain must be extremely low. For example, in the pituitary gland GAD activity is confined to the neurointermediate lobe where GAD immunoreactive processes (from the hypothalamus) terminate (28). Stalk transection drastically reduces GAD activity of the neurointermediate lobe and results in the total loss of GAD immunoreactivity. The residual GAD activity (potentially coming from nonneuronal GAD, but also most likely representing nonspecific decarboxylation) amounts to less than 1% of GAD activity in whole brain homogenate (28). Furthermore, reported GAD enzyme levels in white matter (29) probably reflect at least in part the presence of GAD-containing neuronal fibers, varicosities, and even cell bodies. For example, in rat corpus callosum, ^{3}H-GABA-accumulating neurons (30) and GAD immunoreactive neurons are observed (Wolff, Chronwall, Schmechel, and Oertel, in preparation).

3.3 Methods for Purification

GAD has been purified from mouse brain (7, 8, 12, 31), rat brain (11, 13), bovine brain (32, 33), human brain (34), and catfish brain (35) (Table 1 and Figure 1). However, it is now clear that two forms of GAD exist in brain, a low-molecular-weight form and a high-molecular-weight form (31, 36). When rat brain is homogenized in hypotonic solution without subcellular fractionation, both forms can be separated. With hypotonic release from synaptosomal preparations, the low-molecular-weight form of GAD is preferentially recovered. This suggests that high-molecular-weight GAD is localized in neuronal cell bodies. It is not isolated by the customary method for GAD purification. Currently available antisera to GAD are directed against low-molecular-weight GAD. This may in part explain the high immunoreactivity of axonal endings compared with cell bodies (Section 7.41).

The initial steps of the different purification schemes—homogenization, centrifugation, ammonium sulfate fractionation, molecular sieving, and anion exchange chromatography—are highly reproducible procedures. They yield a partially purified GAD preparation (about 150 times higher

Table 1. Schemes for the Purification of GAD

Mouse Brain (7)	Mouse or Bovine Brain (31)	Catfish Brain (35)
Homogenate (0.25 *M* sucrose)	Homogenate	Homogenate (hypotonic)
Crude extract (supernatant of hypotonically lysed "synaptosomal preparation")	(0.25 *M* sucrose or hypotonic)	Supernatant
$(NH_4)_2SO_4$ (27–62%)	$(NH_4)_2SO_4$	$(NH_4)_2SO_4$ (40–75%)
Sephadex G-200	Sephadex G-200	Ultrogel
$(NH_4)_2SO_4$ (30–68%)	Hydroxyapatite	$(NH_4)_2SO_4$ (40–75%)
Hydroxyapatite	DEAE-Sephadex	Calcium phosphate
$(NH_4)_2SO_4$ (33–70%)	Isoelectric focusing or polyacrylamide gel electrophoresis (nondenaturing conditions)	$(NH_4)_2SO_4$ (40–75%)
DEAE-Sephadex		Ultrogel
$(NH_4)_2SO_4$ (0–75%)		$(NH_4)_2SO_4$ (0–75%)
Sephadex G-200		Polyacrylamide gel electrophoresis (nondenaturing conditions)
$(NH_4)_2SO_4$ (0–75%)		Gel extraction

Mouse Brain (12)	Rat (11) or Human Brain (34)	Rat Brain (13)
Homogenate (hypotonic)	Homogenate (hypotonic)	See Figure 1
Supernatant	Supernatant	
$(NH_4)_2SO_4$ (1–55%)	DEAE-cellulose	
Sepharose-4B-diaminobutyrate-pyridoxal-5′-phosphate	Hydroxyapatite	
	Phenyl-Sepharose	
	QAE-Sepharose	
	Ultrogel	

PRODUCTION OF ANTISERUM S3

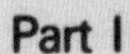

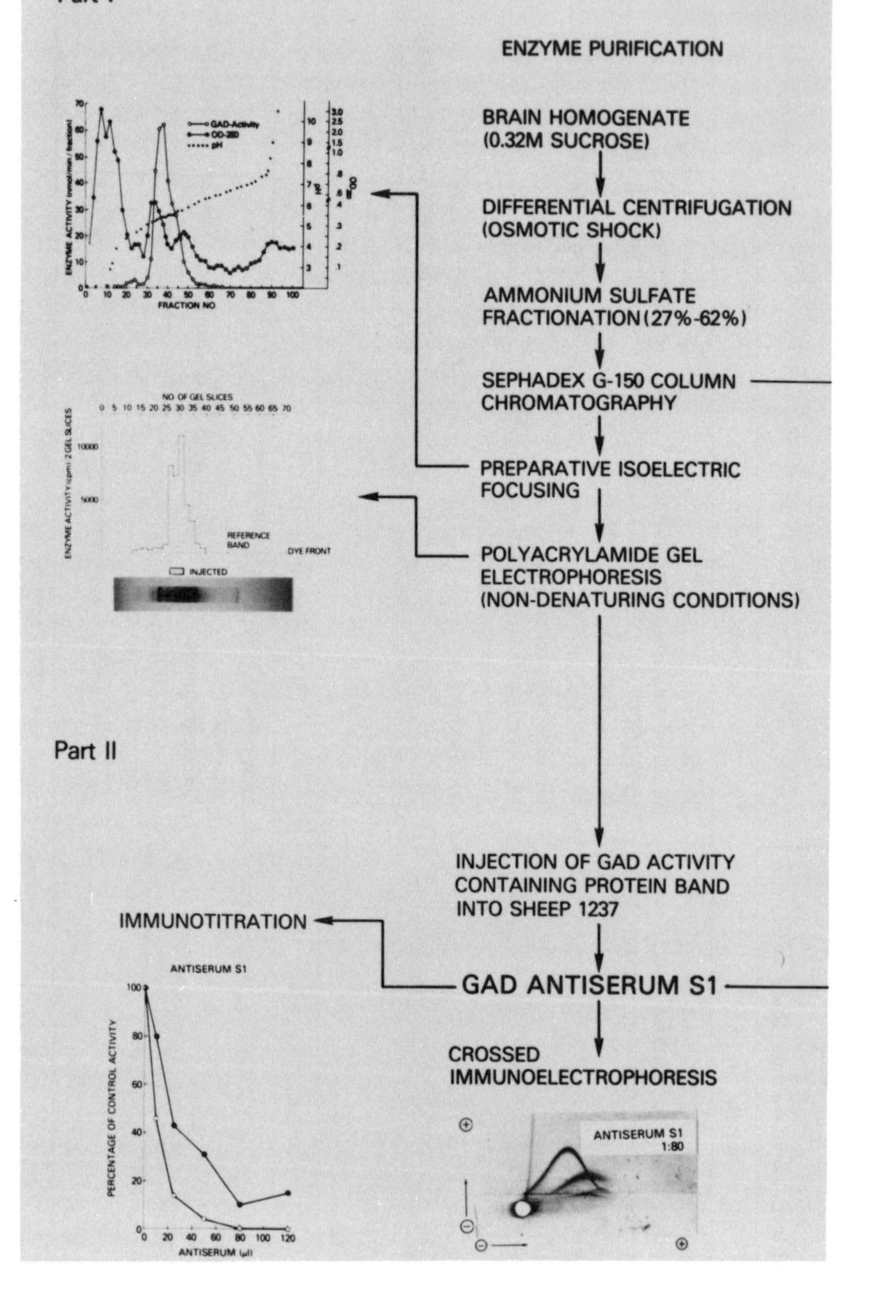

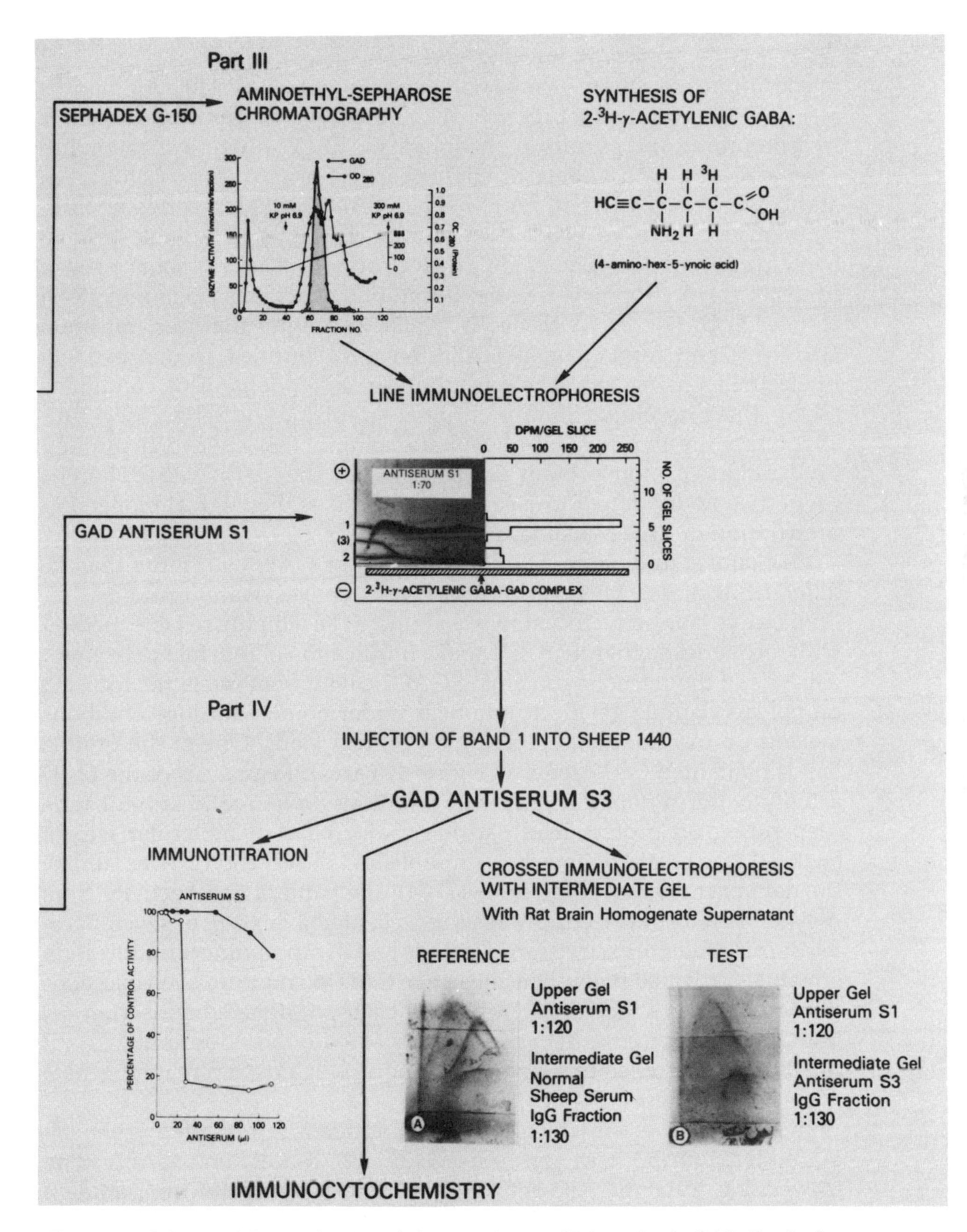

Figure 1. Scheme of the production of sheep antiserum S3 to rat brain GAD. For further explanation, see Section 3.3.

in specific activity than sucrose homogenate), which is stable for months at −20°C to −70°C in the presence of aminoethylisothiouronium bromide hydrobromide and pyridoxal phosphate (7, 13). Further purification has been achieved by a variety of strategies: The original approach by Wu and coworkers (7) carried mouse brain GAD through a series of chromatographic steps to yield a preparation running as a single protein band in routine polyacrylamide gel electrophoresis and as four to seven bands in gel electrophoresis under denaturing conditions (sodium dodecyl sulfate, SDS) (7, 37). With newly available column matrices, rat brain GAD (11) and human brain GAD (34) were purified to one band in isoelectric focusing and gel electrophoresis under denaturing conditions (SDS). More recently, affinity chromatography on a Sepharose-4B-diaminobutyrate-pyridoxal-5′-phosphate column (38) yielded a purified preparation of mouse brain GAD (12) with the same electrophoretic properties under denaturing conditions (SDS) as the original pioneering preparation of Wu, Matsuda, and Roberts (7).

Our efforts to develop affinity columns for GAD with more than 20 different ligands, however not the one above, were unsuccessful. Preparative isoelectric focusing was found to be a highly reliable method with recoveries higher than 85% and a purification of fourfold to fivefold. Analysis of the fractions containing GAD activity from isoelectric focusing on polyacrylamide gel electrophoresis under nondenaturing conditions yielded up to 7 bands (depending on protein load). One of the protein bands contained GAD activity (Figure 1, Part I). Reanalysis of the GAD activity band in denaturing (SDS) conditions gave five to seven bands (depending on protein load), with a major band at molecular weight 58,000 (Tappaz and Schmechel, unpublished observation). Thus, unlike the data of the Strasbourg team (11, 34), the band of GAD activity from polyacrylamide gel electrophoresis represented a mixture of several proteins. Injecting the excised GAD activity band with routine immunization schedules provided polyvalent antisera to GAD after two to seven injections in each of three sheep. The most potent one (designated antiserum S1) immunoprecipitated GAD activity but yielded at least four antigen–antibody arcs in crossed immunoelectrophoresis against brain homogenate supernatant (Figure 1, Part II).

In order to find out which of the four antigens represented GAD, ^{3}H-γ-acetylenic GABA was synthesized (39, 40). γ-acetylenic GABA is an irreversible catalytic site-activated inhibitor of GABA-transaminase (GABA-T), L-ornithine-2-oxoacidaminotransferase, and GAD (40, 41). Rat brain GAD was then partially purified up through aminoethyl-Sepharose column chromatography (Figure 1, Part III). There was no detectable GABA-T at this step of the purification (see subsection below; Sections 5.2, 7.23). When this partly purified GAD preparation was labeled with the irreversible radioligand and analyzed in line immunoelectrophoresis using polyvalent antiserum S1, only one antigen–antibody precipitin line contained radioactivity (Figure 1, Part III). This presumed "GAD–sheep–anti-GAD complex," identified by means of ^{3}H-γ-acetylenic GABA, was therefore used as the source antigen for the production of a new sheep antiserum. The immunopurified GAD preparation elicited an apparently specific antiserum to GAD, termed S3 (Sections 5.2 and 7.3).

The synthesis of radioactive ligand is not necessary for future preparations: neuronal GAD is not present in liver (21), and two of the four antigens, detected by the polyvalent "trapping" antiserum S1, were found in liver (42). Thus separate immunization against the two remaining brain-specific proteins (one of the two being GAD) would have been successful.

Separation of GAD from CSD I and GABA-transaminase

Purified homogeneous preparations of human, bovine, and rat brain GAD demonstrate both GAD and cysteine sulfinate decarboxylase (CSD) activity (Section 4.2). In partly purified preparations, CSD activity can be separated into two distinct peaks, of which the more acidic one, called CSD II, coincides with GAD activity. The less acidic CSD I isoenzyme can be easily separated from GAD (CSD II) with hydroxyapatite column chromatography (33), Sephacryl 200 chromatography (43), or with anion exchange chromatography such as aminoethyl-Sepharose matrix (42) or DEAE matrix with a shallow gradient of 150 m*M* to 300 m*M* phosphate buffer (pH 6.9) (unpublished observation). The possibility of contamination of the GAD preparation by GABA-T is unlikely for several reasons:

1. The initial steps of the subcellular fractionation (7, 8) promote the liberation of GAD from synaptosomes and the retention of most GABA-T in the mitochondrial fraction (8, 44).
2. Rat and mouse brain GABA-T and GAD do have similar molecular weights (8, 45), but the isoelectric point of GABA-T (pI = 6.8) (45) is quite different from that of GAD (pI = 5.4) (11, 13; see Section 4.1).
3. The combination of anion exchange column chromatography and isoelectric focusing probably separates any remaining GABA-T from GAD.
4. The single antigen–antibody line used for producing antiserum S3 reflects additional purification inherent in immunoelectrophoresis and is judged unlikely to contain both GAD and GABA-T (see Section 5.2).

4 CHARACTERIZATION OF GAD

4.1 Isoelectric Point and Molecular Weight

Low-molecular-weight rat brain GAD has an isoelectric point of 5.3 to 5.5 (11, 13) compared with 5.05 for human brain GAD (34).

The molecular weights of both native rat and human brain GAD have been reported to be 140,000 ± 15,000 by electrophoresis on polyacrylamide gradient slab gels (11, 34). The final purified preparation of rat and human brain GAD gave a single band in SDS-gel electrophoresis with a subunit molecular weight of about 67,000 (11, 34). Maitre, Blindermann, and coworkers (11, 34) therefore suggest that rat and human brain GAD are dimers composed of two identical subunits of approximately 60,000 to 70,000 daltons molecular weight. Although our GAD preparation was not purified to homogeneity (13), the catalytic site-bearing subunit, labeled with ^{3}H-γ-acetylenic GABA, was found to have a molecular weight of 58,000 ± 1500 and an isoelectric point of 5.3 to 5.6 (9 *M* urea) in two-dimensional electrophoresis (39). This may correspond to the major band seen during SDS-gel electrophoresis of the GAD activity band used for the production of the polyvalent "trapping" antiserum S1 (see Section 3.3). The apparent subunit size of our rat brain GAD preparation is therefore similar to the above values for rat and human brain GAD (11, 34).

In contrast, the molecular weight of native mouse brain GAD is 85,000 ± 2000, based on sedimentation equilibrium and gel filtration studies. The preparation is apparently homogeneous and consists of two identical subunits of molecular weight 44,000 ± 2000 as demonstrated by its behavior on high-speed sedimentation equilibrium with guanidine-HCl and β-mercaptoethanol (8).

When analyzed by SDS-gel electrophoresis, however, mouse brain GAD shows one major protein band with a molecular weight of 58,000 ± 2000, three minor protein bands with molecular weights of 118,000 ± 7000, 84,000 ± 4000, and 77,000 ± 4000, and five faint bands including one with a molecular weight of 16,000 ± 1000 (8, 37). Mouse brain GAD has been proposed to consist of from two to six (multiple of 15,000) identical subunits (8, 31). Perhaps, the one major band with a molecular weight of 58,000 ("four subunits") corresponds to the apparent subunit of rat and human brain GAD. Mouse brain GAD is reported to have a substrate specificity (27) (see Section 4.2) distinct from bovine, human, and rat brain GAD (33, 34, 42) (Table 2), and yet a close immunological identity with rat brain GAD (46). Further studies with antibody (preferably monoclonal) affinity chromatography or *in vitro* synthesis of

GAD may help to characterize the apparent variation in active-site and subunit composition in the different species.

4.2 Comparison of GAD and Cysteine Sulfinate Decarboxylase Isoenzymes

Hope (4) has suggested that GAD might be identical with cysteine sulfinate decarboxylase (CSD), a proposed biosynthetic enzyme for hypotaurine and the putative inhibitory transmitter taurine (47). The activity of highly purified mouse brain GAD, however, is unaltered by 10 m*M* cysteine sulfinate or cysteic acid, although mouse brain GAD can decarboxylate L-aspartate at 3 to 5% of the rate found with L-glutamate (7, 27). In contrast, highly purified human brain GAD (34), bovine brain GAD (33), and rat brain GAD (42) decarboxylate both cysteine sulfinate and cysteic acid. Total-brain CSD activity from bovine brain or calf brain (33, 43) rat brain GAD (42), and porcine brain GAD (42a) decarboxylate both cysteine sulfinate and cysteic acid. Total-brain CSD activity from bovine brain, calf brain (33, 43) and rat brain (42) can be separated into two isoenzymatic activities, CSD I and CSD II. The more acidic CSD II activity from rat brain coelutes with GAD activity, has a similar pH dependency, and is immunoprecipitated by antiserum S3 in like manner to GAD (42).

CSD I is extremely similar in its biochemical properties to liver CSD and has an about 50 to 100 times higher affinity for cysteine sulfinate than CSD II. Thus, CSD I is likely to be the enzyme responsible for taurine biosynthesis in brain. Neither CSD I nor liver CSD are detected by antiserum S3 to neuronal GAD (CSD II) (42; Table 2). In addition, CSD I is present in anterior and neurointermediate lobe of rat pituitary gland, whereas GAD (CSD II) is present in nerve terminals of the neurointermediate lobe. Pituitary stalk transection abolishes GAD (CSD II) activity in neurointermediate lobe (28) but leaves CSD I activity unchanged (48). These data suggest that CSD I in rat pituitary gland is not localized in (hypothalamic) nerve endings, but rather in neuroendocrine cells or non-neuronal elements. Specific antisera to CSD I will eventually define the precise localization of CSD I in brain and its relation to GAD (CSD II). The identity of liver CSD and brain CSD I also remains to be established.

5 ANTISERUM PRODUCTION

5.1 Antigenicity of GAD

Fifty to 900 μg of purified mouse brain GAD preparation have elicited GAD antiserum production in rabbits (8, 46). Injection of a total of 800

Table 2. Comparison of Properties of GAD and Cysteine Sulfinate Decarboxylase (CSD) Isoenzymes

	Species	GAD	CSD II	CSD I	Liver CSD
Coelution on column chromatography with GAD	rat (42)	+	+	–	–
	bovine (33, 43)	+	+	–	n.t.[a]
K_M (CSA)[b] (CA)[c]	mouse (7, 36)	0.7–2.0 m*M*	–	n.t.	n.t.
	rat (42)	1.5 m*M*	6 m*M* (CSA)	0.050 m*M*–0.100 m*M* (CSA)	0.045–0.050 m*M* (CSA)
	bovine (33)	1.6 m*M*	5.4 m*M* (CA)	0.22 m*M* (CA)	n.t.
	human (34)	1.25 m*M*	4.35 m*M* (CSA)	n.t.	n.t.
pH optimum	mouse (7)	7.0	n.t.	n.t.	n.t.
	rat (42)	6.9–7.1	7.4–7.5	7.4–7.8	7.4–7.8
	bovine (33)	6.8	7.5	7.4	n.t.
	human (34)	6.8	n.t.	n.t.	n.t.
Shape of pH dependency	rat (42)	–	like GAD	like liver CSD	–
Inhibition by high glutamate concentration (8–100 m*M*)	rat (48)	+	+	–	–
	bovine (33)	+	+	n.t.	n.t.
Inhibition by high CSA or CA concentration (10 m*M*)	mouse (25)	–	n.t.	n.t.	n.t.
Immunoprecipitation with GAD-antiserum	rat (42)	+	+	–	–
Inhibition by γ-acetylenic-GABA (tissue homogenate)	rat (42)	+	+	–[d]	–[d]
		brain	brain	brain	liver

[a] Not tested. [b] CSA = cysteine sulfinate. [c] CA = cysteic acid. [d] Partial inhibition in highly diluted tissue preparation (43).

μg of highly purified rat brain GAD preparation into rabbits produced a GAD antiserum that inhibited the pure enzyme by 70% (11). Perez de la Mora et al. (12) produced rabbit anti-mouse GAD sera with 2.5 mg and 5.5 mg of protein, which immunoprecipitated GAD out of crude brain extract by more than 75%. Potent polyvalent antisera were raised in sheep (13) and a single goat (Schmechel and Tappaz, unpublished observation) by six injections of approximately 30 μg protein each containing an unknown amount of GAD protein (13).

Two successive injections of antigen (GAD) antibody precipitin lines (see Section 3.3) at 3-week intervals into a nonimmunized sheep yielded an antiserum that immunoprecipitated GAD activity in brain homogenate supernatant by 70%. After a third and fourth injection at 3-week intervals, the antiserum increased in titer to 85% immunoprecipitation (13). Each injection represented a pool of 12 to 18 precipitin lines obtained in standard line immunoelectrophoresis and was estimated to contain about 15 μg total of pure GAD protein. Thus, an amount of about 30 μg GAD protein, covered with GAD-directed immunoglobulins from "trapping" antiserum S1, was sufficient to produce immunoprecipitating GAD antiserum S3.

In summary, GAD appears to be strongly immunogenic, as only small amounts of low-molecular-weight GAD are necessary to raise polyclonal antisera in rabbit, sheep, and goat.

The above protocol for antigen preparation (see Section 3.3; 49, 50) has rarely been used in neurobiology. Moreover, it is doubtful whether the above-outlined procedure will be employed for raising antibodies against other antigens of neurobiological interest, since successive immunizations are necessary. Efforts are currently directed to the production of monoclonal antibodies (51, 51a) to GAD. This method does not require a homogeneous antigen preparation, although the chances of success increase with the degree of purity. Reproducible methods for GAD purification are published in detail (7, 8, 11–13, 31, 34). Furthermore, new techniques such as protein separation with high-performance liquid chromatography or GAD–antibody affinity columns are also possible ways of preparing partially pure GAD for the purpose of monoclonal antibody production.

5.2 Immunological Characterization of GAD Antisera

The original antiserum to mouse brain GAD detected one antigen in a water extract of crude mitochondrial fraction, according to immunodiffusion and microcomplement fixation (8). The second antiserum to mouse brain GAD yielded a single precipitin line against crude brain extract in double immunodiffusion, in immunoelectrophoretic and counterimmunoelectrophoretic studies (12). Antiserum to purified rat brain GAD gave one precipitin line in immunoelectrophoresis against the purified preparation (11). The corresponding test with crude enzyme preparation, although considered more appropriate for specificity (52), was not reported.

Polyvalent antiserum S1 gave a sharp GAD-containing precipitin line, besides three others, with brain homogenate supernatant in crossed immunoelectrophoresis (Figure 1, part II, second arc) (13). Our sheep antiserum S3 produced *no* precipitin lines under similar conditions, although it immunoprecipitated GAD enzyme activity to 85%. With a partially purified GAD preparation, a single antigen was detected by antiserum S3 in crossed immunoelectrophoresis. Only one of the antigen–antibody precipitin lines produced against brain homogenate supernatant by antiserum S1 was retarded by an intermediate gel containing antiserum S3 (Figure 1, Part IV). No new lines were detected. Antiserum S3 is thus specific by these traditional, although not rigorous, immunochemical criteria. In addition, GAD antiserum S3 immunoprecipitates CSD II activity, but not brain CSD I, liver CSD (Section 4.2), or GABA-T (Weise, Oertel, and Kopin, unpublished observation).

To conclude, immunological tests indicate that sheep antiserum S3 is less potent for immunoprecipitation than antiserum S1 but apparently specific to GAD.

6 BIOCHEMICAL STUDIES WITH GAD ANTISERA

Blindermann et al. (53) have demonstrated that radioimmunoassays give results on GAD levels in brain homogenate comparable to GAD enzyme assays. Hadjian and Stewart (32) employed a radioimmunoassay to investigate the developmental profile of mouse brain GAD protein. To our knowledge no other quantitative assays have been reported using GAD antisera.

7 IMMUNOCYTOCHEMISTRY

7.1 General Remarks: Antisera Specificity

Radioimmunoassays and immunocytochemical techniques are critically dependent on specificity of the particular antiserum for the substance to be quantitated or localized. For an enzyme like GAD, the ability of the antisera to immunoprecipitate enzymatic activity clearly denotes the presence of some antibody response to GAD. The presence of a single antigen–antibody precipitin line during immunodiffusion or immunoelectrophoresis further shows that precipitins to other antigens in the tested (preferably crudest) preparations are absent or below the detection level. Two antisera to mouse brain GAD (8, 12), and one antiserum to rat brain GAD (13) satisfy these criteria.

The methods used in immunocytochemistry reveal *only* immunoglobulin fixation in tissue. The traditional absorption of antisera with a purified preparation of a complex protein such as GAD to generate nonstaining "control sera" reiterates the conclusion and does not necessarily provide further proof of specificity. Nonspecificity of the antiserum may still be present because of

(1) antibodies directed to nonunique determinants of the antigen

(2) the presence of antibodies to other (possibly more immunogenic) trace antigens in the "purified antigen."

Absorption would potentially reduce staining in both of these cases. Thus, the value of absorption in ruling out immunoreactivity unrelated to the "pure antigen" of interest is limited.

Inherent to immunocytochemistry is the possibility that immunoreactivity may be decreased by antigen loss during fixation or by a change of antigenic determinants by "overfixation." Relative immunoreactivity may therefore vary for each antigen. Even a minor serological response to a minor contaminant of the "purified antigen" might be preferentially localized in fixed tissue. For polyclonal antisera, therefore, there is only relative proof of specificity (54).

More direct conclusions about specificity can be derived from comparing staining results of antisera prepared by different methods. Immunocytochemistry using sheep antiserum S3 in fact corresponds to the results published for antimouse GAD antiserum (see Section 7.33). In areas where genetic or anatomical lesions result in a decrease of GAD levels, immunocytochemistry can also be compared with biochemical analysis (see Section 7.3.3d).

The presence of GAD immunoreactivity in cellular elements is presumptive evidence for the presence of GAD (CSD II) and a presumed GABAergic character. The tentative nature of this identification merits the term *GAD like immunoreactivity* (54), but *GAD positivity* or *GAD immunoreactivity* will be used as shorter and more convenient terms.

7.2 Methodological Aspects

7.2.1 *Light Microscopy*

Intracardiac perfusion with 4% paraformaldehyde was recommended as a fixative for GAD immunocytochemistry at the light microscopic level by Saito and coworkers (9, 55). It has subsequently been employed by Ribak and coworkers, as summarized by Roberts (14), and by other groups (12, 56–58). The investigation of different fixatives and fixation protocols showed fixation with periodate, lysine, and 4% paraformaldehyde (59) or the low-high pH formaldehyde fixation (60) to be especially suitable for studies with the indirect immunofluorescence method (61, 62).

For studies with the unlabeled antibody–enzyme method of Sternberger (63) we now employ a modification (58, 62) of the standard procedure (9, 55). The animal is deeply anesthetized with pentobarbital. At a pressure of 90 mm Hg, blood is flushed out of the vascular system with sodium phosphate (0.013 *M*, pH 7.3) buffered saline (0.81%) at 37°C for 1 min, immediately followed by 4% freshly depolymerized paraformaldehyde in sodium phosphate buffer (0.12 *M*, pH 7.3) for 5 min at room temperature and for up to 1 hr at a temperature between 6 and 10 °C. This protocol affords reliable terminal staining and a variable degree of cell body staining. After dissection, the brains can be stored at 4°C in the fixative for a few days and in phosphate-buffered saline (pH 7.4) or Tris-buffered saline (pH 7.6) for several months without substantial loss of immunoreactivity. Tissue blocks can be sectioned by vibratome on the day of perfusion or alternatively sunk in sucrose cryoprotectant for subsequent sectioning with cryostat or freezing microtome. The cytological structure is generally better preserved in vibratome sections, which can be used for both light microscopic (20–25 μm thick sections) and electron microscopic (25–50 μm thick sections) immunocytochemistry. Other fixation protocols are under investigation.

7.2.2 *Electron Microscopy*

To improve the preservation of ultrastructure, a small amount of glutaraldehyde is added to the formaldehyde fixative (10, 58, 62, 64, 65). Areas that contain few myelinated fibers require only a concentration of 0.1% glutaraldehyde for acceptable results, whereas up to 0.4% glutaraldehyde is needed for satisfactory preservation in areas rich in myelin.

Immunoreactivity of nerve terminals is still detectable at concentrations of 0.4% glutaraldehyde.

Immunoreactivity in cell bodies is preserved with up to 0.2% glutaraldehyde in untreated animals and can be detected with a level of up to 0.4% in colchicine-pretreated animals (62, 65, 66).

In brain regions with fenestrated capillaries, such as median eminence or pituitary gland, modifications in perfusion pressure, concentration of glutaraldehyde, and buffer strength provide better preservation of ultrastructure and immunoreactivity (28).

In the absence of detergents in the incubation solution, antibodies penetrate only a few microns into the section. To reliably distinguish between GAD-containing and GAD-negative elements, only the first two microns of a section should be evaluated (Figure 2). Specimens obtained from a deeper level will contain potentially GAD immunoreactive profiles, which are unstained only because of lack of exposure to immunoreagents (Figure 3A; compare with Figure 2).

For further details of fixation procedures, embedding procedures, and the preparation of the embedded specimens for ultrathin sectioning, see refs. 62, 65, and 67.

7.2.3 Variation of Immunocytochemical Methods

Several different immunocytochemical methods have been employed to visualize GAD immunoreactivity. Initial reports described absorption of the antiserum with liver acetone powder and then dilution of 1:6 to 1:8 in conjunction with the indirect immunoperoxidase method (9, 55). In subsequent studies (68–73), modifications of the unlabeled antibody–enzyme method (63) allowed the use of primary antiserum dilutions of up to 1:500 (72).

The second group to prepare mouse brain GAD antiserum carried out staining with primary antibody dilutions of 1:20 to 1:60, after absorption with rat adrenal gland acetone powder, for mapping the distribution of GAD immunoreactive terminals in rat brain by the indirect immunofluorescence method (12, 61). Antiserum to rat brain GAD (11) at a dilution of 1:100 followed by the indirect immunoperoxidase method visualized immunoreactive terminals and somata in the nucleus raphe dorsalis (56, 57).

The sheep antiserum S3 prepared by our group is currently employed in dilutions of 1:1500 to 1:16,000 (see Sections 7.4.1a, 7.4.2) with the unlabeled antibody–enzyme method of Sternberger (63) and in dilutions of 1:500 to 1:2000 with the indirect immunofluorescence method (61). Absorption of antiserum S3 with liver acetone powder did not change the staining characteristics, although it reduced the effective titer by a factor of about four (58). As liver GABA-T is detected by antibodies to

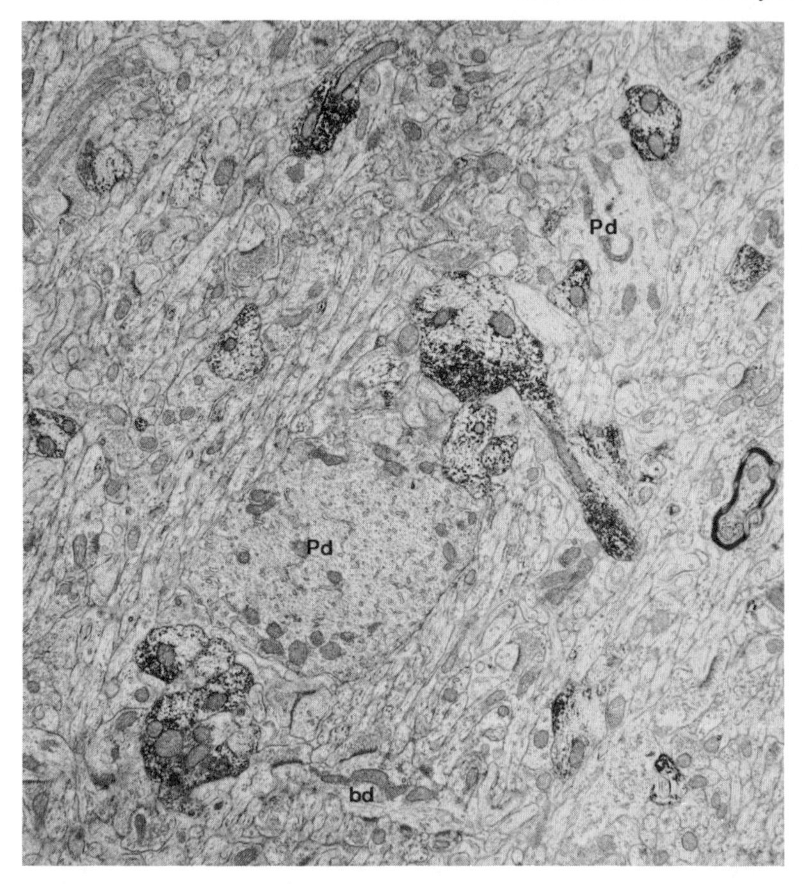

Figure 2. Electron micrograph of GAD immunoreactive terminal staining within 2 μm of surface of Vibratome section of rat cerebellum (see Section 7.4.1b). In molecular layer of cerebellum each axonal profile containing pleomorphic synaptic vesicles is immunoreactive. All other elements of the neuropil are unstained. *Pd*, Purkinje cell dendrite; *bd*, basket cell dendrite. Antiserum S3 (second bleed) 1:2000. Magnification ×13,100. From Oertel et al. (58).

brain GABA-T (46), this preabsorption would have removed or reduced GABA-T-directed antibodies in antiserum S3 (see also subsection in Section 3.3).

7.3 Tests for Immunocytochemical Specificity

The following sections discuss various immunocytochemical tests for specificity of antisera to GAD.

7.3.1 *Control Serum: Absorption Control*

As reference for the various rabbit antisera to mouse brain and rat brain GAD, normal rabbit serum or preimmune rabbit serum was employed (9, 10, 12, 56, 57). For the studies with sheep anti-GAD, preimmune serum was used as control (58). Absorption controls, although a valid procedure for peptide immunocytochemistry, were not performed for antiserum S3, first because we lacked a "pure antigen" and second, as detailed in Section 7.1, because absorption controls only provide evidence that staining is due to immunoglobulins directed against the antigen preparation but are not *independent* proof of antigen purity or antisera specificity (74).

7.3.2 *Serial Dilutions of the Primary Antiserum*

Serial dilution tests were performed with antiserum S3 to minimize possible nonspecific staining properties (62). With high concentrations (1:1000 to 1:1500) and long diaminobenzidine (DAB) development times (20 min), very light staining at the very surface of the tissue is sometimes present in glia and neurons known to use transmitters other than GABA. At higher dilutions of antiserum S3, and with short DAB incubation times, "nonspecific staining" disappears while GAD immunoreactivity is retained. In colchicine-pretreated preparations, the same conditions as above apply with respect to "nonspecific staining." Colchicine-treated animals demonstrate apparent specific cell body staining up to antiserum titers of 1:16,000.

7.3.3 *Test Areas for GAD Immunoreactivity*

7.3.3a Cerebellum. In cerebellar cortex, extensive electrophysiological, biochemical, and pharmacological studies have shown that GABA is the transmitter of Purkinje cells that project to the cerebellar nuclei and lateral vestibular nucleus (75–77). Furthermore, three types of local circuit neurons–Golgi cells, basket cells and stellate cells–use an inhibitory transmitter (75), and the only remaining neuronal cell type of cerebellar cortex, the granule cell, has an excitatory action and is not thought to be GABAergic. Consequently, the cerebellum is an ideal test area for GAD antisera. In the first articles on GAD immunocytochemistry, axon terminals of the four inhibitory neuronal cell classes in cerebellar cortex were immunoreactive at the light microscopic (9) and electron microscopic (10) levels. Subsequent studies on sections of colchicine-pretreated animals demonstrated reaction product in the corresponding neuronal cell bodies (71).

These findings were confirmed and extended in studies on normal rat cerebellum using antiserum S3. In light microscopic preparations, almost all visible members of Purkinje cell, stellate cell, basket cell, and Golgi cell classes are GAD immunoreactive (Figure 4A), whereas granule cells and their parallel fiber axons, mossy fibers, and climbing fibers are unstained (see Section 7.3.3c and Figures 2 and 3A). In ultrathin sections immediately below the surface of randomly selected cerebellar slices, essentially all "inhibitory type" endings are labeled (Figure 2). These data suggest that all inhibitory neurons in the cerebellar cortex use GABA as their transmitter (58, 62).

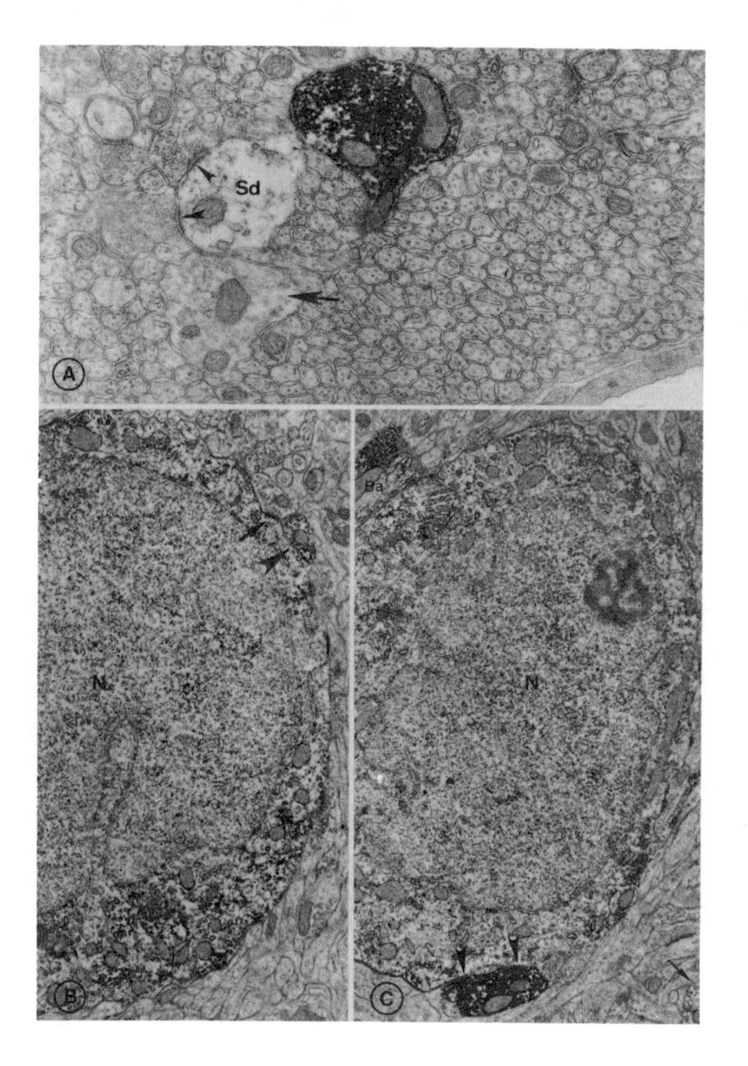

In the cerebellar nuclei, unstained neurons are surrounded by the large GAD immunoreactive terminals of Purkinje cell axons (9, 10, 58).

7.3.3b Reticular nucleus of thalamus. A striking GAD immunoreactive structure in rat brain is the reticular nucleus of thalamus. The vast majority of its neurons (cell bodies, dendrites, and terminals) stain strongly positive with anti-GAD serum (72, 78). Electrophysiological studies suggest, moreover, that reticular neurons of thalamus are inhibitory (79, 80).

The white matter next to the reticular nucleus is essentially nonstaining and thalamic nuclei located medially contain few GAD immunoreactive cell bodies in rat. Because of the prominent immunoreactivity of reticular nucleus and the adjacent relatively nonstaining structures, reticular nucleus is also a good test area for GAD antisera (Figure 4B and C).

7.3.3c Colchicine. Colchicine, a blocker of axoplasmic transport, is used to increase the level of immunoreactive GAD in the cell body above the detection threshold of immunocytochemical methods (71). As transport of other components is also impaired, it has to be ascertained that the antiserum employed still detects only GAD and no other accumulated antigen or antigens. For this reason, cerebellar cortex and reticular nucleus of thalamus are suitable test areas for the use of colchicine to enhance GAD immunoreactivity. In both areas, antiserum S3 already stains GABAergic neurons almost quantitatively even in untreated animals (Figures 4A, 5A; refs. 58, 78). The hope was that colchicine would not produce nonspecific immunoreactivity and would therefore be useful

Figure 3. (A) Electron micrograph of GAD immunoreactive terminal staining approximately 10 μm from the surface of Vibratome slice (see Section 7.4.1b). Sections farther from the surface of Vibratome slice are variable in immunoreactivity (compare with Figure 2). The shaft of a stellate cell dendrite (*Sd*) is in contact (small arrowheads) with two parallel fiber boutons (with round synaptic vesicles) and with two endings of stellate axons (with pleomorphic vesicles). Because of limited immunoreagent penetration only one of the stellate axon terminals is immunoreactive (center); the other is unstained (arrow). Antiserum S3 (second bleed) 1:2000. Magnification ×34,750. From Oertel et al. (58). (B and C) Electron micrographs of colchicine-independent cell body staining in rat cerebellum (see Section 7.4.2). Moderate GAD immunoreactivity (B) is observed in various cytoplasmic regions and plasma membrane of stellate cell body in adult rat cerebellar cortex. *N*, nucleus. Arrowhead indicates labeled axosomatic bouton (example of GABA-GABA-interaction); arrow points to axosomatic bouton of unstained parallel fiber on the stellate cell body. Antiserum S3 (second bleed) 1:1500. Magnification ×14,800. From Oertel et al. (58). (C) GAD-immunoreactive cell body of basket cell in adult rat cerebellar cortex. *N*, nucleus; *Ba*, by-passing basket cell axon. Arrowheads indicate two symmetric synaptic junctions formed by an immunoreactive axosomatic bouton with the basket cell body. Arrow points to immunoreactive basket cell axon. Antiserum S3 (second bleed) 1:1500. Magnification ×15,600. From Oertel et al. (62).

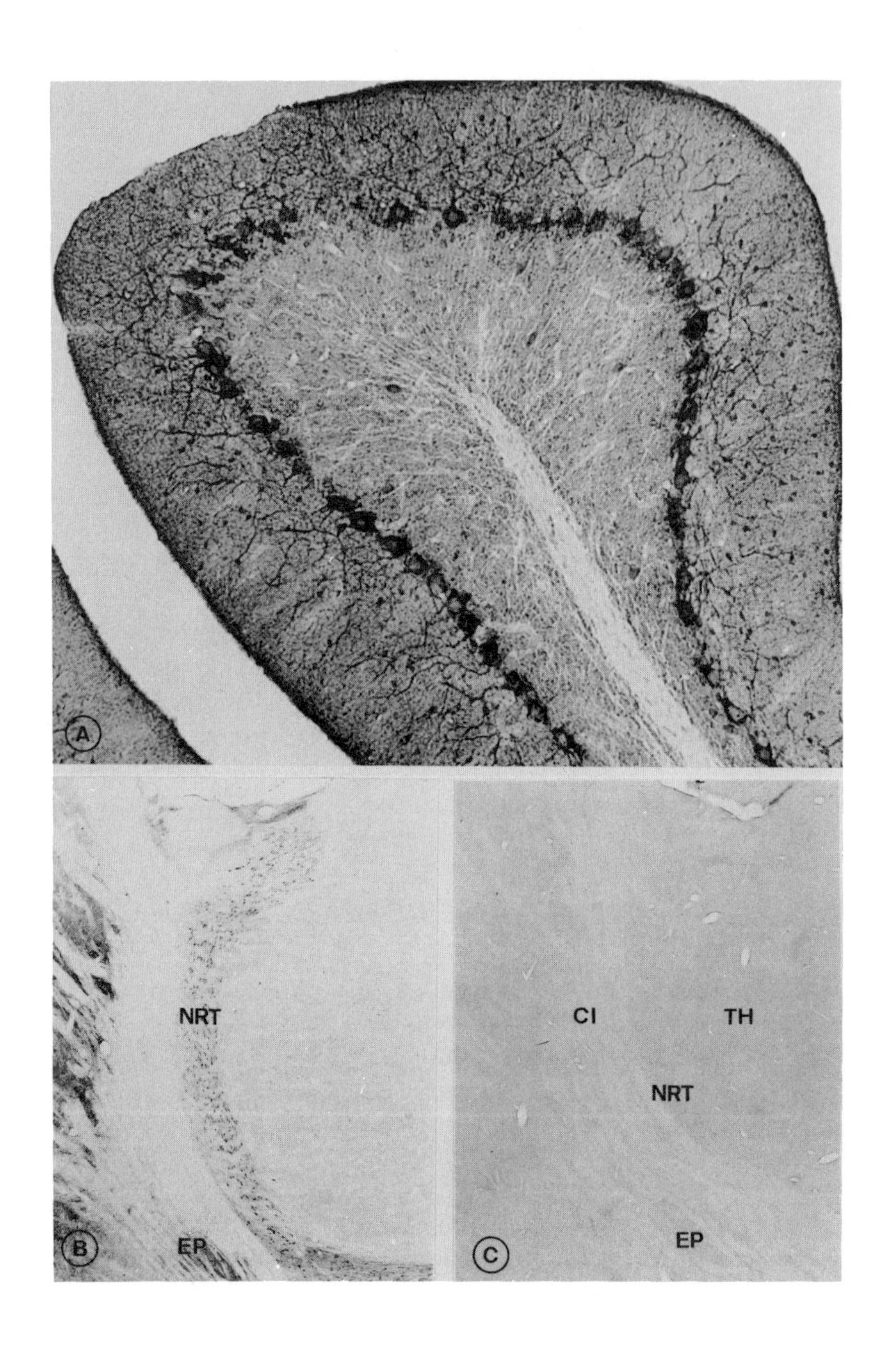
A
B
NRT
EP
C
CI
TH
NRT
EP

for increasing cell body staining of presumed GABAergic neurons in other regions.

After topical colchicine injection into cerebellar cortex, the staining intensity of Golgi and Purkinje cell bodies and, to a lesser extent, basket and stellate cell bodies is enhanced. GABAergic processes are clearly visible, but granule cells, climbing fibers, mossy fibers, and parallel fibers remain negative, as do non-neuronal elements (62, 71). Cell bodies, dendrites, and axons of neurons in reticular nucleus of thalamus are likewise more intensely stained after colchicine treatment (Figure 5B). The adjacent white matter and the vast majority of cells in the rat thalamic relay nuclei remain unstained even in the direct vicinity of the injection site (72, 78). These data demonstrate that antiserum to mouse brain GAD (8) and antiserum S3 (13) do not stain non-neuronal elements or neurons thought to be non-GABAergic following colchicine injection in cerebellum and reticular nucleus of thalamus. It is therefore likely that colchicine can be used with confidence for the enhancement of cell body staining in other areas of brain such as in substantia nigra pars reticulata (see p. 97) (Figure 5C and D).

7.3.3d Lesion studies. It is generally accepted that striatal GABAergic neurons project to substantia nigra (for a review see ref. 81).

Medium-sized GAD-immunoreactive neurons, the presumed cells of origin for the GABAergic striatonigral projection, have been observed

Figure 4. Immunocytochemical localization of GAD immunoreactivity in test areas. (A) Light micrograph of adult rat cerebellar cortex (see Section 7.3.3a). The molecular layer contains numerous immunoreactive punctate profiles. Purkinje cell dendrites, basket cell axons, and stellate and basket cell bodies (see also Figure 7B and C) are lightly labeled. Most of the Purkinje cell bodies are stained. In the granule cell layer, granule cell bodies are unstained whereas Golgi cell bodies are immunoreactive. Rings of labeled punctate profiles are present throughout the granule cell layer, corresponding to immunoreactive Golgi cell terminals around glomeruli. Antiserum S3 (second bleed) 1:1500. Brightfield, magnification ×200. From Oertel et al. (58). (B) Light micrograph of sagittal section through reticular nucleus of thalamus (NRT) (see Section 7.3.3b and 7.3.3c). Rat received a topical injection of 10 μg of colchicine 24 hr prior to fixation into the subthalamus. The vast majority of neurons in the reticular nucleus are GAD-immunoreactive. Note the medium intensity of cell body staining in the dorsal and intermediate aspect of the nucleus and the increased staining intensity ventrally, for example, close to the injection site. Immunoreactive cell bodies are also observed as small dots in the entopeduncular nucleus. The internal capsule and the thalamic relay nuclei are relatively unstained. For explanations, see panel C. Antiserum S3 (second bleed) 1:2000. Brightfield, magnification ×45. From Oertel et al. (78). (C) Control section to panel B, incubated in preimmune serum. *CI*, internal capsule; *EP*, entopeduncular nucleus; *NRT*, reticular nucleus of thalamus; *TH*, thalamic relay nuclei. Preimmune serum 1:1800. Brightfield, magnification ×38. From Oertel et al. (78).

in the striatum of colchicine-pretreated (82) and normal rats (unpublished observation).

In accord with its rich content of GAD by biochemical analysis (83), the substantia nigra is strongly GAD immunoreactive because of numerous terminals forming axodendritic and axosomatic contacts with substantia nigra neurons (69; see also Section 7.6.4b). Biochemical studies have shown a marked decrease of GAD enzyme activity in substantia nigra after surgical interruption of the striatonigral projection (83, 84) or injection of the neurotoxic agent kainic acid (85, 86) into striatum. These experimental

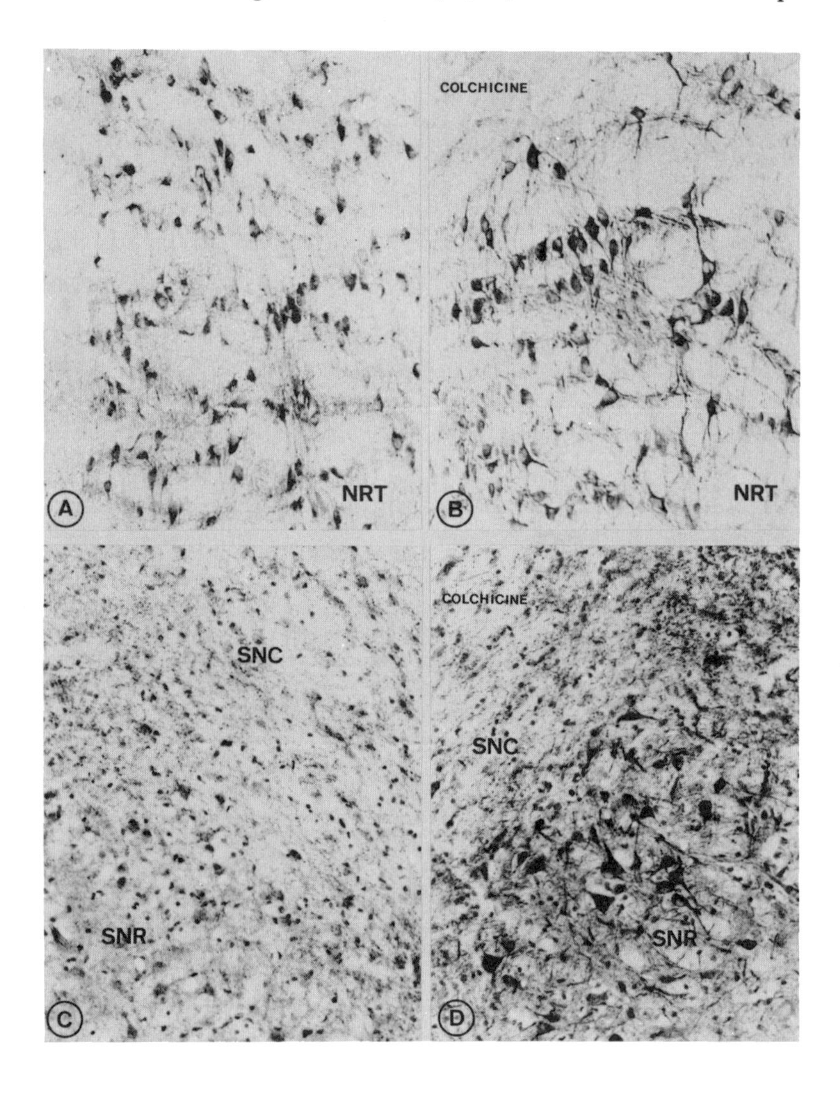

models have therefore been used to test whether GAD immunocytochemistry would faithfully reflect the known loss of GAD in substantia nigra after such lesions: Hemitransection of the striatonigral pathway in rat in fact results in a marked ipsilateral decrease of GAD immunoreactive punctate profiles in the intermediate aspect of pars reticulata of substantia nigra. In contrast, the lateral and medial aspect of pars reticulata had a density of immunoreactive profiles comparable with the control side (87). Essentially identical immunocytochemical data—accompanied by a 65% reduction in GAD enzyme activity—have been reported in rat substantia nigra 7–11 days after an ipsilateral kainic acid-induced striatal lesion (Figure 6; ref. 88).

A dramatic decrease of punctate GAD immunoreactivity is seen in the lateral vestibular nucleus after lesioning the cerebellar cortico-vestibular pathway (89). Decreased immunoreactivity is also observed around deep cerebellar neurons in the mutant mice Pcd (Purkinje cell degeneration) when compared with normals (89a). Thus, staining with rabbit antiserum to mouse brain GAD (8) and sheep antiserum S3 to rat brain GAD (13) reflects the decrease in GAD terminals, in both experimental and genetic models, where there is loss of GABAergic neurons.

7.3.3e Summary. The immunochemical characterization of rabbit antisera to mouse brain GAD (8, 12) and antiserum S3 (13) to rat brain GAD strongly suggest that these antisera are specifically directed against GAD.

In addition, similar immunocytochemical results obtained with two GAD antisera (8, 13), which *differ* in species used for antigen source, means of antigen purification, and antiserum production suggest that

Figure 5. Cell body staining by GAD immunocytochemistry in rat (see Section 7.4.2). Colchicine-independent area (A and B) in the reticular nucleus of thalamus. Colchicine-dependent area (C and D) in the substantia nigra. (A) Processes, punctate profiles and the vast majority of neurons are GAD immunoreactive in reticular nucleus of thalamus of normal rat brain. NRT, reticular nucleus of thalamus. Antiserum S3 (second bleed) 1:2000. Brightfield, magnification ×220. (B) After topical injection of colchicine the staining intensity of cell bodies and processes is markedly enhanced; the number and distribution of immunoreactive neurons remain unchanged. Counterstained with cresyl violet. Antiserum S3 (second bleed) 1:3000. Nomarski interference optic, magnification ×380. (C) No GAD-immunoreactive neurons are seen in colchicine-dependent regions of rat brain such as substantia nigra at the lateral aspect of the border of substantia nigra pars reticulata and pars compacta (indicated in Figure 7B by the arrow pointing to *SNC*, the substantia nigra pars compacta). Immunoreactive punctate profiles are more numerous in pars reticulata than in pars compacta. For explanations, see next panel. (D) Analogous region to (C) from substantia nigra of colchicine-injected side, in contrast, reveals numerous intensively GAD-immunoreactive cell bodies and processes. They are usually fusiform or triangular and are seen in pars reticulata but not in pars compacta. Counterstained with cresyl violet. Antiserum S3 (second bleed) 1:1500. Brightfield, magnification ×220.

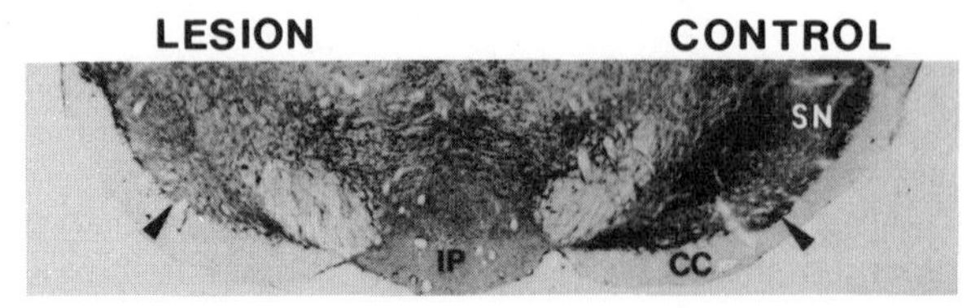

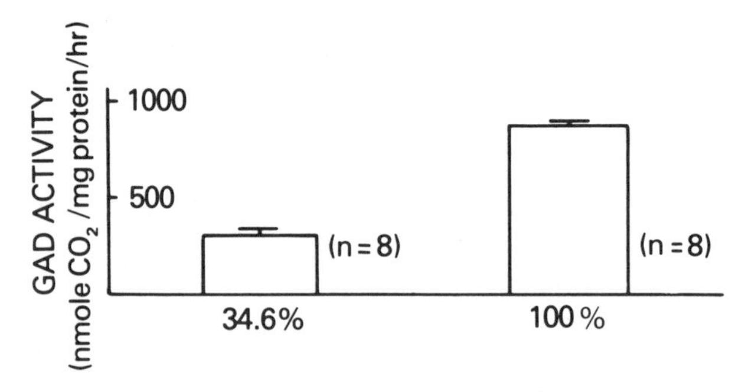

Figure 6. Lesion studies (see Section 7.3.3d). Light micrograph of substantia nigra of a rat with kainic-acid-induced striatal lesion 7 days before fixation (Arrowheads indicate substantia nigra). Substantial reduction in GAD immunoreactivity is observed on the side ipsilateral to the kainic-acid-induced striatal lesion (left) compared with the operated side (right). *CC*, crus cerebri; *IP*, interpeduncular nucleus; *SN*, substantia nigra. Antiserum S3 (second bleed) 1:8000. Brightfield, magnification ×18.5. Nigral GAD enzyme activity on the lesioned side is reduced by 65% compared with control. Values represent means ± SEM. From Oertel et al. (88).

these antisera can be used for the anatomical study of GAD-containing cells (presumed GABAergic neurons).

The above results do not rigorously exclude, however, the possibility that these antisera also detect some non-GABAergic cells either through a common contaminating determinant or through cross-reactivity with other proteins or enzymes (see Section 7.1). Although unlikely, such a possibility makes it advisable to obtain biochemical or electrophysiological support for GAD content or GABAergic function for newly described immunoreactive neurons or cells.

7.4 Characteristics of GAD Immunoreactive Staining

7.4.1 Terminal Staining

GAD is concentrated in synaptosomes, according to biochemical studies (6). So far the synaptosomal (low-molecular-weight) form of GAD has been employed for antibody production and has proved to be very immunogenic (Section 5.1). Thus, available polyclonal GAD antisera may preferentially detect low-molecular-weight GAD in nerve terminals (9, 12, 58) as compared with GAD located in cell bodies. Note, however, that antisera to low-molecular-weight GAD do cross-react with high-molecular-weight GAD (36).

7.4.1a Light Microscopy. GAD immunoreactive punctate profiles corresponding to axon terminals can be classified by aspect into "large" boutons and "small" boutons at the light microscopic level. Large boutons are strongly immunoreactive and clearly seen with dilutions of antiserum S3 of 1:8000 and higher. Smaller, more delicate boutons are better detected with lower dilutions (1:2000–1:4000) or by enhancement of the reaction product (90). These features may correspond to subclasses of GABAergic axon endings with functional differences (see also ref. 12).

In some brain areas, the distribution of large and small terminals follows anatomical borders. For example, the rat caudate nucleus is predominantly characterized by small-type axon profiles, whereas globus pallidus, entopeduncular nucleus, and substantia nigra contain large boutons. Certain thalamic nuclei, such as the ventromedial nucleus, also contain large-type immunoreactive profiles, as do the lateral habenula and subthalamic nucleus (unpublished observation). Other striking examples are the distinction between the large immunoreactive boutons of β-subnucleus of inferior olive compared with the small, fine GAD-immunoreactive profiles of the main nucleus (62). Likewise, the islands of Calleja have large boutons that contrast with the surrounding small GAD boutons of olfactory tubercle (Figure 7C and D; for further details, see ref. 12).

Brain areas can also be classified by the relative density of their GAD immunoreactive profiles. Rat thalamic nuclei, for example, exhibit a low density of GAD terminals, whereas hypothalamus (Figure 7A), substantia nigra (Figure 7B), and globus pallidus are characterized by profuse GAD terminals. These findings correspond in large part to the differences noted in studies of GAD–enzyme activity in these various areas (91).

7.4.1b Electron Microscopy. Systematic studies on the electron microscopic characterization of large vs. small boutons are lacking. GAD immunoreactive terminals form axodendritic (Figure 2; 10, 58, 69), axosomatic (Figure 3B and C; 10, 58, 69), axoaxonic (68), and dendrodendritic (70) synapses. In some instances, such as cerebellar cortex (Figure 3B and C; 58, 71), cerebral cortex (66), substantia nigra pars reticulata (98), and reticular nucleus of thalamus (78), postsynaptic structures have also been shown to contain GAD immunoreactivity (see Section 7.6.4b). These findings provide the morphological correlate for a GABA–GABA mediated disinhibition. In the hypophysis and median eminence, GAD immunoreactive axonal endings are found among unstained neurosecretory endings, in close vicinity to fenestrated capillaries, and in some instances within the perivascular space (28, 92). Whether GABA released from such endings can act as a neurohormone remains to be established (92a).

In most areas of the nervous system, GAD immunoreactive terminals contain pleomorphic vesicles. GAD immunoreactive axon terminals in neurointermediate lobe of hypophysis, however, display predominantly round vesicles, possibly due to the fixation procedure (28). Immunoprecipitate is highly concentrated at the external membranes of vesicles, which themselves are free of precipitate (for further details of the electron microscopic localization see 10, 58, 62, 65). In cerebellar cortex of rat, GAD immunoreactive terminals in general form symmetric synapses (type 2). Improved techniques for pre-embedding staining immunocytochemistry (58, 62) allow the demonstration that a few GAD immunoreactive boutons form asymmetric synaptic junctions (type 1) on Purkinje

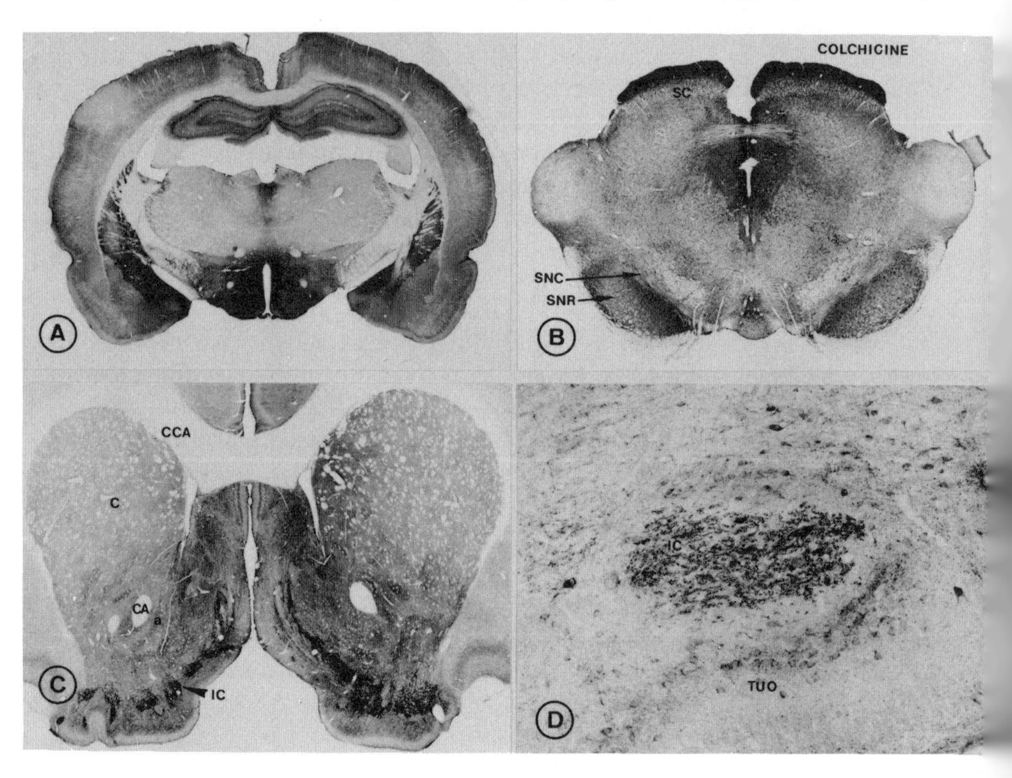

cell spines and granule cell dendrites, as is expected from morphological studies (93, 94).

7.4.2 *Cell Body Staining*

Cell body staining of GABAergic neurons is variable and apparently depends on antiserum, brain region, and fixation.

Using antiserum to mouse brain GAD (8), cell bodies of periglomerular and granule cells in the rat olfactory bulb are consistently stained (70). Numerous GAD immunoreactive cell bodies are observed in cerebral cortex (66), hippocampus (71), and reticular nucleus of thalamus (72). In other areas, however, such as cerebellar cortex (71), striatum, globus pallidus (82), and substantia nigra (87), a consistent demonstration of GAD immunoreactive neurons requires the intraventricular or topical injection of colchicine (Section 7.3.3c). With this procedure, the number of easily detected immunoreactive neurons also dramatically increases in cerebral cortex, hippocampus, and reticular nucleus of thalamus (71, 72).

Figure 7. GAD immunoreactivity in different areas of adult rat brain. Light microscopy (see Section 7.4.1a). (A) Photomicrograph of slightly oblique frontal Vibratome section at the level of the median eminence. Note the high density of terminal staining in the hypothalamus compared with the thalamus. Thalamus is surrounded by a shell of GAD-immunoreactive neurons in the reticular nucleus of thalamus (see also Figure 2B and Figure 5A and B). The internal capsule is relatively unstained except in the region of the entopeduncular nucleus. Other areas of increased staining are the medial aspect of amygdala and globus pallidus. Antiserum S3 (second bleed) 1:4000. Brightfield, magnification ×7.5. (B) Light micrograph of mesencephalon of a rat unilaterally injected with 10 μg colchicine at the medial aspect of the border of substantia nigra pars reticulata and substantia nigra pars compacta. GAD immunoreactivity due to axonal processes and boutons is prominent in substantia nigra pars reticulata, periaqueductal gray and superior colliculus. The staining intensity of the superficial zone in superior colliculus normally resembles that of periaqueductal gray. In this particular section it is artifactually increased due to variation of section thickness. *SC*, superior colliculus; *SNC*, substantia nigra pars compacta; *SNR*, substantia nigra pars reticulata. Arrow pointing to *SNC* indicates the location of Figure 5C. The corresponding area on the colchicine-injected site is shown in Figure 5D. Antiserum S3 (second bleed) 1:1500. Brightfield, magnification ×11.5. From Oertel et al. (98). (C) Light micrograph of a slightly oblique frontal forebrain section. Optic tract has been removed. Note the intense terminal staining in islands of Calleja. *a*, accumbens nucleus; *C*, caudate nucleus; *CA*, anterior commissure; *CCA*, anterior corpus callosum; *IC*, islands of Calleja. Antiserum S3 (second bleed) 1:4000. Brightfield, magnification ×13. (D) Light micrograph of islands of Calleja in a sagittal section of rat brain after topical injection of colchicine (10 μg) in lateral hypothalamus 24 hr prior to fixation. Unstained neuronal processes and cell bodies are outlined by darkly stained large-type punctate profiles. Olfactory tubercle contains a low density of small-type terminals. Immunoreactive cell bodies are present in islands of Calleja and in surrounding areas. *IC*, island of Calleja; *TUO*, olfactory tubercle. Antiserum S3 (fourth bleed) 1:3000. Brightfield, magnification ×130.

Whereas cell body staining has not been obtained with the second rabbit antiserum to mouse brain GAD (12), rabbit antiserum to rat brain GAD (11) visualizes immunoreactive cell bodies in dorsal raphe nuclei of normal rats (56, 57). Finally, the use of sheep antiserum S3 to rat brain GAD at dilutions of 1:1500 to 1:2000 (second bleed) or at dilutions of 1:3000 to 1:4000 (fourth bleed) in rat and cat brain fixed with 4% paraformaldehyde allows reliable light microscopic cell body staining for each of the presumed GABAergic neurons in cerebellar cortex (Figure 4A *i* 58, 62). The vast majority of neurons of reticular nucleus of thalamus of untreated rat (Figure 5A), cat, and monkey are immunoreactive (78), in agreement with the data obtained by Houser and coworkers (72) in colchicine pretreated rats.

The ability of antiserum S3 to yield consistent GAD immunoreactivity in neuronal cell bodies in cerebellum and reticular nucleus of thalamus has been confirmed at the ultrastructural level. In agreement with Ribak and coworkers (65), the highest density of immunoprecipitate is around the cisternae and vesicles of the Golgi apparatus, but immunoprecipitate is found throughout the cytoplasm (Figure 3, B and C) including proximal dendrites (58, 62, 78).

Antiserum S3 obviously has a relatively higher "sensitivity" for detection of GAD immunoreactive neurons than previous antisera. For example, short-axon neurons, a third class of GABAergic neurons in the olfactory bulb (96), and small- to medium-sized neurons in the cerebellar nuclei (97), are GAD immunoreactive with antiserum S3 and represent new additions to the class of known and proposed GABAergic neurons. In contrast to the other three antisera to GAD, antiserum S3 was raised against an antigen–antibody complex. Thus, available antigenic determinants may have been partially occupied by specific immunoglobulins of antiserum S1, which itself was produced against low-molecular-weight GAD (Section 3.3). Perhaps this favored the production of antibodies to antigenic sites common to high- and low-molecular-weight form GAD (36). On the other hand, antiserum S3 may simply contain a higher titer of antibodies with greater detection sensitivity for immunocytochemistry.

Based on data obtained with mouse brain GAD antiserum (8) and antiserum S3 (13), brain areas can be divided into two categories in respect to their cell body staining:

1. In so-called colchicine-independent areas, consistent cell body staining is observed in untreated animals. In some of these areas, detection of presumed GABAergic neurons by means of antiserum S3 appears to be nearly quantitative, since pretreatment with colchicine hardly increases the number of immunoreactive neurons.

2. Other brain areas, called colchicine-dependent, show only occasional and faint staining of cell bodies without prior colchicine injection.

Table 3 presents the staining characteristics and the need for colchicine in a number of regions of rat brain.

The reasons for variation in cell body staining of GABAergic neurons in different regions and cell classes remains unclear. This feature is not exclusively related to the presence of dendrodendritic synapses (70), since not only granule cells and periglomerular cells which form dendrodendritic contacts but also short-axon neurons can be visualized in olfactory bulb with antiserum S3 (96). Of the three GAD immunoreactive local circuit neurons of the cerebellar cortex, the Golgi cells are more intensely stained than basket or stellate cells (58). Some Purkinje cells, the classic GABAergic projection neurons, are nearly as immunoreactive as Golgi cells (58, 62). Based on preliminary studies in other areas than the cerebral cortex, hippocampus, and basal ganglia, both projection *and* local circuit GABAergic neurons of various types may be intensively or faintly stained. Species differences also exist, for example, GAD immunoreactive neurons in thalamic relay nuclei are consistently visualized in cat and monkey but not in rat (Table 3). Other factors may cause these

Table 3. Cell Body Staining of Selected Rat Brain Areas

Colchicine-independent Staining[a]	Colchicine-dependent Staining[a]
Cerebellar cortex[b] (58, 62)	Entopeduncular nucleus (98)
Cerebellar nuclei (97)	Globus pallidus(98)
Cerebral (occipital) cortex[b,c]	Hypothalamus (91a, 92a)
Colliculus inferior (62)	Striatum[e,c]
Dorsal lateral geniculate body[b,c,d]	Substantia nigra (99)
Hippocampus (62, 71)	Thalamic relay nuclei[f,c]
Olfactory bulb[b] (70, 96)	Zona incerta (99)
Reticular nucleus of thalamus[b] (72, 78)	
Retina[c] (73)	
Striatum[c,e]	

[a] With antiserum S3.
[b] No increase or slight increase in number of GAD immunoreactive neurons observed after colchicine pretreatment.
[c] Unpublished observation.
[d] Golgi-like staining.
[e] Substantial increase in number of GAD immunoreactive neurons observed after colchicine pretreatment.
Stain in cat and monkey without colchicine pretreatment.

differences in apparent immunoreactivity, such as the effectiveness of fixation for each brain area and neuronal cell type. Furthermore, functional properties such as pattern and frequency of firing, input resistance, or synaptic relations may be relevant. Efforts at improved fixation protocols are under way, since only a mild increase in the preservation of antigen might raise most if not all GABAergic cells above the threshold of detection.

7.5 Comparison of ^{3}H-GABA Autoradiography and GAD Immunocytochemistry

Only those neurons that use GABA as their transmitter are thought to have a high-affinity GABA uptake (100). ^{3}H-GABA autoradiography has therefore been employed for the visualization of GABA (accumulating) cells (101). ^{3}H-GABA uptake studies *in vivo* and in tissue slabs, however, suffer from certain limitations because of the necessity to apply GABA locally due to the blood-brain barrier (30). In addition to neurons, glial and other non-neuronal cells also take up ^{3}H-GABA, although neuronal uptake may be differentiated from glial ^{3}H-GABA uptake by its time course and by the presence of selective uptake blockers (102, 103).

GAD has not been localized by immunocytochemistry in glial or other non-neuronal cells. To evaluate the reliability of ^{3}H-GABA-autoradiography as a method for detecting GABAergic neurons, a direct comparison was made to GAD immunocytochemistry in dissociated monolayer cell cultures of rat cerebral cortex. This test system avoids the problems related to delivery of the exogenous label in animals. GABA-autoradiographic labeling was found to coexist with GAD immunoreactive staining in about 90% of cells marked with either method in tissue culture (Figure 8A and B). Thus, the use of GABA uptake to localize GABAergic neurons in this particular model is validated by GAD immunocytochemistry (104).

But an indirect comparison of data obtained in animals with ^{3}H-GABA autoradiography and GAD immunocytochemistry reveals a controversial picture. On the one hand, ^{3}H-GABA autoradiography labels neuronal cell types known to be GAD immunoreactive, such as amacrine cells in the retina (73), nonpyramidal (i.e., aspinous and spinous stellate) neurons in the occipital cortex of rat (30, 66), intrinsic neurons in sensory-motor cortex in monkey (105, 106), and some periglomerular cells, granule cells, and short-axon cell bodies in the olfactory bulb (70, 96, 107). On the other hand, ^{3}H-GABA autoradiography fails to label the Purkinje cell, the classical GABAergic projection neuron, although it labels the three cerebellar inhibitory interneurons (101).

In the posterior pituitary lobe ^{3}H-GABA autoradiographic label is reported in non-neuronal profiles (108). In contrast, GAD immunoreactivity

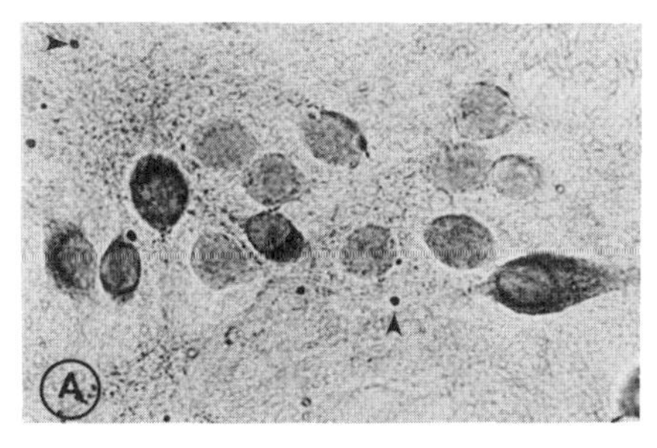

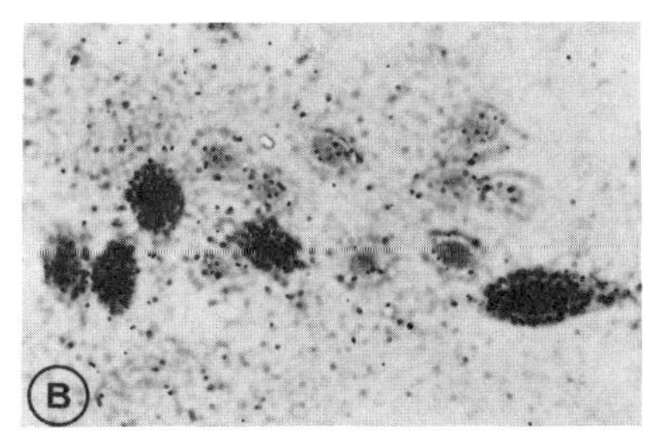

Figure 8. Comparison of GAD immunocytochemistry (A) and high affinity ^{3}H-GABA-uptake autoradiography (B). Primary 22-day-old tissue cultures of rat cerebral cortex were incubated in ^{3}H-GABA, fixed with paraformaldehyde, processed for GAD immunocytochemistry, and subsequently developed for ^{3}H-GABA-autoradiography (3 days of exposure). The GAD-immunoreactive neurons also show accumulation of ^{3}H-GABA. Small arrowheads in (A) mark nonspecifically stained debris, which are also seen in control cultures treated with preimmune serum (not shown). Silver grains are photographed in the focal plane of the emulsion (B), thus blurring the immunoreactive cell bodies. Antiserum S3 (second bleed) 1:2000. Brightfield, magnification $\times 1050$. From Neale et al. (104).

is localized in nerve terminals of central origin in the neurointermediate pituitary lobe; pituicytes and non-neuronal elements are unstained (28). Furthermore, ^{3}H-GABA is selectively accumulated by rat subependymal and supraependymal nerve fibers, the latter of which disappear after 5,7-dihydroxytryptamine treatment (109). GAD immunoreactive profiles intercalated in the ependymal lining of the ventricular system are hardly observed, however, whereas the subependymal layer contains a high density of GAD immunoreactive punctate profiles (12 and unpublished observation). Thus, the capacity of the presumed serotonergic supraependymal fibers to take up ^{3}H-GABA may not be associated with the ability to synthesize GABA. The reasons for the above discrepancies are not yet resolved.

For studies in the living animal or tissue slices it may be advisable to carry out correlative studies with GAD immunocytochemistry before ^{3}H-GABA autoradiography is used as a marker technique for GABAergic neurons.

7.6 GAD Immunocytochemistry in Various Brain Regions: Applications and Perspectives

A detailed review of GAD immunocytochemical studies in basal ganglia, cerebellum, cerebral cortex, hippocampus, retina, and spinal cord has been given by Roberts (14), followed by summaries by Wu and coworkers

(31), Ribak and coworkers (65), Oertel and coworkers (62), and a map of GAD-immunoreactive terminals in rat brain by Pérez de la Mora and coworkers (12). It is not the purpose of this section to review the above reports. Rather, general aspects of GAD immunocytochemical studies are discussed.

7.6.1 *Development*

An important issue in studying GABAergic neurons is to determine their time of neurogenesis and pattern of subsequent development. Biochemical studies suggest that GAD levels are relatively low during neurogenesis and increase thereafter (48, 110, 111). In cerebellum, both early-generated neurons (Purkinje cells) and late-generated neurons (stellate-basket cells) are GABAergic (112). The detection of GAD immunoreactivity in axons occurs as early as at the third postnatal day in rats (113). In rat visual cortex, GABAergic neurons are distributed across all layers of cortex (66). From present concepts of cortical neurogenesis (114, 115), it is expected that GABAergic neurons in the cortex are generated over an extended period of time, as in the cerebellum. Detailed studies of development have not been carried out to date; the need for reliable cell body staining is paramount in such studies.

GABAergic mechanisms are implicated in the development of orientation columns in visual cortex (116). Biochemical evidence suggests that GABA may be closely related to polyamine synthesis during development (117, 118). GABA has an effect on cell-free protein synthesis that is specific to brain (119). The study of developing GABAergic systems may have significance beyond the mere establishment of neuronal networks.

7.6.2 *Intrinsic vs. Extrinsic GABAergic Neurons*

A major issue in GAD immunocytochemistry is the distinction between intrinsic (local circuit) and extrinsic (projection) GABAergic neurons.

Numerous studies have demonstrated that GABA is the neurotransmitter of many local circuit neurons. Examples include: stellate cells, basket cells, and Golgi cells in the cerebellar cortex (9, 68, 71); basket cells in the hippocampus (62, 71); granule cells, medium-sized short-axon neurons, and one type of periglomerular cell in the olfactory bulb (70, 96); amacrine cells in the retina (73); and numerous cells in the spinal cord (120–122).

Only two long pathways involving GABAergic neurons are well established: the projection of Purkinje cells to cerebellar nuclei (Section

7.3.3a) and the lateral vestibular nucleus (Section 7.3.3d), and the striatonigral GABAergic projection (Section 7.3.3d; for a summary, see ref. 81).

There are undoubtedly many other GABAergic projection neurons in the central nervous system. From a biochemical standpoint, the pervasiveness of intrinsic local circuit GABAergic neurons makes the approach of lesioning such proposed pathways a relatively insensitive detection method. GAD immunocytochemistry, however, permits the demonstration of fiber tracts or known projection neurons. Where needed, immunocytochemistry can be combined with anatomical tracing methods for precise delineation of GABAergic projection neurons (see Example below). An example of a newly discovered GABAergic projection is the central GAD immunoreactive innervation of neurointermediate pituitary lobe in rat (28). In addition, numerous GABAergic projection neurons appear to reside in the so-called "tier III" of the telencephalon and basal ganglia (123).

After topical colchicine injection, the vast majority of neurons in rat globus pallidus and entopeduncular nucleus (98, 98a) and the majority of neurons in rat substantia nigra pars reticulata (99) are GAD immunoreactive. These presumed GABAergic neurons may in part represent local circuit neurons and/or they may correspond to the cells of origin for the proposed GABAergic projections from the substantia nigra pars reticulata to the tectum (124–127) and thalamus (125, 128–130), from the entopeduncular nucleus to the thalamus (131, 132) and lateral habenula (133), and from the globus pallidus to the subthalamic nucleus (134, 135). A GABAergic pallidal projection (145) to substantia nigra is discussed (95, 135). Some GABAergic projection neurons may in fact have multiple projections: for example, nigrothalamic and nigrocollicular pathways probably emanate from double-projecting (possibly GABAergic) neurons in pars reticulata of substantia nigra (136).

A number of potential GABAergic projections now proposed by biochemical, electrophysiological, immunocytochemical and morphological data are summarized in Table 4. Based on studies by the groups of Kitai (146) and Hassler (147), Ribak and coworkers (65) have suggested a distinction between projection neurons with profuse axon collaterals, such as medium-sized striatal projection neurons, and neurons with sparse axon collaterals. Thus, GABAergic projection neurons possibly need classification in several subclasses.

Example: Thalamic Relay Nuclei. Like other brain regions, thalamic relay nuclei contain GABAergic local circuit neurons as well as GABAergic projections such as the reticular neurons of thalamus and "tier III"

Table 4. Some Proposed GABAergic Projection Neurons

	Evidence (References)							
Anatomical System	Inhibitory Action	GABA Release	Pharmacological Correlation	Biochemical Studies	Anatomical Tracing	Immunocytochemistry: Cell Bodies in Area of Origin	Nerve Endings in Terminal Field	Immunocytochemical-Morphological Studies
Purkinje cells of cerebellum	75	76	75	77	171	58, 62, 71	9, 10, 12, 58	89
Striatonigral neurons	138	143		84, 137		82	69, 87, 88	87, 88
Striatopallidal neurons	144	143		84, 126, 137			82, 98	
Striatoentopenduncular neurons		143		84, 137	139		82, 98	
Entopedunculothalamic neurons (pallidothalamic neurons)	129, 130 (132, 142)			131	139	98	Section 7.4.1a	
Entopedunculohabenular neurons				133	139	98	Section 7.4.1a	
Nigrothalamic neurons	119, 120		141	118, 125, 128	136	98	Section 7.4.1a	
Nigrotectal neurons	126, 127			124, 125	124, 136	98		
Reticular thalamic neuron "projection" to thalamic nuclei	79, 80					72, 78	12	Section 7.6.2a
Centro- (possibly hypothalamo-) hypophyseal pathway				28			28	

GABAergic neurons (Table 4). The dorsal lateral geniculate body and ventrobasal complex in cat and monkey thalamus have been investigated for presence of local circuit GABAergic neurons by means of a combination of GAD immunocytochemistry and anatomical tracing methods.

Unlike rat thalamus, cat and monkey thalamus are densely populated by GAD immunoreactive cell bodies and terminals. With GAD immunocytochemistry in noncolchicine-treated cat and monkey thalamus, many immunoreactive neurons with relatively small soma are present in dorsal lateral geniculate and ventrobasal complex approximating 20% of total neurons (D. Fitzpatrick, R. Penny and D. Schmechel, unpublished observations). Their dendritic spread is quite extensive, but cell body size is tightly distributed in a low range (150–200 μm^2) compared with all neuronal cell types of each nucleus. To assess whether some of these GABAergic geniculate neurons in thalamic relay nuclei project to cortex, large injections of horseradish peroxidase conjugated to wheat germ agglutinin were made in visual and somatosensory cortex. GAD immunocytochemistry was then combined with demonstration of retrogradely labeled cells using heavy metal-DAB techniques (90, 148). No double-labeled cells have been detected. In contrast, injections into thalamus yield double-labeled reticular neurons of thalamus (GAD-immunoreactive reticular neurons with retrogradely transported horseradish peroxidase conjugate visualized).

Thus, GABAergic neurons in cat and monkey thalamus represent a significant proportion of neurons (approximately 20%) but appear to be predominantly local circuit neurons.

7.6.3 Correlation with Other Cell Characteristics

Staining for GAD in the superficial layers of monkey visual cortex shows regular alternating patches of immunoreactive terminals occurring approximately every 400 μm (149). This pattern correlates exactly to patchy staining of cytochrome oxidase, a mitochondrial enzyme (150). Other investigations suggest that these patches correlate with physiologic columns that are relatively nonorientation-specific and that respond to diffuse light (151).

From the above correlation of GAD staining and cytochrome oxidase activity comes the hypothesis that GABAergic neurons are distinctively rich in mitochondria and iron content (149). There are several regions that show such a correlation, including the cerebellum and basal ganglia (152). However, the areas of relatively lighter cytochrome oxidase activity in superficial layers of monkey visual cortex (149) also contain many GAD immunoreactive terminals and, in fact, respond acutely in their orientation-specific properties to interference with GABAergic transmission (153, 154). In addition, there are certain cell classes, such as deep projection neurons of visual cortex, that are heavily stained by cytochrome oxidase activity (Schmechel, unpublished observation) but are not known to be GABAergic. Furthermore, in the dentate gyrus of the rat hippocampus, granule cell dendrites contain a higher cytochrome oxidase activity than granule cell bodies and axon terminals and pyramidal basket cells (154a), whereas GAD immunoreactivity is present in basket cells (71). Future studies will have to explore whether iron histochemistry or cytochrome oxidase histochemistry may be a reliable alternative or adjunct to GAD immunocytochemistry in a given brain region. Immunocytochemical comparison of GAD antisera with antisera to other neuronal constituents may reveal that GAD immunocytochemistry can be replaced with other marker techniques for certain GABAergic cell types (see also Section 7.5). For example, the calcium-binding protein parvalbumin is found in some of the GABAergic cell classes of cerebellum (155).

7.6.4 GABAergic Neurons and Other Neurotransmitter Systems

Immunocytochemistry with double-staining techniques offers the possibility of studying the interaction of GABAergic neurons with other neurotransmitter systems. In some cases, neurons that are GABAergic (GAD immunoreactive) may also contain other neurotransmitters. For these experiments, brain areas with colchicine-independent GAD immunoreactivity in cell bodies are currently the easiest to investigate (Table 3). Cerebellar cortex is the most suitable since all neuronal cell classes except granule cells appear to be GAD immunoreactive.

7.6.4a Coexistence of GABA with Other Neutrotransmitters. In cerebellum, recent studies have demonstrated methionine enkephalin-like immunoreactivity (156) and ^{3}H-glycine uptake (157) in some Golgi cells and motilin-like immunoreactivity in some Purkinje cells (158, 158a). These data suggest that some GABAergic neurons in cerebellum contain more than one neurotransmitter. Such neurons with multiple neurotransmitters may not include all members of a given GABAergic system, and therefore they may define subclasses.

In reticular nucleus of cat thalamus, GAD immunoreactivity coexists with somatostatin-like immunoreactivity, as shown by a two-color double-immunoperoxidase method (78). The physiological significance of this coexistence remains unclear. Such examples may be limited to certain species, since reticular neurons of rat thalamus are GAD-positive but are not somatostatin-like immunoreactive.

Like many other neurotransmitters, GABA apparently is contained in peripheral neuroendocrine cells. In pancreas, pancreatic β-islet cells are GAD immunoreactive, and biochemical evidence supports a local role for GABA in the regulation of somatostatin release (159). This example also makes the further point that the association of GAD immunoreactivity with other neurotransmitters may not only be species specific but also cell-class specific since pancreatic somatostatin-containing cells do not stain for GAD (159).

7.6.4b Interaction of GABAergic Neurons with Other Cell Classes. As pointed out by Roberts (160), there is an enormous and relatively uncharted area concerning the relationship of GABAergic neurons with other neuronal systems. The widespread distribution of GABAergic neurons and the fundamental role of inhibition in nervous system function suggests that such interactions will be numerous and potentially different from region to region.

One such example is the GABAergic role in processing sensory input in spinal cord, including presynaptic inhibition (68). Another is the presence of synaptoid junctions of GAD immunoreactive axon terminals on neurosecretory axons as well as local neuroendocrine cells in posterior pituitary lobe (28). In these cases, the postsynaptic elements are known to contain, respectively, oxytocin, vasopressin, and α-melanocyte-stimulating hormone (α-MSH). GABAergic interactions with neurons synthesizing neurotransmitters such as dopamine and acetylcholine are known in the nigrostriatal system (161) and thalamus (162). Even in these regions, the relationship of local circuit cholinergic neurons to GABAergic neurons in striatum and the prominent cholinergic innervation of reticular neurons in thalamus (162, 163) remain to be defined more adequately at the ultrastructural level.

The interaction of GABAergic and dopaminergic neurons has considerable relevance because of the involvement of the nigrostriatal and mesolimbic dopaminergic system in human neurological and psychiatric diseases.

The importance of the role of immunocytochemistry in such studies is illustrated by the following examples:

1. In double light microscopic immunocytochemistry with antisera to GAD and tyrosine hydroxylase (TH), a marker for dopaminergic neurons (164), GAD immunoreactive punctate profiles outline TH-positive processes and cell bodies (99).
2. Dopaminergic dendrites in pars reticulata of rat substantia nigra receive striatal axon terminals (165), now considered to contain GABA or the neuropeptide substance P (81).

3. In normal and colchicine-pretreated rats, immunocytochemistry in substantia nigra pars compacta reveals GAD immunoreactive terminals on unstained, presumably dopaminergic cell bodies and dendrites (69, 99). On the other hand, numerous GABAergic cell bodies are present in substantia nigra pars reticulata of colchicine-pretreated rat (87, 89) in agreement with biochemical and electrophysiological data (Section 7.6.2, 125, 128, 129). At the ultrastructural level GAD immunoreactive terminals synapse with both the dendrites and cell bodies of these GAD immunoreactive neurons and unstained profiles (98).

These data show a monosynaptic GABAergic influence on GABAergic neurons of pars reticulata. The finding could correspond to a proposed inhibitory, monosynaptic striatonigral input on nigrothalamic projection neurons in substantia nigra pars reticulata (138, 166). Since GAD immunoreactive terminals in pars reticulata might be of local, striatal (87, 88), or other origin, full confirmation of a GABAergic striatonigral-GABAergic nigro-thalamic connection will require the combination of immunocytochemistry and lesions or anatomical tracers (166).

All studies involving the interaction of GABAergic systems with other neurons may need quantitative double label (pre-embedding) immunocytochemistry at the ultrastructural level. As most polyclonal antisera are raised in *rabbit* and most monoclonal antibodies stem from *mouse* cell lines, *sheep* antiserum S3 in conjunction with non-cross-reacting linking antibodies allows a relatively simple double immunocytochemical technique at the light microscopic level (99). The corresponding technique for the electron microscopic level remains to be established.

SUMMARY

The purpose of this chapter has been to present aspects of (i) GAD purification, (ii) antiserum production and characterization, and (iii) GAD immunocytochemistry.

Three rabbit anti-GAD sera (7, 11, 12) and sheep anti-GAD serum S3, developed at NIMH (13), are compared as specific reagents for the immunohistochemical identification of GAD-containing neurons, presumed to be GABAergic.

The major problems at present in GAD-*like* immunocytochemistry (54) revolve around the staining of cell bodies compared with terminals. This may be due to low concentration of GAD in the cell body, antigen loss during fixation, and/or antigenic difference between high-molecular-weight GAD (cell body) and low-molecular-weight GAD (axonal) (36). The use of colchicine in those brain regions where incomplete staining occurs is at present a tedious but effective answer.

Further investigation of tissue fixation and improved sensitivity of immunocytochemical techniques eventually may circumvent the need for colchicine. In the meantime, there will undoubtedly be additions to the list of known GABAergic projection and local circuit neurons. The study of the ontogeny of GABAergic systems during development, for example, will greatly depend on adequate methods for cell body staining. Analysis of the time of origin of GABAergic neurons, and development of GABAergic local circuit and projection cell classes, will be of critical importance to understanding the neurochemical anatomy of adult nervous system emphasized in recent work (123). Further studies are needed to test the hypothesis that GABAergic neurons are especially metabolically active cells (149) by looking at other indices such as 2-deoxyglucose uptake, cytochrome oxidase staining, or content of glycolytic enzymes.

The relationship of GABAergic neurons to other transmitter systems is a further level of complexity. The solution to some questions may well depend on newly developed techniques, for example, double immunocytochemistry and immunocytochemistry combined with axonal tracing techniques and Golgi ultrastructural analysis (166).

At the simplest level, several GABAergic neurons are now known to contain multiple neurotransmitters (Section 7.6.4a). The presence of other "neurotransmitters" in GABAergic neurons (e.g., somatostatin in reticular neurons of the cat thalamus) may significantly vary from species to species (78) or may be in some cases an indicator of neuronal subclasses.

The widespread occurrence of neurotransmitter receptor subtypes calls for analysis of the relationship of GABAergic neurons to postsynaptic elements and receptor populations. An immediate example is the diazepam-GABA-receptor correspondence (167) in many brain regions.

Application of GAD immunocytochemistry to models of human disease have been limited to date: Experimental models of epilepsy produce a decrease of GAD-positive nerve terminals in seizure foci (168, 169), which may be of relevance to seizure disorders. In respect to human disease in general, the fundamental and widespread nature of GABAergic elements apparently mitigates against clear-cut primary disorders of GABAergic neurons per se (although GABAergic subclasses may be affected). Nevertheless, proposed treatments for several neurological disorders may be dependent on or limited by GABAergic mechanisms (170). The understanding of GABAergic neurons and their interactions with other neuronal classes will be essential to such pharmacological endeavors and will depend in part on GAD immunocytochemistry.

ACKNOWLEDGMENTS

The authors gratefully acknowledge the technical assistance of David H. Ransom, Tony Cridlin, Lester Carmon, Samuel Wright, and Stephanie Smith. We thank Dr. Marcel L. Tappaz and Dr. Joachim R. Wolff for valuable suggestions and Ingeborg Keil, Doris Mitteregger and Ernst Mitteregger for secretarial assistance. This work is supported by Deutsche Forschungsgemeinschaft grant Oe 95/2-1 (WHO). Last, but certainly not least, we wish to dedicate this chapter to Virginia K. Weise, Irwin J. Kopin, Marcel L. Tappaz, J. W. Daly, Anne-Lise Dahl, and our wives.

REFERENCES

1. J. Awapara, A. J. Landua, R. Fuerst, and B. Seale, *J. Biol. Chem.* **187**, 35–39 (1950).
2. E. Roberts and S. Frankel, *J. Biol. Chem.* **187**, 55–63 (1950).
3. S. Udenfriend, *J. Biol. Chem.* **187**, 65–69 (1950).
4. D. B. Hope, *Biochemistry* **59**, 497–500 (1955).
5. E. Roberts, Ed., *Inhibition in the Nervous System and Gamma-Aminobutyric acid*, Pergamon Press, New York, 1960.
6. L. Salganicoff and E. DeRobertis, *J. Neurochem.* **12**, 287–309 (1965).
7. J.-Y. Wu, T. Matsuda, and E. Roberts, *J. Biol. Chem.* **248**, 3029–3034 (1973).
8. J.-Y. Wu, in E. Roberts, T. N. Chase, and D. B. Tower, Eds., *GABA in Nervous System Function*, Raven Press, New York, 1976, pp. 7–55.
9. K. Saito, R. Barber, J.-Y. Wu, T. Matsuda, E. Roberts, and J. E. Vaughn, *Proc. Natl. Acad. Sci. USA* **71**, 269–273 (1974).
10. B. J. McLaughlin, J. G. Wood, K. Saito, R. Barber, J. E. Vaughn, E. Roberts, and J.-Y. Wu, *Brain Res.* **76**, 377–391 (1974).
11. M. Maitre, J. M. Blindermann, L. Ossola, and P. Mandel, *Biochem. Biophys. Res. Commun.* **85**, 885–890 (1978).
12. M. Pérez de la Mora, L. D. Possani, R. Tapia, L. Teran, R. Palacios, K. Fuxe, T. Hökfelt, and Å. Ljungdahl, *Neuroscience* **6**, 875–895 (1981).
13. W. H. Oertel, D. E. Schmechel, M. L. Tappaz, and I. J. Kopin, *Neuroscience* **6**, 2689–2700 (1981).
14. E. Roberts, in P. Krogsgaard-Larsen, J. Scheel-Krüger, and H. Kofod, Eds., *GABA-Neurotransmitters*, Alfred Benzon Symposium 12, Academic Press, New York, 1979, pp. 28–45.
15. P. L. McGeer, J. C. Eccles, and E. G. McGeer, *Molecular Neurobiology of the Mammalian Brain*, Plenum Press, New York, 1978.
16. R. W. Albers and R. Brady, *J. Biol. Chem.* **234**, 926–928 (1959).
17. J. Storm-Mathisen and F. Fonnum, *J. Neurochem.* **18**, 1105–1111 (1971).
18. O. Chude and J.-Y. Wu, *J. Neurochem.* **27**, 83–86 (1976).
19. E. Hamel, I. E. Goetz, and E. Roberts, *J. Neurochem.* **37**, 1032–1038 (1981).
20. I. Kanazawa, L. L. Iversen, and J. S. Kelly, *J. Neurochem.* **27**, 1267–1269 (1976).

21. J.-Y. Wu, O. Chude, J. Wein, E. Roberts, K. Saito, and E. Wong, *J. Neurochem.* **30**, 849–857 (1978).
22. B. Haber, K. Kuriyama, and E. Roberts, *Science* **168**, 598–599 (1970).
23. B. Haber, K. Kuriyama, and E. Roberts, *Biochem. Pharmacol.* **19**, 1119–1136 (1970).
24. L. P. Miller and D. L. Martin, *Life Sci.* **13**, 1023–1032 (1973).
25. P. H. Wu, D. A. Durden, and L. Hertz, *J. Neurochem.* **32**, 379–390 (1979).
26. E. Hamel, D. N. Krause, and E. Roberts, *Brain Res.* **223**, 199–204 (1981).
27. J.-Y. Wu, *J. Neurochem.* **28**, 1359–1367 (1977).
28. W. H. Oertel, E. Mugnaini, M. L. Tappaz, V. K. Weise, A. L. Dahl, D. E. Schmechel, and I. J. Kopin, *Proc. Natl. Acad. Sci. USA* **79**, 675–679 (1982).
29. J. S. Enna, M. J. Kuhar, and S. H. Snyder, *Brain Res.* **93**, 168–174 (1975).
30. B. Cronwall and J. R. Wolff, *J. Comp. Neurol.* **190**, 187–208 (1980).
31. J.-Y. Wu, Y. Y. T. Su, M. K. Lam, and C. Brandon, in F. V. DeFeudis and P. Mandel, Eds., *Advances in Biochemical Psychopharmacology, Vol. 29: Amino Acid Neurotransmitters*, Raven Press, New York, 1981, pp. 499–508.
32. R. A. Hadjian and J. A. Stewart, *J. Neurochem.* **28**, 1249–1257 (1977).
33. J.-Y. Wu, *Proc. Natl. Acad. Sci. U.S.A.*, **73**, 4270–4274 (1982).
34. J. M. Blindermann, M. Maitre, L. Ossola, and P. Mandel, *Eur. J. Biochem.* **86**, 143–152 (1978).
35. Y. Y. T. Su, J.-Y. Wu, and D. M. K. Lam, *J. Neurochem.* **33**, 169–179 (1979).
36. J.-Y. Wu, E. Wong, K. Saito, E. Roberts, and A. Schousboe, *J. Neurochem.* **27**, 653–659 (1976).
37. T. Matsuda, J.-Y. Wu, and E. Roberts, *J. Neurochem.* **21**, 167–172 (1973).
38. L. D. Possani, A. Bayon, and R. Tapia, *Neurochem. Res.* **2**, 51–57 (1977).
39. W. H. Oertel, D. E. Schmechel, J. W. Daly, M. L. Tappaz, and I. J. Kopin, *Life Sci.* **27**, 2133–2141 (1980).
40. M. J. Jung, B. Lippert, B. W. Metcalf, P. J. Schechter, P. Böhlen, and A. Sjoerdsma, *J. Neurochem.* **28**, 717–723 (1977).
41. M. J. Jung and N. Seiler, *J. Biol. Chem.* **253**, 7431–7439 (1978).
42. W. H. Oertel, D. E. Schmechel, V. K. Weise, D. H. Ransom, M. L. Tappaz, H. C. Krutzsch, and I. J. Kopin, *Neuroscience* **6**, 2701–2714 (1981).
42a. R. M. Spears and D. L. Martin, *J. Neuro Chem.*, **38**, 985–991 (1982).
43. A. A. Heinamaki and R. S. Piha, *Trans. Int. Soc. Neurochem.* 375 (1979).
44. I. Schousboe, B. Bro, and A. Schousboe, *J. Biochem.* **162**, 303–307 (1977).
45. M. Maitre, L. Ciesielski, C. Casil, and P. Mandel, *J. Biochem.* **52**, 157–169 (1975).
46. K. Saito, in E. Roberts, T. N. Chase, and D. B. Tower, Eds., *GABA in Nervous System Function*, Raven Press, New York, 1976, pp. 103–111.
47. J. G. Jacobsen and L. H. Smith, Jr., *Physiol. Review* **48**, 424–511 (1968).
48. W. H. Oertel, V. K. Weise, A. H. Bruckner, and I. J. Kopin, "Cysteine Sulfinate Decarboxylase Form I Present in Denervated Rat Neurointermediate Pituitary Lobe," in preparation.
49. A. J. Crowle, A. J. Revis, and K. Jarrett, *Immunol. Commun.* **1**, 325–336 (1972).
50. H. D. Caldwell, C.-C. Kuo, and G. E. Kenny, *J. Immunol.* **115**, 969–975 (1975).
51. A. Köhler and C. Milstein, *Nature* **258**, 495–497 (1975).

51a. J.-Y. Wu, C.-T. Lin, C. Brandon, T.-S. Chan, H. Möhler, and J. G. Richards, in V. Chan-Palay and S. L. Palay, Eds., *Cytochemical Methods in Neuroanatomy*, Alan R.Liss, New York, 1982, pp. 279–296.

52. E. G. Jones and B. K. Hartmann, *Ann. Rev. Neurosci.* **1**, 215–296 (1978).

53. J. M. Blindermann, M. Maitre, and P. Mandel, *J. Neurochem.* **32**, 245–246 (1979).

54. J. Rossier, *Neuroscience* **6**, 989–991 (1981).

55. R. Barber and K. Saito, in E. Roberts, T. N. Chase, and D. B. Tower, Eds., *GABA in Nervous System Function*, Raven Press, New York, 1976, pp. 113–132.

56. D. Nanopoulos, M.-F. Belin, M. Maitre, and F.-J. Pujol, *C. R. Acad. Sci., Paris, Ser. D*, **290**, 1153–1156 (1980).

57. D. Nanopoulos, M.-F. Belin, M. Maitre, G. Vincendon and J. F. Pujol, *Brain Res.* **232,** 375–389 (1982).

58. W. H. Oertel, D. E. Schmechel, E. Mugnaini, M. L. Tappaz, and I. J. Kopin, *Neuroscience* **6**, 2715–2735 (1981).

59. I. W. McLean and P. K. Nakane, *J. Histochem. Cytochem.* **22**, 1077–1083 (1974).

60. A. Berod, B. K. Hartmann, and J. F. Pujol, *J. Histochem. Cytochem.* **29**, 844–850 (1981).

61. A. H. Coons, in J. F. Danielli, Ed., *General Cytochemical Methods*, Academic Press, New York, 1958, pp. 399–422.

62. W. H. Oertel, E. Mugnaini, D. E. Schmechel, M. L. Tappaz, and I. J. Kopin, in V. Chan-Palay and S. L. Palay, Eds., *Cytochemical Methods in Neuroanatomy*, Alan R. Liss, New York, 1982, pp. 297–329.

63. L. A. Sternberger, *Immunocytochemistry*, 2nd ed., Wiley, New York (1979).

64. J. G. Wood, B. J. McLaughlin, and J. E. Vaughn, in E. Roberts, T. N. Chase, and D. B. Tower, Eds., *GABA in Nervous System Function*, Raven Press, New York, 1976, pp. 133–148.

65. C. E. Ribak, J. E. Vaughn, and R. P. Barber, *Histochem. J.* **13**, 555–582 (1981).

66. C. E. Ribak, *J. Neurocytol.* **7**, 461–478 (1978).

67. V. L. Friedrich, Jr., and E. Mugnaini, in L. Heimer and M. J. Robards, Eds., *A Handbook of Neuroanatomical Tract Tracing Techniques*, Plenum Press, New York, 1981, pp. 345–375.

68. R. P. Barber, J. E. Vaughn, K. Saito, B. J. McLaughlin, and E. Roberts, *Brain Res.* **141**, 35–55 (1978).

69. C. E. Ribak, J. E. Vaughn, K. Saito, R. Barber, and E. Roberts, *Brain Res.* **116**, 287–298 (1976).

70. C. E. Ribak, J. E. Vaughn, K. Saito, R. Barber, and E. Roberts, *Brain Res.* **126**, 1–18 (1977).

71. C. E. Ribak, J. E. Vaughn, and K. Saito, *Brain Res.* **140**, 315–332 (1978).

72. C. R. Houser, J. E. Vaughn, R. P. Barber, and E. Roberts, *Brain Res.* **200**, 341–354 (1980).

73. C. Brandon, D. M. K. Lam, and J.-Y. Wu, *Proc. Natl. Acad. Sci. USA* **76**, 3557–3561 (1979).

74. F. Vandesande, *J. Neurosci. Methods* **1**, 3–23 (1979).

75. M. Ito and M. Yoshida, *Experientia* **20**, 515–516 (1964).

76. K. Obata and K. Takeda, *J. Neurochem.* **16**, 1043–1047 (1969).

77. F. Fonnum, J. Storm-Mathisen, and F. Walberg, *Brain Res.* **20**, 259–275 (1970).

78. W. H. Oertel, A. M. Graybiel, E. Mugnaini, R. P. Elde, D. E. Schmechel, and I. J. Kopin, *J. Neuroscience*, in press (1983).

79. J. Schlag and M. Waszak, *Brain Res.* **21**, 286–288 (1970).

80. Y. Lamarre, M. Filion, and J. P. Cordeau, *Exp. Brain Res.* **12**, 480–498 (1971).

81. A. Dray, *Neuroscience* **4**: 1407–1439 (1979).

82. C. E. Ribak, J. E. Vaughn, and E. Roberts, *J. Comp. Neurol.* **187**, 261–284 (1979).

83. M. J. Brownstein, E. A. Mroz, M. L. Tappaz, and S. E. Leeman, *Brain Res.* **135**, 315–323 (1977).

84. F. Fonnum, Z. Gottesfeld, and I. Grovofa, *Brain Res.* **143**, 125–138 (1978).

85. J. T. Coyle and R. Schwarcz, *Nature* **263**, 244–246 (1976).

86. E. G. McGeer and P. L. McGeer, *Nature* **263**, 517–519 (1976).

87. C. E. Ribak, J. E. Vaughn, and E. Roberts, *Brain Res.* **192**, 413–420 (1980).

88. W. H. Oertel, D. E. Schmechel, J. M. Brownstein, M. L. Tappaz, D. H. Ransom, and I. J. Kopin, *J. Histochem. Cytochem.* **29**, 977–980 (1981).

89. C. R. Houser, J. E. Vaughn, and R. P. Barber, *Soc. Neurosci. Abstr.* **7**, 40 (1981).

89a. M. Wassef, M. L. Tappaz, W. H. Oertel, L. Paût and C. Sotelo, *Neurosci. Lett. Suppl.* S513 (1982).

90. J. C. Adams, *J. Histochem. Cytochem.* **29**, 775 (1981).

91. M. L. Tappaz, M. J. Brownstein, and M. Palkovits, *Brain Res.* **108**, 371–379 (1976).

91a. S. R. Vincent, T. Hökfelt, and J.-Y. Wu, *Neuroendocrinology* **34**, 117–125, 1982.

92. M. L. Tappaz, M. Wassef, W. H. Oertel, L. Paut and J. F. Pujol, *Neuroscience*, in press (1983).

92a. M. L. Tappaz, W. H. Oertel, M. Wassef and E. Mugnaini, in R. M. Buijs, P. Pévet and D. F. Swaab, Eds., Chemical Transmission in the brain. The role of amines, amino acids and peptides. Elsevier Biomedical, Amsterdam. *Progress in Brain Res.* **55**, 77–96 (1982).

93. E. Mugnaini, *Brain Res.* **17**, 169–179 (1970).

94. E. Mugnaini, in O. Larsell and J. Jansen, Eds., *The Comparative Anatomy and Histology of the Cerebellum: The Human Cerebellum, Cerebellar Connections, and Cerebellar Cortex*, University of Minnesota Press, Minneapolis, 1972, pp. 201–332.

95. G. DiChiara, M. Morelli, M. L. Porceddu, M. Mulas, and M. DelFiacco, *Brain Res.* **189**, 193–208 (1980).

96. E. Mugnaini and W. H. Oertel, "Three Classes of Glutamate Decarboxylase Immunoreactive Neurons in the Olfactory Bulb of the Rat," in preparation.

97. E. Mugnaini and W. H. Oertel, *Soc. Neurosci. Abstr.* **7**, 112 (1981).

98. W. H. Oertel, E. Mugnaini, C. Nitsch, D. E. Schmechel, and I. J. Kopin, *Soc. Neurosci. Abstr.* **8**, 508 (1982).

98a. W. H. Oertel, C.Nitsch and E. Mugnaini, in R. Hassler, Ed., *Advances in Neurology*, Raven Press, New York, in press (1983).

99. W. H. Oertel, M. L. Tappaz, A. Berod, and E. Mugnaini, *Brain Res. Bull.* **9**, 463–474 (1982).

100. M. J. Neal and L. L. Iversen, *J. Neurochem.* **16**, 1245–1252 (1969).

101. T. Hökfelt and Å. Ljungdahl, *Exp. Brain Res.* **14**, 354–362 (1972).

102. N. G. Bowery, G. P. Jones and M. J. Neal, *Nature* **264**, 281–284 (1976).

103. M. Tappaz, M. Aguera, M. F. Belin, and J. F. Pujol, *Brain Res.* **186**, 379–391 (1980).

104. E. A. Neale, W. H. Oertel, L. M. Bowers, and V. K. Weise, *J. Neurosci.* In press (1983).

105. C. Houser, J. E. Vaughn, E. G. Jones, and S. H. C. Hendry, *Soc. Neurosci. Abstr.* **6**, 159 (1980).

106. S. H. C. Hendry and E. G. Jones, *J. Neurosci.* **1**, 390–408 (1981).

107. N. Halasz, Å. Ljungdahl and T. Hökfelt, *Brain Res.* **167**, 221–240 (1979).

108. P. M. Beart, J. S. Kelly, and F. Schon, *Biochem. Soc. Trans.* **2**, 266–268 (1974).

109. M. F. Belin, H. Gamrani, M. Aguera, A. Calas, and J. F. Pujol, *Neuroscience* **5**, 241–254 (1980).

110. M. F. Diebler, E. Farkas-Bargeton, and R. Wehrlé, *J. Neurochem.* **32**, 429–435 (1979).

111. B. W. L. Brooksbank, D. J. Atkinson, and R. Balázs, *Dev. Neurosci.* **4**, 188–200 (1981).

112. M. Jacobson, *Developmental Neurobiology*, 2nd ed., Plenum Press, New York, 1978.

113. B. J. McLaughlin, J. G. Wood, K. Saito, E. Roberts, and J.-Y. Wu, *Brain Res.* **85**, 355–371 (1975).

114. P. Rakic, *Science* **183**, 425–427 (1974).

115. J. R. Wolff, B. M. Chronwall, and M. Rickmann, in V. Neuhoff, Ed., *Proceedings of the European Society for Neurochemistry*, Vol. 1, Verlag Chemie, Weinheim, New York, 1978, pp. 158–173.

116. W. Singer, *Exp. Brain Res.* **30**, 25–41 (1977).

117. N. Seiler, G. Bink, and J. Grove, *Neurochem. Res.* **4**, 425–447 (1979).

118. N. Seiler, S. Sharhan, and B. F. Roth-Schechter, *Dev. Neurosci.* **4**, 181–187 (1981).

119. B. Goertz, *Exp. Brain Res.* **34**, 365–372 (1979).

120. R. P. Barber, J. E. Vaughn, and E. Roberts, *Brain Res.* **238**, 305–328 (1982).

121. A. I. Basbaum, E. J. Glazer, and W. H. Oertel, *Soc. Neurosci. Abstr.* **7**, 528 (1981).

122. S. P. Hunt, J. S. Kelly, P. C. Emson, J. R. Kimmel, R. J. Miller, and J.-Y. Wu, *Neuroscience* **6**, 1883–1898 (1981).

123. H. J. W. Nauta, *Neuroscience* **4**, 1875–1881 (1979).

124. S. R. Vincent, T. Hattori, and E. G. McGeer, *Brain Res.* **151**, 159–164 (1978).

125. G. DiChiara, M. L. Porceddu, M. Morelli, M. L. Mulas, and G. L. Gessa, *Brain Res.* **176**, 273–284 (1979).

126. G. Chevalier, A. M. Thierry, T. Shibazaki, and J. Féger, *Neurosci. Lett.* **21**, 67–70 (1981).

127. G. Chevalier, J. M. Deniau, A. M. Thierry, and J. Féger, *Brain Res.* **213**, 253–263 (1981).

128. M. S. Starr and I. C. Kilpatrick, *Neuroscience* **6**, 1095–1104 (1981).

129. A. Ueki, M. Uno, M. Anderson, and M. Yoshida, *Experientia* **33**, 1480–1482 (1977).

130. J. M. Deniau, D. Lackner, and J. Féger, *Brain Res.* **145**, 27–35 (1978).

131. J. B. Penney, Jr., and A. B. Young, *Brain Res.* **207**, 195–199 (1981).

132. M. Uno and M. Yoshida, *Brain Res.* **99**, 377–380 (1975).

133. J. I. Nagy, D. A. Carter, J. Lehmann, and H. C. Fibiger, *Brain Res.* **145**, 360–364 (1978).

134. F. Fonnum, I. Grofová, and E. Rinvik, *Brain Res.* **153**, 370–374 (1978).

135. D. van der Kooy, T. Hattori, K. Shannak, and O. Hornykiewicz, *Brain Res.* **204**, 253–268 (1981).

136. M. Bentivoglio, D. van der Kooy, and H. J. M. Kuypers, *Brain Res.* **174**, 1–17 (1979).

137. J. I. Nagy, D. A. Carter, and H. C. Fibiger, *Brain Res.* **158**, 15–29 (1978).

138. J. M. Deniau, J. Féger, and C. LeGuyader, *Brain Res.* **104**, 152–156 (1976).

139. D. van der Kooy and D. A. Carter, *Brain Res.* **211**, 15–36 (1981).

140. A. M. Graybiel and C. W. Ragsdale, in M. Cŭenod, G. W. Kreutzberg, and F. E. Bloom, Eds., Development and chemical specificity of neurons, Elsevier Biomedical, Amsterdam, *Prog. Brain Res.* **51**, 239–283 (1979).

141. M. Yoshida and S. Omata, *Experientia* **35**, 794 (1979).

142. M. Uno, N. Ozawa, and M. Yoshida, *Exp. Brain Res.* **33**, 493–507 (1978).

143. C. Gauchy, M. L. Kemel, J. Glowinski, and M. J. Besson, *Brain Res.* **193**, 129–141 (1980).

144. M. Yoshida, A. Rabin, and M. Anderson, *Exp. Brain Res.* **15**, 333–347 (1974).

145. T. Hattori, P. L. McGeer, H. C. Fibiger, and E. G. McGeer, *Brain Res.* **54**, 103–114 (1973).

146. S. I. Kitai, R. J. Preston, G. A. Bishop, and J. D. Kocsis, in L.-J. Poirier, T. L. Sourkes, and P. J. Bédard, Eds., *Advances in Neurology*, Vol. 24, *The Extrapyramidal System and Its Disorders*, Raven Press, New York, 1979, pp. 45–51.

147. R. Hassler, J. W. Chung, A. Wagner, and U. Rinne, *Neurosci. Lett.* **5**, 117–121 (1977).

148. R. M. Bowker, H. W. M. Steinbusch, and J. D. Coulter, *Brain Res.* **211**, 412–417 (1981).

149. A. E. Hendrickson, S. P. Hunt, and J.-Y. Wu, *Nature* **292**, 605–607 (1981).

150. M. Wong-Riley, *Brain Res.* **171**, 11–28 (1979).

151. D. H. Hubel and M. S. Livingstone, *Soc. Neurosci. Abstr.* **7**, 367 (1981).

152. L. Heimer, R. D. Switzer, and G. W. van Hoesen, *Trends Neurosci.* **5**, 83–87 (1982).

153. T. Tsumoto, W. Eckart, and O. D. Creutzfeldt, *Exp. Brain Res.* **34**, 351–363 (1979).

154. A. M. Sillito, J. A. Kemp, and C. Blakemore, *Nature* **291**, 318–320 (1981).

154a. C. E. Ribak, *Brain Res.* **212**, 169–174 (1981).

155. M. R. Celio and C. W. Heizmann, *Nature* **293**, 300–302 (1981).

156. J. A. Schulmann, T. E. Finger, N. Brecha, and H. J. Karten, *Neurosci.* **6**, 2407–2416 (1981).

157. R. Balázs, J. Cohen, J. Garthwaite, and P. L. Woodhams, in F. Fonnum, Ed., *Amino Acids As Chemical Transmitters*, Plenum Press, New York, 1977, pp. 629–651.

158. G. Nilaver, R. Defendini, E. A. Zimmermann, M. C. Beinfeld, and T. L. O'Donohue, *Nature* **295**, 597–598 (1982).

158a. V. Chan-Palay, C. Nilauer, S. L. Palay, M. C. Beinfeld, E. A. Zimmermann, J.-Y. Wu, and T. L. O'Donohue, *Proc. Natl. Acad. Sci. U.S.A.* **78**, 7787–7791, 1981.

159. M. S. Robbins, R. L. Sorenson, R. P. Elde, D. E. Schmechel, and W. H. Oertel, in S. Raptis and J. Gerich, Eds., *Proceedings of the 2nd International Symposium on Somatostatin*, Academic Press, New York, in press (1983).

160. E. Roberts, in P. Krogsgaard-Larsen, J. Scheel-Krüger, and H. Kofod, Eds., *GABA-Neurotransmitters*, Alfred Benzon Symposium 12, Academic Press, New York, 1979, pp. 533–543.

161. G. Bartholini, *Trends Pharmacol. Sci.*, 138–141 (1980).

162. Y. Ben-Ari, R. Dingledine, I. Kanazawa, and J. S. Kelly, *J. Physiol.* **261**, 647–671 (1976).

163. H. Kimura, P. L. McGeer, J. H. Peng, and E. G. McGeer, *J. Comp. Neurol.* **200**, 151–201 (1981).

164. V. M. Pickel, T. H. Joh, and D. J. Reis, *Brain Res.* **85**, 295–300 (1975).

165. M. Wassef, A. Berod, and C. Sotelo, *Neuroscience* **6**, 2125–2139 (1981).

166. P. Somogyi, A. J. Hodgson, and A. D. Smith, Neuroscience **4**, 1805–1852 (1979).

167. H. Möhler, J. G. Richards, and J.-Y. Wu, *Proc. Natl. Acad. Sci. U.S.A.* **78**, 1935–1938 (1981).

168. C. E. Ribak, A. B. Harris, J. E. Vaughn, and E. Roberts, *Science* **205**, 211–214 (1979).

169. C. E. Ribak, R. M. Bradburne, and A. B. Harris, *Soc. Neurosci. Abstr.* **7**, 628 (1981).

170. P. Krogsgaard-Larsen, J. Scheel-Krüger, and H. Kofod, Eds., *GABA-Neurotransmitters*, Alfred Benzon Symposium 12, Academic Press, New York, 1979.

171. P. L. McGeer, T. Hattori, and E. G. McGeer, *Exp. Neurol.* **47**, 26–41 (1975).

Chapter 3

Immunocytochemistry of Peptides

Stanley J. Watson and Huda Akil

Department of Psychiatry and Mental Health Research Institute
University of Michigan, Ann Arbor

1 INTRODUCTION

The ability to localize a specific protein or peptide sequence to a particular cellular population or to a specific organelle within a cell has proved to be an enormously powerful tool. By far the most specific and efficient method for this localization involves the use of immunocytochemistry (ICC) (Figure 1) (1). This method is an interesting combination of strategies from anatomy, biochemistry, and immunology (2). Anatomy provides the context for localizing the molecule. Sectioning, fixation methods, and anatomical relationships are examples of the technical and logical base provided from anatomy. Biochemistry is often drawn on to supply

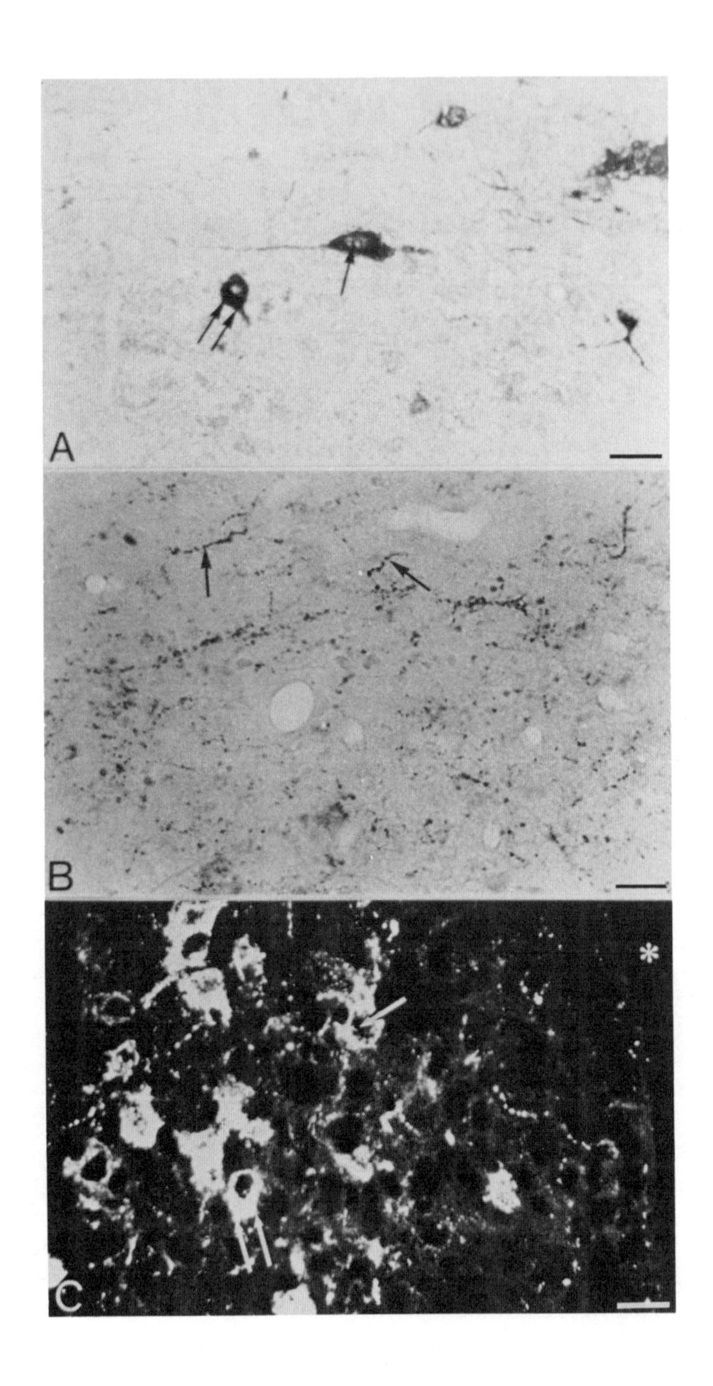
A
B
C

a "pure" protein, or for the synthesis of the substance in question, not to mention the primary logical context for studying the molecule. Finally, immunological tools are needed for the production of antisera against all or a critical part of the molecule to be localized. For example, it is often of great interest to study a biologically active peptide sequence in several tissue types using antisera against two or three different regions of the peptide molecule. With the methods properly applied from the main areas mentioned above, information can be obtained about the molecule in question. That is, it can be localized to a particular cell type and subcellular organelle, and the general structure and biochemical identity of the molecule can be estimated by the use of convergent antisera and different competition paradigms. Often critical inferences can flow from this type of data. For example, if a mythical protein X is found in secretory granules in an endocrine cell, one is inclined to think of it as a potentially releasable substance (and should perhaps consider studies in body fluids and of its endocrine effects, etc.). In contrast, if protein X is located in the cell membrane of fibroblasts a very different type of physiology is suspected.

Under the proper conditions, and with some luck, the types of inference described above can be validly drawn from a careful series of immunocytochemical studies. Unfortunately, several logical and technical barriers can produce major interpretive flaws in such work. In this chapter we address some of the clearer immunocytochemical methods, some common problems, and a few solutions to these problems, while attempting to illustrate the logic of the method throughout.

2 THE LOGIC OF IMMUNOCYTOCHEMISTRY

In later sections we present specific techniques for the several methods involved in ICC. In this section we discuss the logical framework involved in the application of those methods. For the reader not familiar with immunocytochemical methods or results, the technique section lists re-

Figure 1. Typical immunocytochemical demonstrations in the central nervous system of the rat. Calibration bars, 20 μm. (A) Peroxidase-antiperoxidase (PAP) stained beta-endorphin (β-END) cells in the rat hypothalamus. Single arrow indicates the unstained nucleus, double arrows show the immunoreactive cytoplasm. In the PAP technique the horseradish peroxidase enzyme acts on its substrate to produce a dark precipitate as an indication of a positive stain. The light area is unstained. (B) β-END stained fibers in the hypothalamus (arrows). Note the beaded appearance. PAP method as in panel A. (C) β-END cells in hypothalamus stained with the fluorescein–double antibody techniques. In this system the positively stained cytoplasm (double arrows) is green fluorescent (white in photograph) and the unstained nucleus (single arrow) is dark.

agents, methods, and photographic examples of immunocytochemically stained cells and fibers.

The goal of an immunocytochemical demonstration is the use of antisera to localize a specific protein or peptide in its normal cellular environment. Several confluent methods and factors must be managed properly to reach this end point.

To begin with, the tissue to be studied must be chemically fixed so that the protein or peptide in question (the antigen) is in its native environment (i.e., it does not diffuse) and is not degraded. This fixation must be adequate for good localization but not be so strong as to "overfix" the tissue and prevent the antigen from being bound by the antibody. The nature of the fixatives often needs to be changed in order to match the chemistry of the antigen. For example, alcohol-containing fixatives are often poor for peptide localization because many peptides can be extracted from the tissue by alcohols. On the other hand, acid alcohol fixatives may work well for certain types of proteins.

The next critical element in the series is a specific antiserum against the antigen. The specificity of the binding to the antigen depends on the antigenic determinant or determinants of the antiserum and the existence of one or more subpopulations of antibodies of varying affinities and possibly a wide range of specificities as well. The method of preparing the antigen to elicit and harvest antibody is therefore critical. In the case of a peptide or protein *extracted* from tissue one must be concerned about other substances contaminating the "pure" extracted antigen. The immune systems of the rabbit may be a better chemist than the investigator. Put more seriously, antibodies may be produced against the antigen of interest as well as against the contaminants. Where such a "dirty" antiserum is used, it may well allow the visualization of many molecular species, including the one of interest. If only an impure antigen is available for central studies, it may not be possible to block the anatomical demonstration with any degree of specificity (both the antibody of interest and those against contaminants will be blocked by the impure antigen leading to the circular conclusion of a pure antibody).

If a pure protein or synthetic peptide is available and a "clean" antiserum is raised, then that antiserum can be used for ICC. The key to a good immunocytochemical stain is both a specific antiserum and a clear signal (a clear stain with a low background).

The logic of ICC involves the proper use of two parts of the immunoglobulin G (IgG) molecule (see Figure 2). Initially the variable region of the molecule (the "receptor" site) binds to the antigen in the tissue (hopefully the antigen against which the rabbit was immunized). After the antibody has bound to the antigen, its constant region continues to

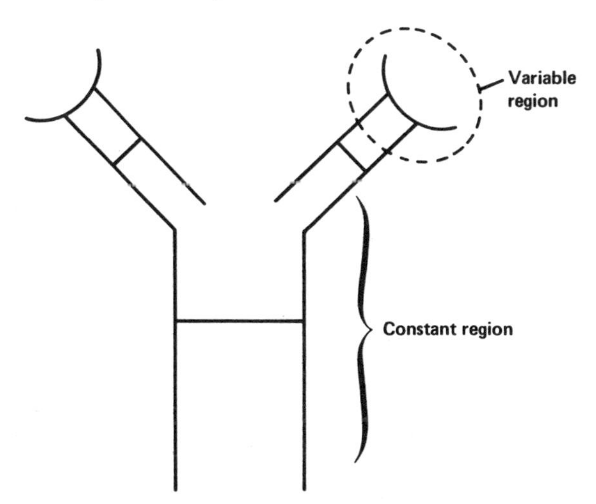

Figure 2. Protein domains of a model IgG molecule.

be exposed. The constant region is that protein piece which is invariant for all IgG molecules within the species in question (i.e., rabbit). A second antibody binds to the constant region of the first antibody (i.e., goat IgG raised against rabbit IgG) and is subsequently visualized (Figures 3 and 4). Thus the first antibody serves two purposes in this paradigm, taking a specific binding role and then becoming an antigen for the second antibodies.

When the tissue is fixed and sectioned and a good antiserum secured, the two major problems are solved. But a multitude of technical issues are yet to be managed. For example, the time, temperature, and dilution of the primary antiserum (i.e., the antiserum directed against the antigen of interest) and the nature of the marker or detection system (fluorescence or peroxidase-antiperoxidase, etc.) need to be determined. Even if a good, clear, visible signal has been obtained in a usable anatomical context, a host of biochemical issues are still to be resolved. Is the antibody binding the exact antigen-peptide or a structurally similar peptide? Is it cross-reacting with a "family" of similar structures? Or a repeated sequence in an unrelated protein? How are these questions to be approached? In the biochemistry laboratory, gel permeation chromatography, polyacrylamide gel electrophoresis systems, high performance liquid chromatography (HPLC), and particularly amino acid sequencing can help answer these questions. But these methods are not available in an immunocytochemical preparation. Some questions can be approached and others cannot. For these reasons it is important to characterize immunoreactive species biochemically whenever possible.

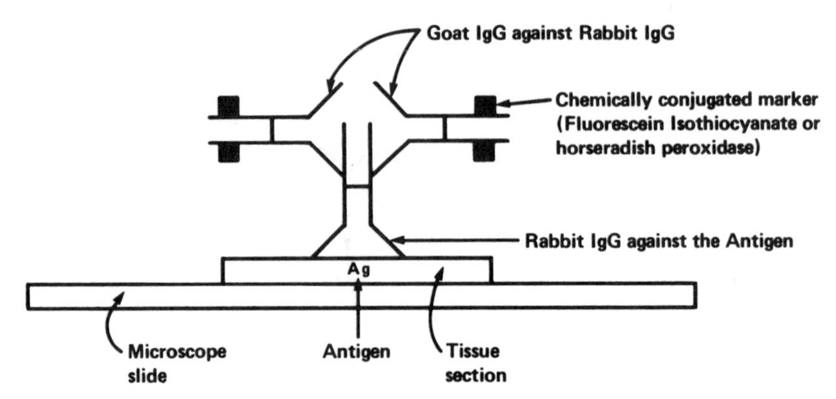

Figure 3. Both the fluorescein and conjugated horseradish peroxidase (HRP) indirect immunocytochemical techniques use two antibody steps. After the tissue–antigen complex is fixed and mounted on the microscope slide, the first antibody (rabbit antiantigen IgG) is bound to the antigen via the variable (receptor) region. The second antibody (goat IgG against rabbit IgG constant region) binds to the constant region (see Figure 2) of the rabbit antibody. The goat IgG is usually obtained commercially conjugated to fluorescein isothiocyonate (FITC) or the enzyme HRP. With FITC-labeled antiserum, special excitation and emission filters must be used to visualize the stained tissue. With the HRP-linked method, an enzyme substrate must be used (see Section 3.3.1) in order to produce a visible colored precipitate.

2.1 Immunocytochemistry and Immunoprecipitation vs. Radioimmunoassay

The use of antisera to identify or bind to a particular molecular species is a very powerful tool. Yet the interpretation of results from this basic antigen–antibody interaction is fraught with potential error. There are two approaches to the issue of antigen–antibody specificity, seen in the context of radioimmunoassay (RIA) on the one hand and ICC on the other. In an RIA, a labeled (radioactive) antigen is mixed with the antiserum and used to "preselect" a certain population of antibodies. Only those IgG molecules (antibody) with an affinity for the labeled antigen will be detected by counting the radioactivity. Displacement of the labeled antigen will be detected by a decrease in counts because of competition from unlabeled antigen for the same IgG sites. Thus, in the RIA paradigm only the IgG molecules against a specific antigen are detected. Antibodies directed against other antigens do not bind the labeled antigen and are not detected. If we inject a rabbit with a peptide, for example beta-endorphin (β-END) and a carrier protein, bovine serum albumin (BSA), the animal will produce both anti-β-END and anti-BSA antibodies. How-

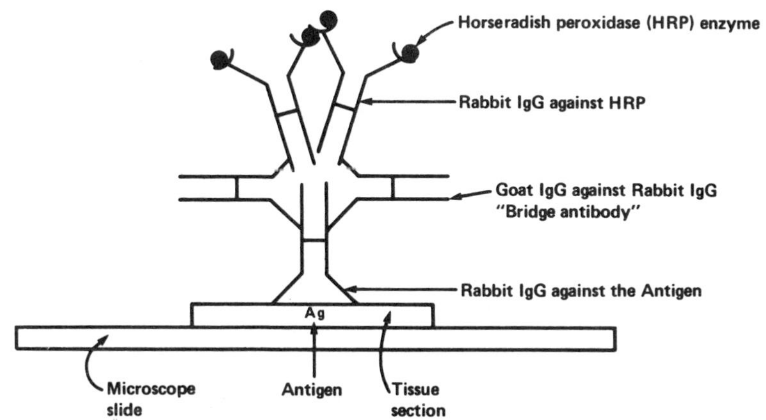

Figure 4. Sternberger's peroxidase-antiperoxidase (PAP) technique is thoroughly described in his text (2). In essence it involves a three-antibody sandwich. The first two layers (rabbit anti-antigen and goat anti-rabbit) are as described in Figure 2. There are two differences between this two-antibody method and the PAP method. First, in the PAP method the goat anti-rabbit (GAR) antibody is not linked to a marker substance (HRP or FITC). Second, the concentration of the GAR or "bridging antibody" is higher in the PAP technique in order to leave half the GAR receptor sites free to bind another type of rabbit IgG molecule, rabbit IgG, against HRP. This rabbit anti-HRP molecule plus the HRP enzyme itself forms a PAP complex in solution. The PAP reagent is usually purchased as an antigen–antibody complex. It is applied to the section after the bridging GAR antibody, the GAR antibody binds the rabbit PAP complex thereby producing a three-antibody (and enzyme) sandwich. The enzyme is visualized using substrate to produce a visible colored precipitate.

ever, under RIA conditions, the use of radioactive β-END will allow binding only to β-END antibodies. We could use BSA in the buffer, but the BSA–anti-BSA interactions will not be detected by the RIA.

In contrast, ICC and immunoprecipitation paradigms may detect *any* antigen in the tissue to which the antiserum is directed. For example, in our mythical antiserum, we would precipitate both β-END and BSA. Or if the rabbit serum had both anti-substance P and anti-β-END antibodies (because of impure antigen), we might find that both substance P and β-END cell types would be shown in an immunocytochemical study. The RIA method would only detect the antibodies against the labeled substance P (if that is what we added) and show no signs of β-END. From the perspective of ICC, the antiserum to be used should contain antibodies against only one antigen in the tissue of interest, whereas RIA could theoretically have an infinite number of antibodies against several antigens—although this may begin to create some practical problems even under RIA conditions.

2.2 Families of Antigens

Even if one has an excellent antiserum for ICC, there is another major problem, that of families of proteins and peptides with similar structures. In the neuropeptide field, the family of opioid peptides exemplifies this principle (and immunocytochemical problem) very well. Currently all opioid peptides are all known to have either methionine-enkephalin (Tyr-Gly-Gly-Phe-Met) or leucine-enkephalin (Tyr-Gly-Gly-Phe-Leu) at their NH_2 terminus (β-END, methionine-enkephalin, leucine-enkephalin, dynorphin, a-neoendorphin, not to mention the six to eight peptides found in the precursor to the enkephalins) (3–9). Antisera directed against the NH_2 terminus of one of these peptides may well cross-react with some or all of the others (10). When that antiserum is used in immunocytochemical studies, major questions about the specificity of the peptide in the cells visualized must be raised. For example, if we used an anti-dynorphin antiserum we would stain dynorphin cells in the supraoptic nucleus (10), but we might also stain enkephalin (or its precursor) in enkephalin cells in the caudate or even β-END in its cells. Blocking studies using excess β-END or enkephalin, or even fragments of the enkephalin precursor, might help—but each of them has that troublesome common NH_2-terminus "enkephalin" sequence.

2.3 Convergent Antisera

A rational (and more direct) approach to the common antigen sequence problem is to attempt to develop antisera against the unique sequences of the members of the antigen family (Figure 5). For example, the 17-amino-acid peptide dynorphin (11) has a COOH terminal region that is not repeated in other opioid peptides. Thus the approach of choice is to synthesize that region and develop antibodies against it.

We have an interesting example of this approach with peptide F (an enkephalin precursor fragment) (cf. 9). The two enkephalin sequences at each end and the total length of the peptide were our motivations for choosing a midportion for antiserum production. We chose the sequence (. . . Asp-Glu-Leu-Tyr-Pro-Leu-Glu-Val-Glu . . .) because it did not include the enkephalin structures, contained no complex amino acids, and had a tyrosine (for iodination). This peptide was coupled via glutaraldehyde to thyroglobulin and injected into rabbits. The resulting antiserum bound to the cells of the system in question (enkephalin neurons) as well as to β-END-containing cells in brain and pituitary! After a considerable effort the cross-reactivity problem was partly solved. To put it succinctly, the F antiserum could bind to F-containing proteins (i.e., enkephalin-syn-

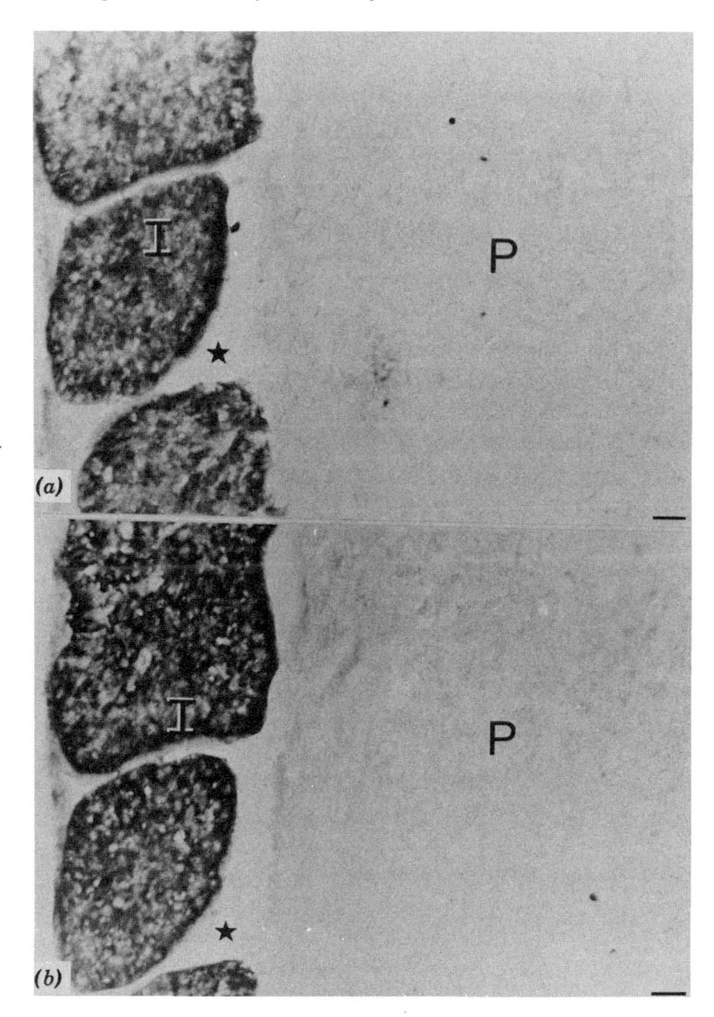

Figure 5. Rat pituitary, stained with β-END antiserum (panel A) and α-MSH antiserum (Panel B) both via the PAP technique. Serial 5 μm sections were used. *I*, intermediate lobe; *P*, posterior lobe. All the cells of the intermediate lobe were stained with both antisera. Star notes a common point for reference. Calibration bar, 20 μm.

thesizing cells). It also bound to brain and pituitary ACTH/β-END cells. How could that be interpreted? There was no enkephalin pentapeptide in F to cross-react with β-END. After comparing the structures of F with the sequence of the ACTH/β-END precursor we found that there was a region of five identical or very similar amino acids in sequence with one substitution in the COOH terminus of ACTH (see underlined amino acids in the F fragment). We therefore used $ACTH_{1-39}$ and the F peptide fragment for competition studies against the antibody. In unblocked sections of brain we could see both cell types, enkephalin and β-END. Peptide F could block demonstrations in both enkephalin and β-END/ACTH cells, whereas ACTH blocked only the β-END/ACTH cells but not the enkephalin cells (Figure 6A, B, and C). Thus we concluded that there were really two populations of IgG molecules, those that bound true F-containing sequences and those that cross-reacted with ACTH. In the pituitary (which has only ACTH/β-END cells), all cellular immunoreactivity was blocked by ACTH and the F fragment (Figure 6D and E).

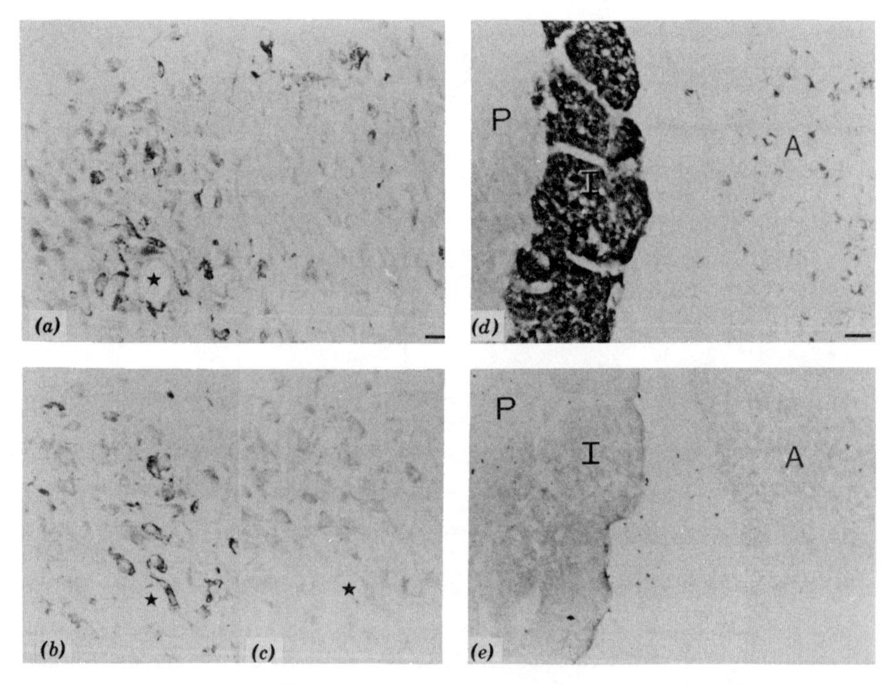

Figure 6. Peptide F-like immunoreactivity in hypothalamus (A to C) and pituitary (D and E). Calibration bars, 20 μm for (A), (B), and (C), 50 μm for (D) and (E). (A) F-like immunoreactive cells. (B) Same area (note star in vessel) of F-like immunoreactivity; unsuccessfully blocked by 50 μ*M* $ACTH_{1-39}$. (C) Same as (A) (see starred vessel) but blocked by 10 μ*M* peptide F fragment. (D) F-like immunoreactivity in rat pituitary. *A*, anterior lobe; *I*, intermediate lobe; *P*, posterior lobe. F antisera stain scattered the anterior lobe cells (corticotrophs) and all intermediate-lobe cells (β-END/α-MSH cells). (E) Same as panel D except F-fragment antibody was blocked by $ACTH_{1-39}$, suggesting a different antibody specificity than that seen in panel B.

2.4 Biochemical Characterization of Immunoreactivity

The problem of characterizing the F immunoreactivity in hypothalamus is best resolved with biochemical tools. As a general principle we attempt to validate any new immunoreactivity, antibody, or tissue with biochemical methods. While immunocytochemical methods can localize antigens with much higher resolution than biochemical methods, the latter have much greater precision in revealing the nature of the antigen: its size, chromatographic behavior, hydrophilic, or hydrophobic character, and its specific amino acid sequence. Perhaps the most sensitive and useful system for characterizing the nature of the immunoreactivity involves molecular sizing (gel permeation chromatography) followed by HPLC-C_{18} reverse-phase chromatography coupled with RIA (12). This type of analysis on the tissue in question can yield surprising results. For example, using antisera against dynorphin$_{1-13}$ it is possible to immunocytochemically stain the bovine adrenal medulla (10). Yet when that tissue is analyzed by HPLC/RIA with both dynorphin and enkephalin RIAs, there is 400-fold more enkephalin than dynorphin. The dynorphin RIA was carried out at a much greater antibody dilution than was ICC (1/60,000 vs. 1/500) and therefore each method depends on different IgG subpopulations. The differences in antibody subpopulations and the number of antigens each technique can detect (see ICC vs. RIA above) make it very difficult to compare results from these two methods. Perhaps the best that one can achieve is a statement that the tissue contains the antigen of interest and it is in the form hypothesized. Equally revealing is the complex result—that is, the tissue has the desired antigens plus a slightly different form, a "large" form (precursor) or a related form. The issues of multiple forms and precursor–product relationships are currently critical in neuropeptide analysis.

2.5 Controls

There are two major types of controls to be carried out in ICC, technical and specificity.

The *technical controls* are logical evaluations of the effect of the marker antibody system used (2). That is, are the secondary or tertiary antibodies binding to the section? If these antibodies are applied alone, do they alone account for the demonstration? The method here is exactly the same as that used with the primary antibody except that a "nonimmune" serum is employed as the first step. Each serum (see the Technique section below) and treatment condition (DAB and H_2O_2, O_sO_4, etc.) can be evaluated for its contribution to nonspecific background and "specific signal."

The *specificity control* deals with the primary antibody and the structures it binds to. In that sense, these studies are analogous to RIA cross-reactivity studies. The key issue is to demonstrate that the antigen is the only substance capable of inhibiting (or blocking) the primary antibody (ref. 10 and Figure 6). For example, when we use a β-END antibody we block with 10 n*M* of β-END (see Section 3.5 for peptide handling) and unsuccessfully block with a series of related substances (methionine-enkephalin, dynorphin, etc.). The family of similar antigens is the main potential culprit. This type of study depends on a great excess of peptide to block (or occupy) the antibody sites and prevent them from binding to the antigen in the tissue.

Any immunocytochemical demonstration from a crude antiserum depends on a variety of IgG populations. Thus, a loss of signal in the tissue after blocking is a reflection of the general antigen requirements of these several populations. It does not mean that only one antibody population is present.

2.6 New Antibody Calibration

The use of a new serum requires an orderly strategy, especially for the study of a new substance. The key is to try a broad concentration range of the primary serum with a known marker-antibody system—such as peroxidase-antiperoxidase (PAP) or fluorescein—in a tissue with the antigen in it (as indicated by RIA or other techniques). Too high a background staining can easily obscure a subtle, specific stain. Upon dilution, the background (nonspecific staining) decreases rapidly whereas the specific stain often does not drop until a much higher dilution. Using the lowest amount of antibody necessary to get a particular signal saves serum and blocking peptide while favoring the binding of higher affinity subpopulations over low-affinity subpopulations.

3 TECHNIQUES

The following is a list of successful methods used in our laboratory for ICC across a broad range of antigens ranging from pentapeptides to 300,000-dalton proteins.

3.1 Antibody Preparation

3.1.1 *Peptide Coupling to a Carrier Protein*

Peptides are generally too small to induce a large immune response. The peptide must be conjugated to a large protein by either the carbodiimide (cf. 13) or glutaraldehyde techniques (14).

The *carbodiimide reaction,* a two-step reaction, creates a new amide–peptide bond between the peptide and protein. The carbodiimide reacts with either a carboxyl or primary amino group to form *O*-acyl-urea or *N*-acyl-urea intermediates. These intermediates react with a free amino or carboxyl group to create the amide bond. Peptides may be conjugated to large proteins or polymerized to themselves by this process.

The conjugation of peptides to thyroglobulin has been a very successful procedure in our laboratory. For example, 8 mg of a peptide (in this case with a molecular weight of 1300) are dissolved in 1.9 mL of HCl (pH 5.0) and added to the peptide solution. Then 85 mg of 1-ethyl-3-(3-dimethy-aminopropyl) carbodiimide-HCl are dissolved in 0.7 mL of HCl (pH 5.0) and added dropwise (while stirring) to the peptide-thyroglobulin mixture. The reaction mixture is stirred for 48 hr. A white precipitate appears in the vial. The carbodiimide conjugate does not need to be dialyzed before injection. The conjugated peptide is divided into 15 tubes (based on 50% yield), lyophilized, and stored at −20°C.

The *glutaraldehyde reaction* also conjugates peptides at a primary amino group. In this method 8 mg of the same peptide (molecular weight 1300) and 40 milligrams of thyroglobulin are dissolved in 3.10 mL of 0.1 *M* potassium phosphate buffer (pH 7.5). The reaction mixture is cooled to 0°C in an ice bath and 1.6 mL of a cooled 25% glutaraldehyde:H_2O solution (1:100) is added dropwise to the mixture. The reaction mixture is stirred at 0°C for 30 min and then at room temperature for 2 hr. The conjugated peptide is dialyzed against saline and double-distilled water. A yellow precipitate may appear in the dialysis bag.

The efficiency of both the carbodiimide and glutaraldehyde reactions can be calculated by adding iodinated peptide (chloramine-T method) to the reaction vial. The conjugated peptide is separated from the uncoupled peptide and counted to evaluate efficiency of coupling. The glutaraldehyde reaction yields 50–80%. The carbodiimide reaction has a lower efficiency, 25–50%. This low efficiency may be due to the competition of peptide with peptide instead of peptide with protein.

3.1.2 Immunizations and Bleeding

Rabbits are immunized with 100 to 500 μg of *peptide* (as estimated from ^{125}I-peptide coupling of the peptide carrier complex). The smaller the peptide the more we use for immunization. One mL of antigen (peptide plus carrier) in aqueous buffer plus 1 cc of Freund's complete adjuvant is mixed until uniformly viscous and subcutaneously injected in multiple sites on the backs of 3–4 kg New Zealand rabbits. The immunization schedule involves this "hyperimmunizing" once a week for 3 weeks followed by the first bleed 3 weeks later. Subsequently, the animals are boosted with half the amount of peptide and Freund's incomplete adjuvant 1 week after the last bleed and rebled 2 weeks later. Typically a 10- or 12-amino-acid peptide might yield 1/100 to 1/300 immunocytochemical titer on the first bleed (variable across the 3 or 4 rabbits immunized), peaking at boost 6–8 at 1/1000–1/10,000 titer. We follow each bleed in each rabbit with a usable-dilution study in a standard tissue. When the titer peaks (or certainly if it begins to drop) the rabbit is "bled out." Each bleed yields 25–30 cc of serum, with the final bleed yielding 50–70 cc.

3.1.3 Affinity Purification

Affinity purification of an antiserum involves using the peptide linked to a solid matrix to bind the specific IgG population directed at the peptide while discarding the nonspecific IgG. This allows the preselection of relevant populations of antibodies and the elimination of unwanted populations. It often results in a better demonstration, with lower background (15).

From several commercially available affinity chromatography gels, Cyanogen Bromide-Activated Sepharose-4B was chosen for its simplicity and efficiency in peptide immobilization. Under mild conditions, Cyanogen Bromide-Activated Sepharose-4B reacts with primary amino groups (NH_2-terminus or basic R groups).

To immobilize the peptide, 3 mg of highly purified peptide are dissolved in 1.7 mL of sodium bicarbonate buffer (0.1 *M* $NaHCO_3$ with 0.5 *M* NaCl, pH 8.3). Next .450 mg of Cyanogen Bromide-Activated Sepharose-4B are swelled and washed in 100 m*M* HCl and transferred to the peptide solution. The peptide-gel mixture is gently stirred at room temperature for 2 hr and then transferred to a Bio-Rad disposable minicolumn and washed with 10 column volumes of sodium bicarbonate buffer.

In order to inactivate the remaining cyanogen groups, the gel bed is reequilibrated with 1 *M* ethanolamine, Tris, or glycine in 0.1 *M* sodium bicarbonate buffer (pH 8.3). The gel is incubated for an additional 2 hr and then washed with sodium acetate buffer (0.1 *M* sodium acetate with 0.5 *M* NaCl, pH 4.2) followed by sodium borate buffer (0.1 *M* sodium borate with 0.5 *M* NaCl, pH 8.2). This washing cycle is repeated 3 times (3 × three volumes) to dissociate any noncovalently bound peptide. The column is washed with 6 *M* guanidine-HCl and then stored at 4°C in 0.2% sodium azide.

To purify the specific IgG, the stored peptide–Sepharose-4B column is washed with 10 column-volumes of double-distilled water and reequilibrated in a sodium phosphate buffer (0.150 *M* sodium phosphate with 1% NaCl, 0.3% BSA, and 0.1% Bacitracin, pH 8.25). The gel is transferred to a vial containing 12.0 mL of sodium phosphate and 0.250 mL of crude antiserum. The reaction mixture is shaken for 2–3 hr at room temperature. At that stage, the antipeptide antibodies are coupled to the peptide on the gel. The gel is transferred back to the Bio-Rad disposable minicolumn and washed with 10 volumes of sodium phosphate buffer to eliminate the unbound IgG molecules and various serum constituents.

The specific IgG molecules are eluted off the gel with 5 volumes of 0.1 *M* glycine-HCl (pH 2.3). The eluate is collected and mixed into 3.0 mL of 3 *M* Tris buffer (pH 8.6) containing 5 mg BSA. The elution is completed, the gel bed is washed with 6 *M* guanidine, and the column is stored at 4°C in 0.2% sodium azide.

The purified IgG is transferred to a dialysis membrane and dialyzed against 100 m*M* Tris (pH 7.4) with 2 *M* sodium chloride. The dialysis buffer is diluted 2/3 times every hour (4 times) with double-distilled water. The dialysis bag is transferred to phosphate buffered saline (pH 7.4) for overnight dialysis. The purified, dialyzed antibody is lyophilized, reconstituted in a smaller volume of double-distilled water, and stored at −20°C. It can be used for immunohistochemistry as well as for immunoassay or immunoprecipitation. However, some of the characteristics of the affinity purified antiserum may have changed compared with the original serum. Therefore, recalibration of optimal dilution as well as cross-reactivities is essential.

3.2 Preparing the Tissue

3.2.1 Formaldehyde Perfusion

1. The rat is anesthesized with 1 cc (50 mg/cc) Pentobarbital and placed on ice.
2. The abdominal cavity is cut open.

3. The diaphragm is removed and the chest cavity opened.
4. A suture is placed *loosely* around the aortic arch (use curved tweezers to cut through the pericardium).
5. The right atrium is cut open, and 50–100 cc of cold normal saline solution is injected via the left ventricle to flush the vascular tree.
6. The tip of both ventricles is removed and a flexible plastic cannula is inserted into the aorta and tied in place.
7. Approximately 750 mL of 4% formaldehyde (FA) (made from paraformaldehyde; see *Buffers and Solution*, Section 3.10) in 0.1 *M* Na phosphate buffer at 4°C are pumped through the rat for 30 min at 140 mm Hg.
8. The tissue is removed and placed into 50 cc of 4% FA for 1 hr at 4°C and then into cold 15% sucrose in phosphate buffered saline (see Section 3.10) for 1 to 2 days.
9. The tissue is then frozen with O.C.T. (a mounting medium) on brass chucks in liquid nitrogen. The tissue is stored at −60°C.

3.2.2 *Cryostat Sectioning*

1. The tissue is frozen on cryostat chucks with O.C.T. mounting medium in liquid nitrogen and the temperature is allowed to equilibrate with the temperature of the cryostat (−15 to −20°C).
2. 5 to 10 μm serial sections are cut and melted onto glass slides at room temperature directly from the cryostat knife. (The slides are subbed beforehand with a solution of sodium azide and gelatin; see Section 3.10.)
3. The sections (on slides) are stored at −60°C until ready for use (which may be up to several months later).

Many other fixation methods can be used. The *Manual of Histologic Staining Methods* of the Armed Forces Institute of Pathology (16) is an excellent general reference for fixation and embedding.

An unusual method combining catecholamine histofluorescence and neuropeptide ICC has been reported (and well used) by McNeill and Sladek (17).

Electron microscopic sectioning is not be discussed in this chapter, nor is electron microscopic ICC. The reader is referred to the excellent papers of V. Pickel (18, 19) and G. Pelletier (20) for pre- and post-embedment methods and to Meek's excellent text (21) for general electron microscopic methods.

3.3 Visualizing the Antigen in Tissue

3.3.1 The Peroxidase-Antiperoxidase Procedure

The primary antibody, normal goat serum (NGS), 0.3% Triton-phosphate buffered saline (T-PBS) (cf. Figure 1, A and B; ref. 2), goat anti-rabbit (GAR), and peroxidase-antiperoxidase (PAP) used below should be kept frozen until ready for use.

Day 1

1. Air-dry the sections (on slides) and ring with fingernail polish to maintain fluid level.
2. Dilute a working stock of NGS with 0.3% T-PBS to 1/300. Add 500 μg/mL Bacitracin to NGS. Apply 100 μL NGS to each section in a 37°C moist box (hot box) for 5–10 min.
3. Drain the NGS off of the slide onto a paper towel. Do not allow the section to dry. Apply 100–200 μL of the primary antibody (the dilution must be empirically determined for each bleed of each antibody). Incubate the section in the hot box for 1 hr, put it in "humid" slide boxes (we use wet foam rubber) and into the refrigerator overnight (24 hr).

Day 2

1. Wash the section for 30 min (3 washes × 10 min) in PBS.
2. Incubate the section with NGS for 5 to 10 min in a hot box (37°C).
3. Drain the NGS off the slide onto a paper towel.
4. Apply the bridging antibody (GAR) diluted 1/100 with 0.3% T-PBS. Incubate it for 30 min in a hot box, then overnight at 4°C.

Day 3

1. Wash the slides for 3 × 10 min in PBS.
2. Incubate with NGS for 5 to 10 min in a hot box.
3. Incubate the section with PAP (Sternberger-Meyer, Inc.) diluted 1/80 with 0.3% T-PBS for 1 hr in a hot box (37°C).
4. Wash for 30 min (3 washes × 10 min each) in PBS. Remove the fingernail-polish circle with a razor blade.

5. Prepare diaminobenzidine (DAB) when the slides begin to wash. Add 25 mg DAB to 200 mL PBS, stir to dissolve, then filter through a Buchner funnel with Whatman No. 2 filters. Refilter the solution, using Millipore filters (0.45 μm). Use the solution immediately to prevent "clumping" on the sections.
6. Immerse the slides in DAB solution. Add 2 mL 3% H_2O_2. Allow the tissue to incubate for 15 min while stirring.
7. Wash 3 × 10 min in distilled H_2O.
8. Apply filtered OsO_4 dropwise (25 μL) to each section for 5 sec to enhance the DAB staining.
9. Wash 3 × 10 min in distilled H_2O.
10. Dehydrate stepwise in 50%, 70%, 90%, 95%, 100% (× 2) ethanol and 2 times in xylene for 3 min each.
11. Coverslip in permount.

3.3.2 The Fluorescence Technique

See Figures 1C, 2, and 3. The procedure is as follows:

Day 1

Same as PAP.

Day 2

1. Wash 3 × 10 min in PBS.
2. Dilute the fluorescein-isothiocyanate (FITC)–GAR complex (Cappel Laboratories, Dowington, PA) to 1/100 with 0.3% T-PBS.
3. Filter the solution with a millipore filter (0.22 μm) and apply it *immediately* to the sections on slides (to prevent fluorescent clumping).
4. After 1 hr incubation at 37°C, wash the sections in PBS for 3 × 10 min.
5. Coverslip the sections, using KPO_4-buffered glycerol (see below).
6. Observe and photograph immediately.

3.3.3 The Conjugated HRP Technique

The *conjugated horseradish peroxidase (HRP) technique* (22) is a double antibody method. It is similar to the fluorescence method up to Day 2, where GAR-HRP is added. It is "developed" with DAB and H_2O_2 as in the PAP procedure (see Day 3, steps 5–10).

3.4 Colchicine Pretreatment

This method enhances peptide content in cell bodies by inhibiting axonal transport, thereby allowing visualization of cells of origin.

1. Dissolve the colchicine (1 mg/mL) in normal saline.
2. Lightly etherize the rat.
3. Make a longitudinal cut with scalpel on top of skull.
4. Cut away the periosteum over the skull.
5. Use clean gauze and apply pressure to prevent bleeding.
6. Drill a hole in the skull 1 mm lateral to the midline and 1 mm behind bregma.
7. Inject 50 μl of colchicine solution deep into the lateral ventricle (3.7 mm deep).
8. Seal the injection site with bone wax.
9. Clip the skin closed with wound clips.
10. Sacrifice the animal by perfusion 48 hr after colchicine injection.

3.5 Peptide Controls

1. To avoid nonspecific absorption of peptide to walls, plastic tubes and pipette tips are precoated with "3X" protein buffer (10 mL 0.5 *M* Tris HCl with 400 mg BSA, 20 mg lima bean trypsin inhibitor, 10 mg polylysine made up to 100 mL with double-distilled H_2O).
2. Coat the walls with the 3X buffer and allow them to air dry.
3. Weigh enough peptide (μg quantities) to make a 500 to 200 μ*M* stock.
4. Dissolve the peptide in the appropriate solution (varies with the peptide in question).
5. Add the peptide to antibody diluted to its final working concentration with precoated pipette tips to make a 10 μ to 50 μM blocking condition.
6. Be sure to dilute the unblocked antibody with an equal volume of 0.3% T-PBS or peptide buffer.
7. Allow the blocked antiserum to incubate 30 min to 1 hr at 4°C before applying it to the section.
8. Process according to the PAP technique.

3.6 Buffers and Solutions

This is a list of the solutions referred to above.

Buffered Glycerol for Fluorescent Section Coverslips

1. 0.4 *M* Potassium bicarbonate can be made with 8.01 g potassium bicarbonate plus 190 mL H_2O. Adjust pH to 8.6 with KOH (0.1 *M*) and bring the solution to a total volume of 200 mL.
2. Add 50 mL of 0.4 *M* potassium bicarbonate to 50 mL glycerol to make the final solution.

Phosphate-Buffered Saline (PBS)

1. 1 *M* KPO_4 can be made with K_2HPO_4 (143.2 g) plus KH_2PO_4 (g) added to 1 liter of 40°C water.
2. PBS is made with 80 mL of 1 *M* KPO_4 plus 37.4 g NaCl added to 4 liters of H_2O.

Slide Subbing Solution

1. Add a stock gelatin solution of 1 g Na azide plus 5 g gelatin to 100 mL H_2O.
2. Divide the solution into 30 mL fractions and freeze.
3. Add 30 mL gelatin solution to 1 liter H_2. Dip clean microscope slides into the diluted gelatin. Dry the slides in an 80°C oven.

Formaldehyde

1. 1 *M* Na Phosphate buffer is composed of NaH_2PO_4 H_2O (11.04 g) plus Na_2HPO_4-anhydrous (45.44 g) dissolved in 4 liters hot distilled H_2O (80°C).
2. Add 160 g paraformaldehyde powder to the hot buffer.
3. When solution clears, filter the buffered formaldehyde with Whatman No. 2 filter and cool to 4°C for use (final pH 7.4).

Nissl Stain for Frozen Sections

1. Add 0.135 g basic fuchsin to 0.365 methylene blue.
2. Dissolve in 50 mL of 30% ETOH.
3. Filter the solution through a millipore filter before use.

ACKNOWLEDGMENTS

This work was supported in part by NIDA Grant #DA02265 to H. A. and S. W., NIDA Center Grant #DA00154 to S. W. and H. A., and NSF Grant #BNS8004512 to S. W. and H. A.

REFERENCES

1. A. H. Coons and M. H. Kaplan, *J. Exp. Med.* **91**, 1 (1950).
2. L. A. Sternberger, *Immunocytochemistry*, Wiley, New York, 1979.
3. J. Hughes, T. W. Smith, H. W. Kosterlitz, L. A. Fothergill, B. A. Morgan, and H. R. Morris, *Nature* **258**, 577 (1975).
4. C. H. Li and D. Chung, *Proc. Natl. Acad. Sci. USA* **73**, 1145 (1976).
5. R.Guillémin, N. Ling, and R. Burgus, *C. R. Hebd. Séances Acad. Sci., Ser. D* **282**, 783 (1976).
6. A. F. Bradbury, W. F. Feldberg, D. G. Smyth, and C. Snell, in H. W. Kosterlitz, Ed., *Opiates and Endogenous Opioid Peptides*, Elsevier/North-Holland, Amsterdam, 1976, p. 9.
7. A. Goldstein, S. Tachibana, L. I. Lowney, M. Hunkapiller, and L. Hood, *Proc. Natl. Acad. Sci. USA* **76**, 6666 (1979).
8. K. Mizuno, N. Minamino, K. Kangawa and H. Matsuo, *Biochem. Biophys. Res. Commun.* **97**(4), 1283 (1980).
9. D. L. Kilpatrick, T. Taniguchi, B. N. Jones, A. S. Stern, J. E. Shively, J. Hullihan, S. Kimura, S. Stein, and S. Udenfriend, *Proc. Natl. Acad. Sci. USA* **78**(4), 3265 (1981).
10. S. J. Watson, H. Akil, V. E. Ghazarossian, and A. Goldstein, *Proc. Natl. Acad. Sci. USA* **78**(2), 1260, (1981).
11. A. Goldstein, W. Fischli, L. I. Lowney, M. Hunkapiller, and L. Hood, *Proc. Natl. Acad. Sci., USA* in press (1981).
12. H. Akil, Y. Ueda, H.-L. Lin, and S. J. Watson, *Neuropeptides* **1**, 429 (1981).
13. S. Sullivan, H. Akil, S. J. Watson, and J. D. Barchas, *Commun. Psychopharmacol.* **1**, 605 (1977).
14. V. E. Ghazarossian, C. Chavkin, and A. Goldstein, *Life Sci.* **27**, 75 (1980).
15. R. Axen, J. Porath, and S. Emback, *Nature* **214**, 1302 (1967).
16. *Manual of Histologic Staining Methods of the Armed Forces Institute*, McGraw-Hill, New York, 1968.
17. T. H. McNeill and J. R. Sladek, *Brain Res. Bull.* **5**, 599 (1980).
18. V. M. Pickel, T. H. Joh, D. J. Reis, J. E. Leeman, and R. J. Miller, *Brain Res. Bull.* **160**, 387 (1979).
19. V. M. Pickel, K. K. Sumai, S. C. Beckley, R. J. Miller, and D. J. Reis, *J. Comp. Neurol.* **189**, 721 (1980).
20. G. Pelletier, R. Puviani, O. Bosler, and L. Descarries, *J. Histochem. Cytochem.* **29**(6), 759–764 (1981).
21. *Principles and Techniques of Electron Microscopy*, Van Nostrand Reinhold, New York, 1970.
22. P. Nakane, *J. Histochem. Cytochem.* **16**, 590 (1968).

Chapter 4

Immunoelectron Microscopy of Neuropeptides: Theoretical and Technical Considerations

Gerald P. Kozlowski

Department of Physiology
Southwestern Medical School
University of Texas, Dallas

Gajanan Nilaver

Department of Neurology
College of Physicians and Surgeons
Columbia University, New York

1 INTRODUCTION

Localization studies for neuronal substances at the ultrastructural level not only include all the procedural considerations due to light microscopic immunocytochemistry (ICC) but also encompass additional problems associated with the preservation of tissue structure and antigenicity at the subcellular level. Remarkably rapid progress has been made in gaining useful information employing electron microscopic ICC for a great variety of proteins and peptides. Within recent years, a number of monographs and reviews have appeared: Childs (12), Feldman et al. (22), Feţeanu (23), Larsson (51), Pickel (76), Sternberger (95), Swaab and Boer (98), Williams (108), and Livett (53). There has especially been an explosive increase in the number of papers dealing with the localization of neuropeptides in the central nervous system (CNS) using immunoelectron microscopy. In this regard, Swaab's (98) statement is particularly apt:

> *Now that the first period of pure excitement about the enormous potential of these staining techniques is over, it seems appropriate to ask ourselves what exactly we are staining, and what methods we should select for a given problem.*

It is the purpose of this chapter to address these questions as they pertain to immunoelectron microscopy of neuropeptides.

2 ANTIBODY STRUCTURE

The success of immunochemical techniques is based on the highly specific interactions that can occur between an antibody and an antigen. Of course, other factors are important to the reaction, such as affinity and avidity of the antibody. The primary reaction of an antibody with its antigen and the resultant antigen–antibody complex is represented by the equilibrium equation:

$$\text{Ab} + \text{Ag} \underset{\text{Avidity}}{\overset{\text{Affinity}}{\rightleftharpoons}} \text{Ag–Ab}$$

The power of association for an antibody to react with its respective antigen is termed affinity, and it drives the equation to the right. Avidity is a measure of the strength of the bond forming the complex. Antibodies with weak avidity drive the equation to the left. In the ICC reaction, greater association and less dissociation of the antigen–antibody complex occurs. Long incubation times of highly dilute antisera drive the equation to the right, allowing greater opportunity for the antibodies with weak affinities to bind. Under these conditions background staining is dimin-

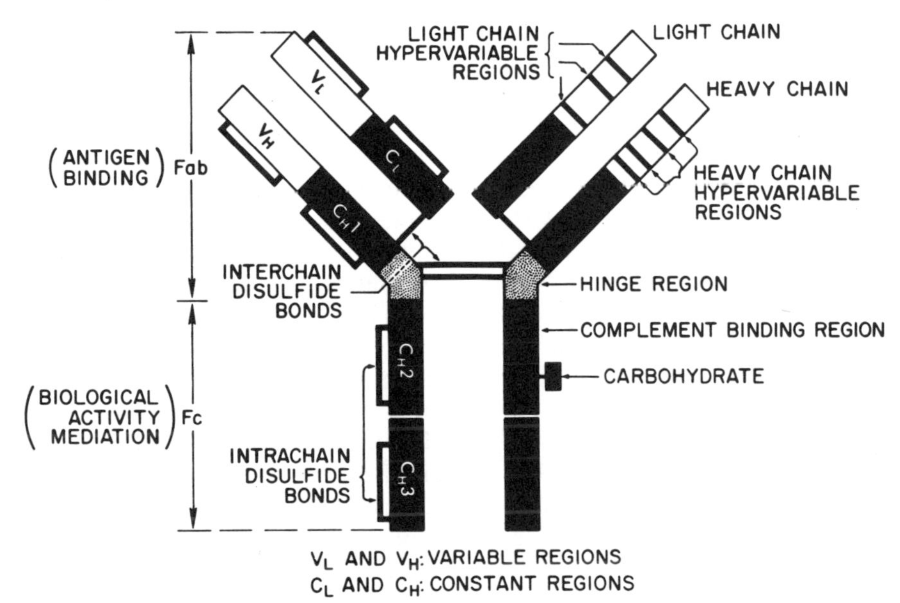

Figure 1. Topology and functional architecture of the IgG molecule. From Wasserman and Capra (105), courtesy of Dr. J. Donald Capra.

ished because many cross-reacting antibodies are in such low concentration. Such long incubation times are also necessary with fixed tissue because the antigen is cross-linked to cellular structures and may be in an altered state, whereas with radioimmunoassay (RIA) the antigen is free and unaltered. Each antiserum must be tested at different dilutions and incubation times in order to determine optimal staining conditions.

Although antibodies can be immensely diverse in their reactivities, they still share common structural features. The basic molecular structure of an antibody of the immunoglobulin (IgG) class (Figure 1) consists of two light and two heavy chains joined by disulfide bonds.

Each molecule of this four-chain unit has a molecular weight of about 150,000. Enzymatic digestion with papain at the hinge yields several fragments (F), mainly Fab and Fc. The variable region of the Fab fragment is the antigen-binding site; the Fc fragment (c = crystalline) is a species-specific polypeptide and constant.

3 ANTIBODY PRODUCTION

Artificial active immunization of animals with an antigen results in the production of antibodies to parts of the antigen called the antigenic determinant, or epitope. Usually, antibodies recognize a sequence of four to seven amino acids of the terminal portion of a peptide (95). Most peptides are weak immunogens or haptens, and it is necessary to couple the hapten to a carrier such as bovine serum albumin (BSA) or thyroglobulin (Tg). The latter procedure, as suggested by Skowsky and Fisher (91), has become popular. However, our earlier experience (Kozlowski et al., 47; Zimmerman et al., 111) with antisera generated against hapten-carrier conjugates in light microscopic ICC points to the danger of false positives when using such antisera. When using Nett–Niswender anti-LHRH* (luteinizing hormone-releasing hormone) No. 42 to stain rat and sheep brains, the final reaction product appeared not only in specific structures but also nonspecifically as high background in the median eminence ependymal cells and neurons of the supraoptic and paraventricular nuclei. Indeed, the latter localizations for LHRH were reported by Naik (61) to be specific for the rat. However, rabbit No. 42 not only produced anti-LHRH antibodies but also anti-BSA antibodies to the BSA carrier that cross-reacted to albuminoid substances of the magnocellular system. The anti-BSA reactivity in tissues could be progressively inhibited by the addition of increasing amounts of BSA to the whole antiserum before use for ICC. The affinity chromatographic technique in which the carrier is conjugated to Sepharose 4B is our method of choice for eliminating antibodies made to carrier proteins.

False positives such as the ones described here are not a danger when using RIA techniques, this points to the vast difference between RIA and ICC. With RIA the antigen is unfixed and competes with radiolabeled antigen for binding to the antibody; the data are expressed in terms of bound vs. unbound antigen. With ICC we are dealing with chemically or mechanically fixed antigens in tissues that can react with a variety of antibodies within a particular antiserum, which in turn form a final reaction product (Figure 3). Although RIA can be useful in determining titer and specificity of a particular antibody, this information does not necessarily prove specificity under the conditions of incubation for ICC. The techniques of RIA and ICC are two different systems and their use should be viewed as providing corroborative evidence rather than definite proof of the relative presence or absence of a particular antigen. A case in point is the report by Schultzberg et al. (87), in which capsaicin-depleted cholecystokinin-8-like immunoreactivity was determined with ICC but not with RIA. In another instance, reported by Sokol et al. (93) and Zimmerman et al. (111), an anti-oxytocin was discovered in antiserum to vasopressin with ICC but not with RIA. Indeed, Moriarity et al. (59)

* The terms gonadotropin-releasing hormone (Gn-RH), luteinizing hormone-releasing factor (LRF), and luliberin refer to the same decapeptide here referred to as LHRH.

have shown that ICC is a more sensitive technique than RIA, able to detect anti-ACTH antibodies at dilutions 100 to 1000 times the limits of the RIA.

Ideally, antigens should be available in pure form. Of course this is nearly impossible for many proteins, especially enzymes. But even in the case of synthetic peptides, antibodies formed against acetylated antigens can provide false positive immunostaining (106). Unfortunately, most peptides are sold in this form.

The development of antibodies for use in ICC requires rigorous characterization procedures since their potency for staining can vary from bleeding to bleeding of the rabbit. Swaab (99) found that potency of immunofluorescent staining for oxytocin was at its highest after nine weeks of immunization and subsequently decreased despite maintenance of high titers and maximal binding throughout the course of immunization. This provides another example of the differences in results that can be obtained with ICC and RIA techniques.

4 THE LABELED ANTIBODY TECHNIQUE

The direct labeling of antibody was first developed by Coons (15), using immunofluorescent techniques. Avrameas (1) and Nakane and Pierce (62) were pioneers in developing the methodology for using horseradish peroxidase (HRP) conjugated antibodies for use in light microscopic and electron microscopic ICC. Gonatas et al. (30) used radiolabeled Fab fragments versus intact sheep anti-rat IgG antibody to compare any differences that might exist between the two in achieving intracellular penetration. They found that in the plasma cell there was excellent intracytoplasmic localization of ^{125}I-Fab of sheep anti-rat IgG, whereas the intact iodinated antibody was unable to penetrate into the cell. On that basis it appeared that the size of the conjugate was directly responsible for the degree of penetration of the immunolabel. However, Sternberger (95) stated that it is still an open question whether smaller conjugates penetrate better than larger ones, and that perhaps effects of the conjugates on polarity of protein are more important limitations to penetration than size. Sternberger further discussed the advantages and disadvantages of direct labeling of antibody. In many instances, a major disadvantage has been the unavailability of primary antisera in large enough amounts for extensive studies. The labeled antibody technique requires a large volume of purified antisera and yields only about 2.5% (95) of peroxidase-conjugated immunologically active antibody even under the best of conditions.

5 THE UNLABELED ANTIBODY TECHNIQUE

The labeled antibody technique requires a biochemical reaction for conjugating a marker molecule such as HRP for conventional light microscopy or fluorescein isothiocyanate (FITC) for fluorescence microscopy. As shown in Figure 2 (*1A* and *B*, *2*), the marker may be attached to the entire IgG molecule, to an Fab fragment, or to a second antibody directed against the primary antibody.

The unlabeled antibody technique utilizes an antibody made against the antigen to be localized (usually generated in rabbits), a "bridging" antibody from a different species (like sheep or goat, usually made against the IgG fraction of the rabbit), and a third antibody made in a rabbit

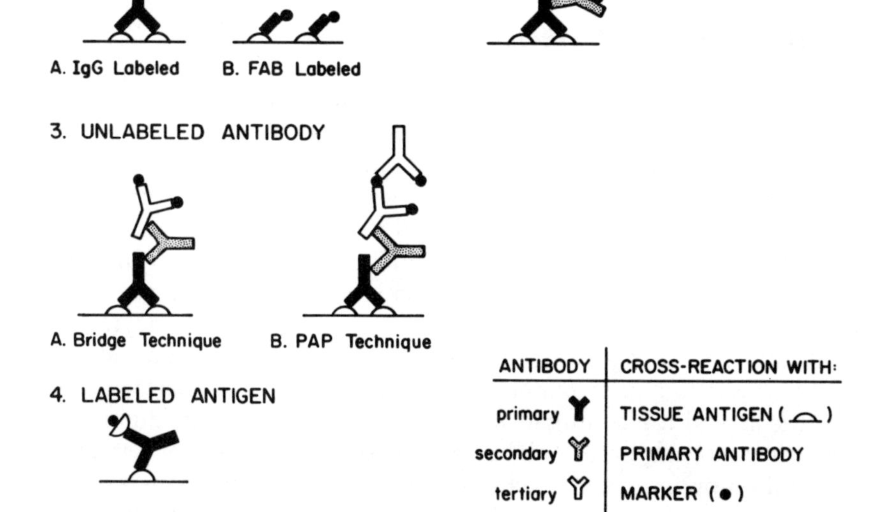

Figure 2. Some of the popular methods for detecting antigens are illustrated. A common marker is horseradish peroxidase, which in the presence of DAB and H_2O_2 forms a brown final reaction product. Among others, some markers are fluorescent, ferritin, and radiolabeled compounds. With direct labeled antibody techniques either whole IgG molecules (*1A*) or their Fab fragments (*1B*) react with the antigen. With indirect labeled antibody methods (*2*), a second labeled antibody is used. The peroxidase-antiperoxidase (PAP) technique (*3B*) is an improvement on the bridge technique (*3A*), both of which utilize a third antibody that is usually generated against horseradish peroxidase. The RICH technique (52) utilizes the bivalent properties of IgG molecules; one Fab arm binds to the antigen to be localized, while the other arm binds to a radiolabeled antigen (*4*).

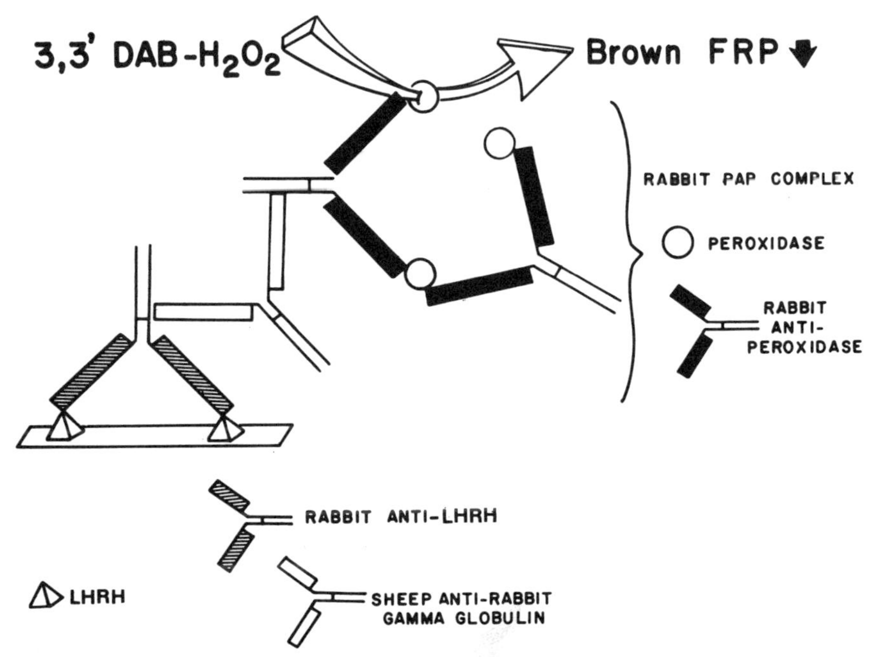

Figure 3. The detection of LHRH in tissue sites using the peroxidase-antiperoxidase (PAP) technique is depicted. Fab fragments of the rabbit anti-LHRH antisera cross-react with its antigen, LHRH, or with some substance having similar antigenic determinants that would be LHRH-like. PAP consists of two antiperoxidase molecules in complex with three horseradish peroxidase (HRP) molecules. Sheep anti-rabbit gamma globulin serves as the so-called "bridging" antibody. The IgG fraction of rabbit serum is used as an immunogen in a series other than the rabbit; it cross-reacts with the Fc fragment of the rabbit anti-LHRH and the rabbit anti-HRP of the PAP complex, thereby forming the "bridge." A precipitable brown final reaction product (FRP) is developed in the presence of 3,3'-diaminobenzidine (DAB) and H_2O_2. The colored final reaction product is visible with light microscopy and osmiophilic for use with electron microscopy.

against the marker protein-HRP (see Figure 2, *3A*). The last steps consist of incubating the section in diaminobenzidene (DAB) and H_2O_2, which form an insoluble brown final reaction product at the site of the antigen. DAB serves as the electron donor to the HRP-hydrogen peroxide complex, which undergoes oxidative polymerization and cyclization during the reaction to form the phenazine polymer or final reaction product (88). This technique is based on the method published by Mason et al. (54) in 1969; it is sometimes called the "bridging" or "sandwich" technique, since an intermediate antibody is used to "bridge" the primary antibody with a third antibody. Usually this is a sheep or goat anti-rabbit IgG that binds to the Fc portion of the primary rabbit antisera and the Fc portion of the rabbit anti-HRP. A significant improvement was achieved by Sternberger et al. (96), who reacted HRP with rabbit anti-HRP to form a peroxidase-antiperoxidase (PAP) complex. This had the advantage of not only reducing the number of reagents used but also increasing the ratio of HRP molecules to the antigenic site. The amplification effect is illustrated in Figure 3.

6 METHODS OF PROCEDURE: PRE-EMBEDDING IMMUNOCYTOCHEMISTRY (ICC)

The procedures described here are those used in our laboratory for the past eight years and are based on our experience with a wide range of fixatives, reagents, and buffers at different times and concentrations. There are several suppliers of materials and reagents other than those indicated here, and at times it may be necessary to make substitutions of reagents from other companies. The reagents and methods presented here can serve as a guide to those without previous experience with ICC techniques and should provide a working set of directions capable of producing positive results, hopefully upon the first effort. One of the most important messages anyone could pass on to the novice is that ICC should not be regarded as if it were some type of sophisticated conventional staining technique. First of all, the antigen–antibody reaction does not follow the law of mass action and is not stoichiometric. Therefore, one should expect to use an organized trial of dilutions and incubation times of the immunoreagents in order to determine maximum staining intensity in relation to background. It is important to structure such an approach with the least number of variables so that true comparisons can be made of the variant sections, allowing meaningful conclusions about the efficacy of the reagents used. With the ever increasing number of laboratories now employing light microscopic and electron microscopic ICC as primary tools of research, there has been a concomitant increase in the number of improvements in the technique, some of which are critical for particular applications. Thus, one should not regard the protocol presented here as a recipe to be religiously used; but one should be open to making adjustments according to results achieved for specific applications. Such matters as improvements in fixation, the use of different antisera, and development of the final reaction product with various intensification procedures, if tried and found successful, can sometimes provide exciting and dramatic differences in obtaining improved experimental data. Of course each experimental run should have a primary objective intended to provide experimental data, but many times it is enjoyable to include a few extra wells of tissue treated differently for the purpose of attempting to improve one's procedures.

6.1 Procedure for Light Microscopic ICC

6.1.1 *Zamboni's Fixative (94)*

0.05 M Phosphate Buffer

3.31 g $NaH_2PO_4 \cdot H_2O$
33.77 g $Na_2HPO_4 \cdot 7\ H_2O$
Dissolve in 1 liter distilled H_2O.
pH to 7.3 (use few drops of 10 *N* NaOH to adjust).

In a separate beaker, add 20 g paraformaldehyde (SPI-Chem, Structure Probe, Westchester, PA) to 150 mL of saturated picric acid that has been filtered. Heat to 60°C and add several drops of 10 *N* NaOH until solution clears. Allow to cool. Add phosphate buffer and q.s. to about 900 mL. Adjust to pH 7.3, using 10 *N* HCl and q.s. to a final volume of 1000 mL.

6.1.2 *0.05 M Phosphate Buffered Saline (PBS)*

28.36 g Na_2HPO_4
27.24 g KH_2PO_4
35.04 g NaCl
Dissolve in 4 liters H_2O; adjust pH to 7.4 with 10 *N* NaOH.

6.1.3 *Perfusion Procedure*

We routinely use male, 200–250 g, Long-Evans rats purchased from Simonson Laboratories, Gilroy, CA. The animal is given i.p. injections of:

0.2 mL of 50 mg/mL Nembutal (Abbot Laboratories, North Chicago, IL)
0.1 mL of Heparin, 1,000 units/mL (Elkinns-Sinn, Cherry Hill, NJ)
0.5 mL of 1% sodium nitrite (vasodilator)

Place a grate over the sink and on it place the anesthetized rat on its back; tie the limbs to the grate. Excise a V-shaped skin flap over the thorax. Make two parallel cuts into both sides of the rib cage. Cut away the diaphragm. The heart is now exposed. Hold ribs back with a heavy hemostat. Cut away the thymus. With a small hemostat, pull a piece of thread under the superior aorta and make a loose knot but do not tie off.

We use a perfusion system consisting of 2 i.v. drip bottles; one for the phosphate-buffered saline (PBS) flush solution, the other for the fixative. They are hung on a stand about three feet above the animal. Each has tubing leading to a "Y" connector which in turn leads to a peristaltic pump. There is a "T" connector on the tubing from the pump with an attached mercury pressure gauge, allowing for measurement of fluid pressure which is continuously adjusted to read 90 mm Hg during perfusion. It is important to eliminate any bubbles in the tubing. The amount of pressure is dependent on the size of the cannula, constrictions in the tubing, and other variables (26). Adjustments in pressure used for a particular delivery system should be made according to quality of ultrastructural preservation achieved.

Make a cut in the left ventricle and insert a small cannula into the aorta. Tighten the knot over the aorta. Clip the right atrium to allow fluid to drain. Slowly flush the heart with PBS solution. Use a hemostat to clamp the descending aorta and vena cava at the base of the thoracic cavity next to the diaphragm, superior to the liver and ventral to the spine. After the return flow from the atrium becomes clear, switch to fixative solution. Perfuse with 800 mL of fixative for 15 min. Remove the head with a guillotine. Cut away skin, exposing the skull. Using a pair of rongeurs and starting with rostral skull, carefully expose the dorsal and lateral surface of brain. As you gently lift the rostral part of brain, cut the olfactory lobes, optic nerves, pituitary stalk, and other cranial nerves. The final cut is of the spinal cord.

Place the brain in a capped bottle with 100 mL fixative at room temperature for one day. Then wash three times daily for two days with PBS buffer in order to remove the picric acid.

6.1.4 Sectioning with the Vibrating Microtome

The Lancer Vibratome Series 1000 is currently available from Ted Pella, Inc., Tustin, CA, and has replaced the Oxford Vibratome.

1. Fill the trough with PBS.
2. Break a double-edged Schick Super Chromium razor blade in two and place one half in the blade holder adjusted to an angle of 15–20°.
3. Block the tissue to a height of no more than 8 mm, preferably less.

4. Dry the bottom of the piece of brain tissue on a paper towel and fix to the specimen stage with half a drop of Super Glue or Eastman 910 adhesive. (We prepare our own specimen stages by cutting a 1 cm^2 rectangular aluminum rod into 1 cm lengths and then gluing a 2 cm^2 piece of a microscope slide to the surface of the block.) While waiting for the adhesive to dry, about 3 or 4 min, keep the top and sides of the tissue block moist with a paper towel soaked in buffer.
5. When the adhesive is dry enough so that the block remains secure on the stage when it is pressed gently, the stage can be mounted in the vibrating microtome.
6. Raise the stage until the tissue is just below the cutting edge of the blade. Section at 100 μm until the brain is level.
7. After leveling, cut sections at 50 μm, collect with a camel's hair paintbrush, and store in the refrigerator in buffer.
8. Every vibrating microtome is different but the speed should be set in the 3–6 range and the amplitude in the 5–7 range. If a block is not cutting well, varying the speed and/or the amplitude can improve sectioning.
9. The tissue should remain in buffer at all times and there should always be sufficient buffer in the trough to cover the razor blade.

6.1.5 Light Microscopic ICC Staining Procedure

Sections are free-floating and are stained in reusable plastic boxes with compartments holding approximately 2.5 mL fluid. All incubations and washings are done on a Junior Orbital Shaker (Lab-Line Instruments, Melrose Park, IL) with speed set so that sections are free-swirling (about 100 rpm). We use two forms of PBS buffer.

PBS A: PBS buffer + 0.1% Triton X-100 (Grand Island Biological Co., Grand Island, NY)

PBS B: PBS A + Millipore-filtered 1% NGS (normal goat serum) (Cappel Laboratories, Cochranville, PA)

Sections are transferred from well to well with a camel's hair paintbrush. The sections are placed in:

1. 1% H_2O_2 in PBS for 30 min to eliminate endogenous peroxidase. (30 mL tot. vol. = 10 mL of 3% H_2O_2/20 mL PBS)

2. PBS A, 15 min wash. (50 mL tot. vol. = 0.05 mL Triton/50 mL PBS A)
3. PBS B, 15 min wash.
4. 24-hr incubation at 4°C in primary antibody generated in rabbit. Antiserum is diluted in PBS B to 1:1,000. (2.5 mL tot. vol. = 2.5 μl antiserum/2.5 mL well). This dilution is commonly used, but optimal results with a particular antiserum should be determined by trying varying concentrations.
5. PBS A, 15 min wash.
6. PBS B, 15 min wash.
7. 1:100 goat anti-rabbit IgG (Cappel Laboratories) for 30 min (2 mL tot. vol. = 20 μl/2 mL PBS B)
8. PBS A, 15 min wash.
9. PBS B, 15 min wash.
10. 1:500 rabbit PAP (Dako, Accurate Chemical Scientific Corp., Westbury, NY) in PBS buffer for 1 hr.
11. PBS buffer, 3 × 10 min washes.
12. 3,3′ diaminobenzidene tetrahydrochloride (DAB) 15 mg% (Aldrich, Milwaukee, WI) + .003% H_2O_2 in PBS buffer (for 50 mL tot. vol., then 7.5 mg DAB + 50 μL 3% H_2O_2/50 mL buffer). DAB is a suspected carcinogen. It is light-sensitive and should be dissolved under a hood. Protect from light by covering the beaker with aluminum foil. Just before use, filter the DAB with a 0.45 μm Nalgene Membrane Filter unit (Nalgene Labware, Division of Sybron Co., Rochester, NY. Nalgene No. 245-0045). Unused DAB solutions should be placed in a 10% Clorox™ bleach solution in order to eliminate harmful effects to the environment. Sections are incubated with shaking until a visible reaction product is noted in the region of interest as viewed with a stereoscopic microscope. When an intense reaction is observed, sections are immediately placed in buffer. If sections remain in the DAB solution too long, increased background can result, which in turn reduces the contrast of staining achieved.

6.1.6 *Mounting Sections*

1. Rinse sections in PBS buffer × 15 min.
2. Float the section in a staining dish filled with 50:50 PBS and distilled water. A subbed slide is partially immersed in the liquid and the section is maneuvered onto the slide with a brush. The subbing solution affixes the section to the slide.

Subbing solution

a. Dissolve 1 g gelatin in 50 mL heated distilled H_2O.
b. Bring volume to 1 liter with distilled H_2O.
c. Cool and add 0.1 g chromium potassium sulfate ($CrK(SO_4)_2 \cdot 12H_2O$).
d. Place clean microscope slides in staining racks. Dip slides 3 times in solution, dry under heat lamp between dips. Dry completely after last dip. Store subbed slides in refrigerator.

3. If a nuclear counterstain is desired, proceed to 6.1.7. If not, allow slides to air dry in staining rack for about 20 min under a heat lamp.
4. Transfer the slides to staining dishes:

   a. 70% ethanol 1 × 2 min,
   b. Absolute ethanol 1 × 2 min.
   c. Xylene 1 × 5 min.

5. Coverslip with Permount (Fisher Scientific Co., Pittsburgh, PA).

6.1.7 *Hematoxylin Counterstain*

1. Place mounted sections in slide holder in a staining dish that has just the bottom of the dish covered with 10% formalin; place in 55-60°C oven for 1 hr.
2. Rinse with tap water for 10 min. Place a baffle between running water and slides so that water rinses gently.
3. Filter and add 3 to 4 mL glacial acetic acid per 100 mL hematoxylin. Place slides in Harris hematoxylin (Anderson Laboratories, Ft. Worth, TX) for 15 min.
4. Rinse in tap water 2 × 10 min.
5. Differentiate in acid alcohol for 30 sec. Acid alcohol is 1 liter of 70% ethanol plus 10 mL concentrated HCl. The time of differentiation may vary, so check sections with a microscope.
6. Rinse in tap water 1 × 3 min.
7. Blue sections in ammonia water: 1 liter of tap water with 3 mL ammonium hydroxide. Three to five dips.
8. Rinse in distilled H_2O 2 × 10 min.
9. Dehydrate 3 min each in 60%, 80%, 95%, 100%, 100% ethanol, xylene, xylene.
10. Permount and coverslip.

6.2 Procedure for Electron Microscopic ICC

6.2.1 Preparation of Fixative

0.1 M Phosphate Buffer

Solution A: 13.9 g sodium phosphate monobasic (Na $H_2PO_4 \cdot H_2O$) plus 500 mL distilled H_2O.

Solution B: 28.82 g sodium phosphate dibasic (Na $HPO_4 \cdot 7H_2O$) plus 500 mL distilled H_2O.

Add 405 mL of Solution B to 95 mL of Solution A.

Add 500 mL distilled H_2O.

pH should be 7.4.

600 mL fixative is enough to perfuse one 250–300 g rat with enough left over for two days of postfixation of the brain.

3% Paraformaldehyde + 2% Glutaraldehyde

This preparation makes 600 mL. Heat 200 mL distilled H_2O to 60°C. Do not exceed 60°C, even though the paraformaldehyde dissolves slowly.

Add 18 g paraformaldehyde; the solution will be milky white.

Add 0.1 *M* NaOH dropwise until solution clears. Approximately 2 mL of NaOH is usually necessary to get all of the paraformaldehyde into solution.

Add 48 mL of 25% glutaraldehyde (if using 50% stock, dilute with 0.1 *M* phosphate buffer 1:1).

Bring this volume to 300 mL with 0.1 *M* phosphate buffer, q.s. to 600 mL with distilled water.

pH should be 7.2. If basic, adjust with HCl; if acid, adjust with 1 *N* NaOH.

If solution is refrigerated before use, some paraformaldehyde may precipitate out; this will clear when heated to room temperature. The pH varies with temperature and should be adjusted before use. The perfusion and sectioning procedures have been described in Sections 6.1.3 and 6.1.4.

6.2.2 Electron Microscopic ICC Staining Procedure

Sections are processed essentially the same as described in Section 6.1.5, with the notable exception of substituting Tris-buffered solutions for

PBS. In our hands, Tris buffer rather than PBS yields a consistently superior preservation of structure and reaction product at the subcellular level.

0.1 M Tris buffer (to make 1 liter, pH 7.6 at 25°C)

12.12 g Trizma HCl (Sigma Chemical Co., St. Louis, MO)
2.78 g Trizma Base
1.00 g gelatin dissolved in 50 mL heated H_2O

Check pH, using a glass calomel electrode (Fisher Scientific Co., Pittsburgh, PA).

Tris A: Tris buffer + 0.1% Triton
Tris B: Tris A + 1% NGS (normal goat serum)

After step 12 of 6.1.5, the sections are further processed, still using the plastic compartment boxes:

1. 0.1 *M* phosphate buffer, wash 1 × 10 min.
2. 2.5% glutaraldehyde in 0.1 *M* phosphate buffer, 1 hr. This step serves to fix the reaction product.
3. 0.1 *M* phosphate buffer, wash 1 × 5 min.
4. 1% OsO_4 in 0.1 *M* phosphate buffer 1 hr. This step is done under the hood; gloves and eye protection must be worn when handling osmium tetroxide.
5. 0.1 *M* phosphate buffer, wash 3 × 2 min.
6. 50% ethanol 1 × 10 min.
7. 70% ethanol 1 × 10 min.
8. 95% ethanol 2 × 10 min.
9. 100% ethanol 3 × 10 min.
10. 1:1 propylene oxide:absolute ethanol, 1 × 15 min. Starting with this step, sections are transferred to 50 mL Tri-Pour beakers, since propylene oxide will dissolve the plastic compartment boxes.
11. Propylene oxide, 3 × 10 min each.
12. 1:1 propylene oxide: Araldite 6005 (Ladd Research Industries, Burlington, VT), overnight under the hood with no cover. Other embedding medias may be used. Next morning, the sections will be mounted in a pool of resin on microscopic slides coated with 1% dimethyl dichlorosilane in benzene.

Coated Slides

a. Wash glass slides in soap and water.
b. Rinse well.
c. Dry in oven.
d. Dip in a solution of 1% dimethyl dichlorosilane (Sigma Chemical Co., St. Louis, MO) 5–10 min under the laboratory hood. Discard solutions after use.
e. Dry in 60°C oven.
f. Wash in tap water for 10 min; dry slides in oven. Alternative methods include using Teflon-coated or silicone-coated slides, or slides coated with evaporated carbon.

13. Place a few drops of Araldite 6005 on a coated slide, place section in the pool of embedding resin, coverslip with another coated slide, and place, lying flat, in a 60°C oven overnight.

Araldite 6005 Solution

a. 1:1 (by weight) Epoxy 6005:DDSA. Add 5% (by volume) plasticizer, DBP.
b. Mix well 5–10 min with glass stirring rod.
c. Add accelerator, 0.1 mL DMP/10 mL stock plastic.
d. Mix 5–10 min with glass rod. Unused resin can be stored frozen in plastic syringes.

When slides are cooled, they are pried apart with a single-edged razor blade. Sections are examined under a stereoscopic microscope and areas of interest are cut away with a single-edged razor blade. Place a small drop of Super Glue on the top of a blank resin Beem capsule, and affix section to capsule. After drying, the tissue is ready for ultramicrotomy.

14. Ultrathin sections are stained as follows:

a. 4% uranyl acetate in 50% ethanol for 5 min.
b. 50% ethanol wash; blot dry with filter paper.
c. Lead citrate for 2.5 min
   1.33 g lead nitrate
   1.76 g sodium citrate
   30 mL of boiled distilled H_2O
   Shake vigorously for 1 min.
   8.0 mL 1.0 *N* NaOH
   pH of final solution, 12.0.
d. 50% ethanol wash; blot dry with filter paper.

Sections are now ready for examination with the electron microscope.

7 POST-EMBEDDING ICC

The hypothalamic magnocellular neurosecretory system serves as a useful model for studying neuropeptides. Large cell bodies found in the paraventricular, supraoptic, and accessory nuclei send their fibers via the hypothalamo-hypophysial tract through the zona interna of the median eminence to terminate in the pars nervosa of the neurohypophysis (Figure 4).

These cells synthesize the hormones vasopressin and oxytocin, and their associated neurophysins. Since the advent of ICC our appreciation of the complexity of this system at the morphological, biochemical, electrophysiological, and functional level has greatly increased (10, 27, 64, 100, 102, 104, 109, 111). With post-embedding ICC for neurophysins, (45), visualization of the final reaction product is largely limited to neurosecretory granules within the cell bodies, fibers, and their terminals (Figure 5).

Silverman and Zimmerman (90) have provided some examples of extragranular staining for neurophysins in the neural lobe, using freeze-

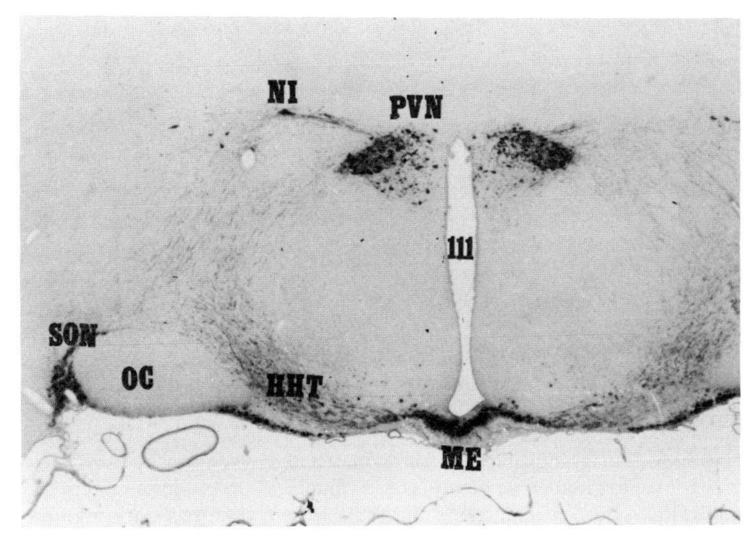

Figure 4. A paraffin-embedded section of the anterior hypothalamus of the rat immunostained with rabbit anti-bovine neurophysin II showing components of the magnocellular neurosecretory system. The paraventricular nucleus (*PVN*), cells of intermediate nuclear groups (*NI* or periforniceal cells), and supraoptic nucleus (*SON*) cells give rise to fibers that sweep ventromedially to form the hypothalamo-hypophysial tract (*HHT*). At the level of the median eminence (*ME*), these fibers pass through the zona interna and project to the pars nervosa of the neurohypophysis, but some fibers arising from the PVN project to the zona externa (not shown at this magnification) and deliver their peptides to the pars distalis of the adenohypophysis. *OC*, optic chiasm; *III*, third ventricle.

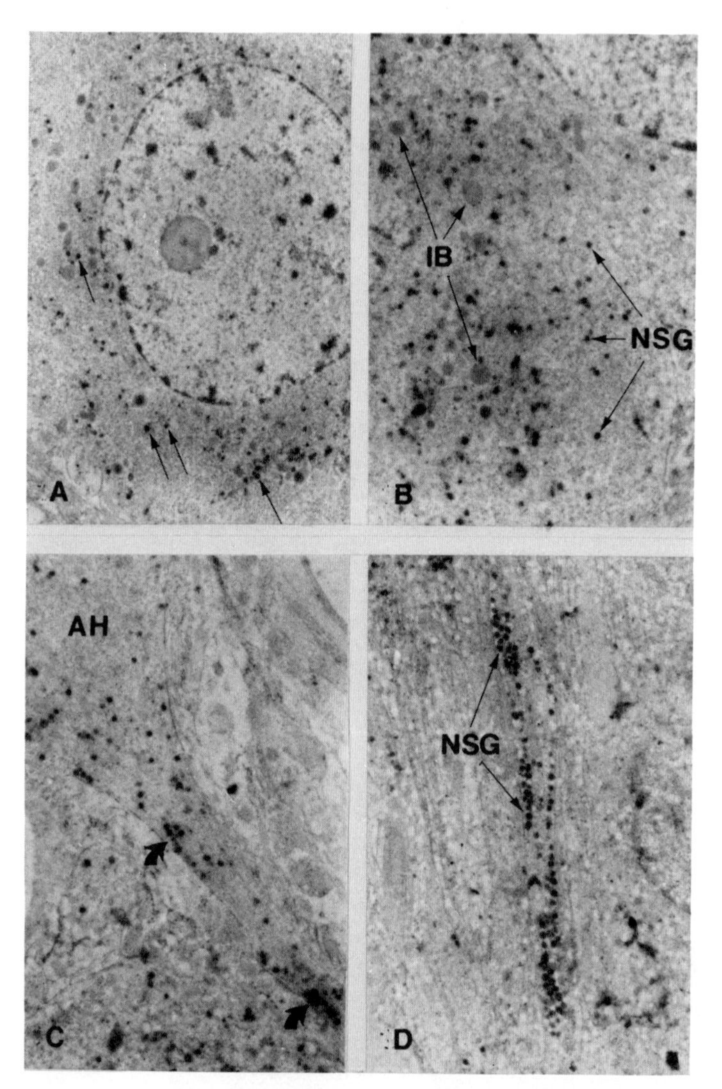

Figure 5. (A) High-power magnification of a magnocellular neurosecretory neuron of the supraoptic nucleus, containing numerous neurosecretory granules (arrows) immunopositive for neurophysin. (B) High-power magnification of cytoplasm with neurosecretory granules (*NSG*). The large inclusion bodies (*IB*) are not immunostained. (C) The neurosecretory granules tend to converge in a region resembling the axon hillock and form aggregates (arrows). (D) Longitudinal section of a neurosecretory axon in which the neurosecretory granules accumulate as linear aggregates. These accumulations provide the ultrastructural correlate for the beaded appearance of the fibers as seen with light microscopy. Postembedding technique. From Kozlowski et al. (45).

substitution techniques. Krisch (48) reported that post-embedding ICC is capable of demonstrating vasopressin as it is released in a nongranular form from the supraoptic nucleus perikarya into the intercellular cleft. Buijs and Swaab (10) found that specificity of staining using the post-embedding technique for vasopressin was lacking even when antivasopressin was immunoabsorbed with solid-phase procedures. With this treated antiserum they were able to detect staining in the brains of animals lacking vasopressin.

Childs and Ellison (13) and Childs (12) provide details of the post-embedding procedure as well as criteria necessary to prove staining specificity. It is clear from their papers that high dilutions of primary antisera (e.g., anti-ACTH at 1:100,000) with long incubation times (24–48 hr) are preferable. In this regard, Bigbee et al. (5) demonstrated that antigen-rich tissue required low concentrations of primary antisera to minimize the binding of both Fab fragments of the primary antibody in order to ensure that the bridge antibody molecules will be only singly bound and thus free to attach to the PAP complex. With post-embedding staining the final reaction product can appear as grapelike clusters in association with antigen-rich areas such as neurosecretory granules containing LHRH (Figure 6). In other cases, the cyclic or pentagonal nature of single PAP complexes is preserved (18, 95).

Post-embedding ICC techniques for neuropeptides probably demonstrate only 5 to 10% of immunoreactive peptide originally present in brain tissue. RIA studies of the amounts of peptide found after fixation and subsequent dehydration in alcohol showed that 90 to 95% of LHRH (29, 39), somatostatin (16), or vasopressin (58) was extracted. Recent advances in freeze-fixation technology hold great promise for improved retention and localization of neuropeptides with post-embedding ICC. Coulter and Elde (17) have used frozen-dried posterior pituitary tissue fixed with OsO_4 vapor to demonstrate neurophysin at the ultrastructural level, and they have further shown that little or no peptide is lost with this procedure. Recent application of the Coulter–Terracio freeze-dry apparatus (Ladd Research Industries, Burlington, VT) to ICC studies with other tissues such as the pancreas, has introduced an improved dimension for visualizing the final reaction product both in granules and within well-defined intracellular membranous compartments such as GERL (*G*olgi-associated *e*ndoplasmic *r*eticulum giving rise to *l*ysosomes) (21). This technique also holds great promise for successfully localizing hormone receptors that are susceptible to the effects of conventional fixation techniques (86). And since post-embedding ICC allows for the localization of multiple antigens in serially sectioned cells or granules (12), another likely application for the future would be to extend light

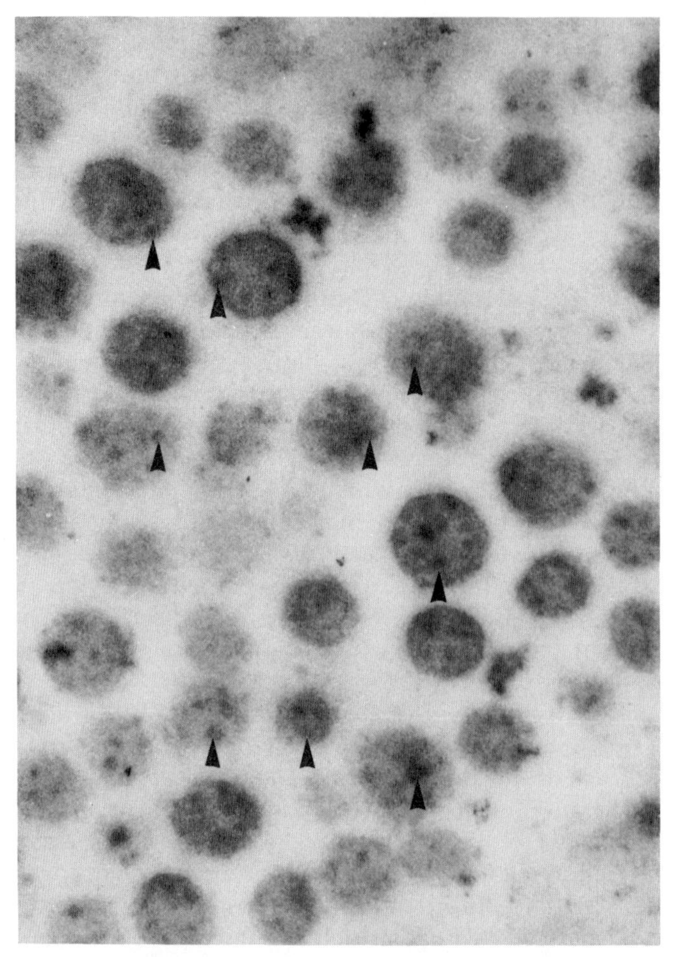

Figure 6. High-power magnification of several neurosecretory granules immunopositive for LHRH, showing the appearance of immunoprecipitate (arrows) as seen with post-embedding immunocytochemistry. From Kozlowski (43).

microscopic ICC studies to demonstrate the coexistence in single neurons of neurotransmitters and/or neuropeptides (3, 37) to the ultrastructural level, providing much greater resolution even to the point of resolving whether these substances coexist in the same granule.

8 PRE-EMBEDDING ICC

Our studies have concentrated on the comparison of ultrastructural similarities and differences that exist between a representative parvocellular neuroendocrine cell (i.e., the LHRH cell) and a magnocellular neuron immunolabeled for either neurophysin or vasopressin (41–46). Localization of neurophysin in magnocellular cells confirms their glandular nature for the production and processing of neurosecretory products. As predicted from the general sequence of events that protein synthesis follows, the

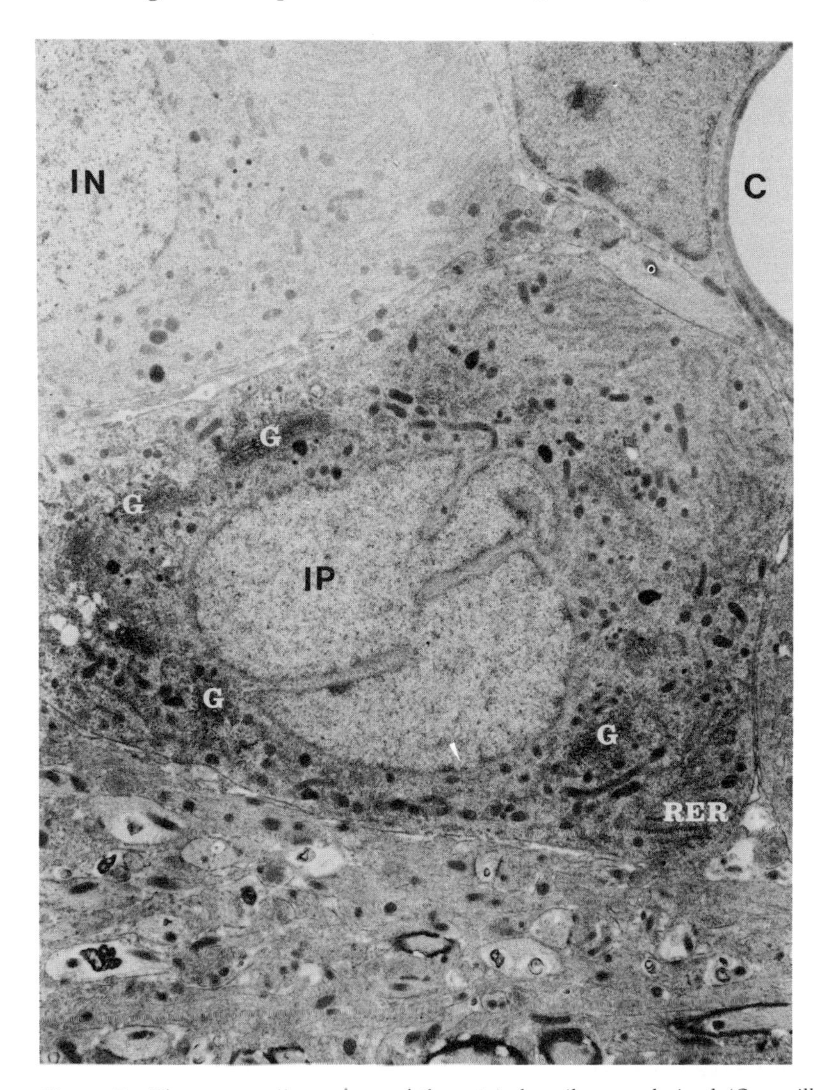

Figure 7. The supraoptic nucleus of the rat is heavily vascularized (*C*, capillary). An immunonegative cell (*IN*) is compared with an immunopositive cell (*IP*) for neurophysin. There is a general osmiophilic density to the cytoplasm, but the Golgi apparatus (*G*) and patches of rough endoplasmic reticulum (*RER*) are especially prominent. Pre-embedding technique.

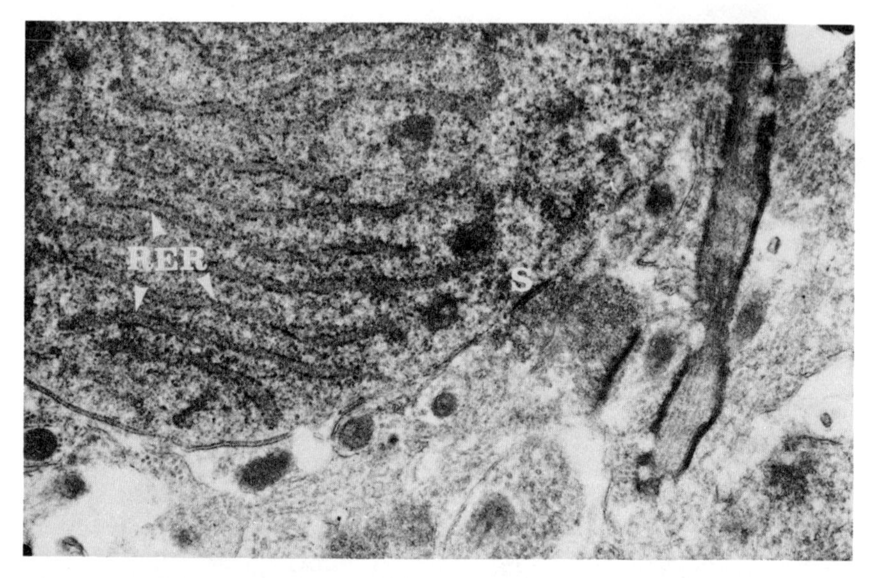

Figure 8. High-power magnification of a neuron of the rat paraventricular nucleus stained with rabbit anti-rat neurophysin IV. Cisternae of the rough endoplasmic reticulum (*RER*) are filled with final reaction product. *S*, synapse. Pre-embedding technique.

final reaction product is associated with the rough endoplasmic reticulum, Golgi complex, shuttle vesicles, and neurosecretory granules (Figures 7–10).

We have also observed that the final reaction product can impart a general background density to the cytoplasm of immunopositive cells vs. immunonegative cells stained for either neurophysin (Figure 7) or LHRH (Figure 14). High-voltage electron microscopy of LHRH-containing fibers demonstrated that the final reaction product is not only associated with neurosecretory granules but is also interspersed between them as extragranular matter (Figure 11).

A similar situation exists for magnocellular neurons containing neurophysin. The general background staining seen in the cytoplasm of neurophysin-containing cells (Figures 12 and 13) relates to several technical factors, including diffusion problems and the development of DAB, that will be discussed later.

Parvocellular neurons containing LHRH also have final reaction product associated with subcellular structures involved in the synthesis, packaging, and transport of secretory product (Figures 14 and 15). In contrast to magnocellular neurons, LHRH neurons have smaller and fewer neurosecretory granules. However, there are numerous examples of immunopositive saccules of smooth endoplasmic reticulum (Figure 15) that most probably transport considerable amounts of LHRH along cellular processes (20, 41, 44). This compartmentalized immunoreactivity is to be distinguished from final reaction product associated with microtubules (Figure 15).

Interpretation of results of the subcellular distribution of final reaction product seen with pre-embedding ICC involves considerations of the effects of fixation, detergents (used to aid in the penetrability of reagents), and development of the PAP complex with DAB/H_2O_2.

Fixation should be regarded as one of the most important factors involved in successfully localizing antigens. Varying both the type and method of fixation for electron microscopic ICC is a necessary prerequisite for its use in any ICC study. Childs (12) has shown that demonstrating different anterior pituitary hormones with ICC requires the use of different fixatives. For conventional electron microscopy of CNS tissues, considerable effort has been spent by many investigators in attempting to achieve what would be regarded as excellent morphological preservation of cells and their organelles. Superimpose upon these efforts the demand

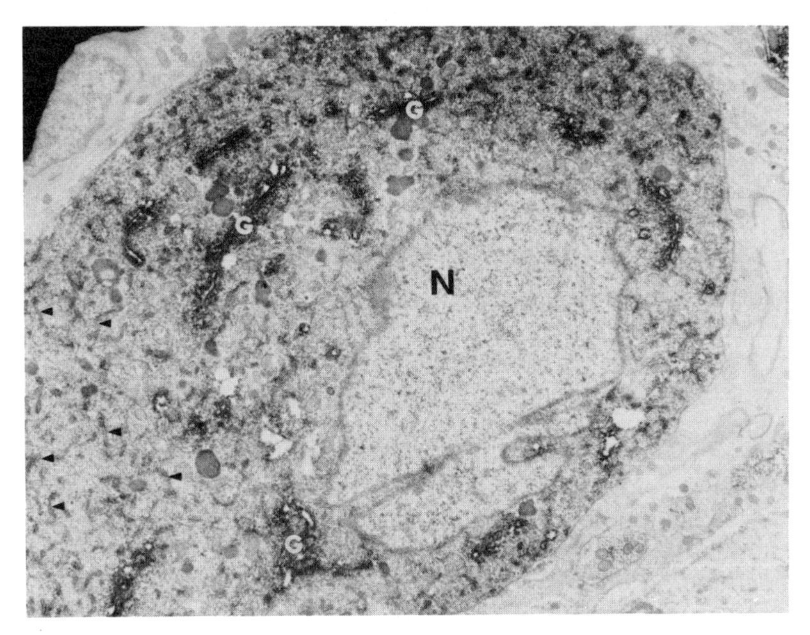

Figure 9. Low-power magnification of a neuron of the rat paraventricular nucleus stained with rabbit anti-rat neurophysin IV. These large cell bodies contain an indented nucleus (*N*), numerous arrays of highly immunoreactive Golgi complexes (*G*), and rough endoplasmic reticulum (small arrowheads). Pre-embedding technique.

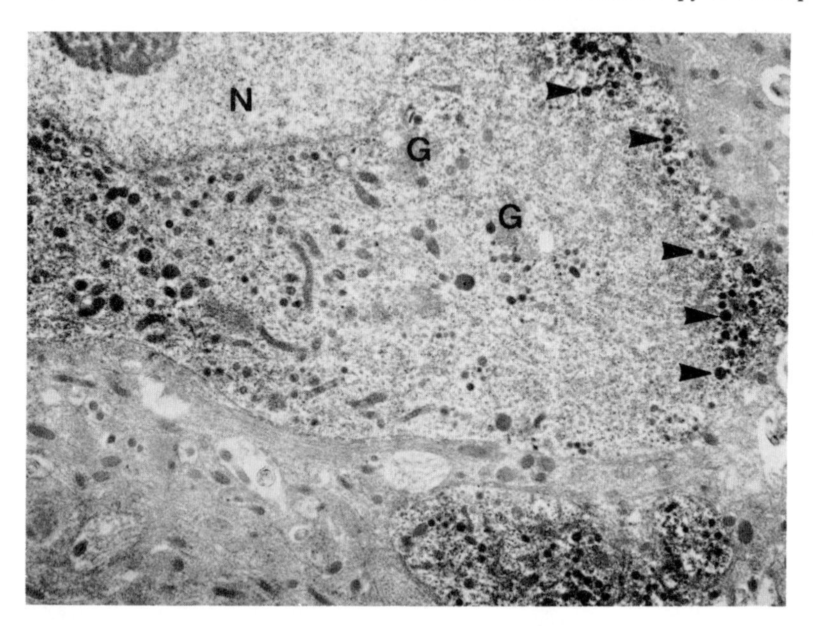

Figure 10. Low-power magnification of a neuron of the rat supraoptic nucleus stained with rabbit anti-rat neurophysin IV. The nucleus (*N*) is centrally located. Golgi complexes (*G*) are not immunoreactive. Numerous immunopositive neurosecretory granules are distributed along the periphery of the cell cytoplasm. Pre-embedding technique.

of retaining antigenicity of neuropeptides as they occur *in situ*, and the problem becomes even more acute. It has almost become an accepted maxim that it is impossible to simultaneously obtain excellent antigenicity and preservation of cellular morphology. Actually, this may only apply to particular antigens, antisera, and chemical fixation techniques. Indeed, our earlier attempts were unsuccessful in obtaining localization of antigens in tissues fixed with OsO_4, an experience shared by many colleagues at the time. Recent studies have now shown that using OsO_4 can significantly improve the preservation of cellular morphology while still retaining antigenicity of neuropeptides (2, 17, 71) and pituitary hormones (for review, see 12). Also, most workers advocate the use of low concentrations (0.1–0.2%) of glutaraldehyde because higher concentrations of glutaraldehyde inhibited immunoreactivity of the antigen (9, 11, 76), allowed nonspecific binding of antibody to free aldehyde groups (25, 107), or interfered with the penetration of antibodies (38). Pretreatment of the tissue sections with a dilute solution of lysine can be used to bind free aldehyde groups if their presence seems to pose a problem. However, in our studies, 3% paraformaldehyde and 2% glutaraldehyde as a fixative

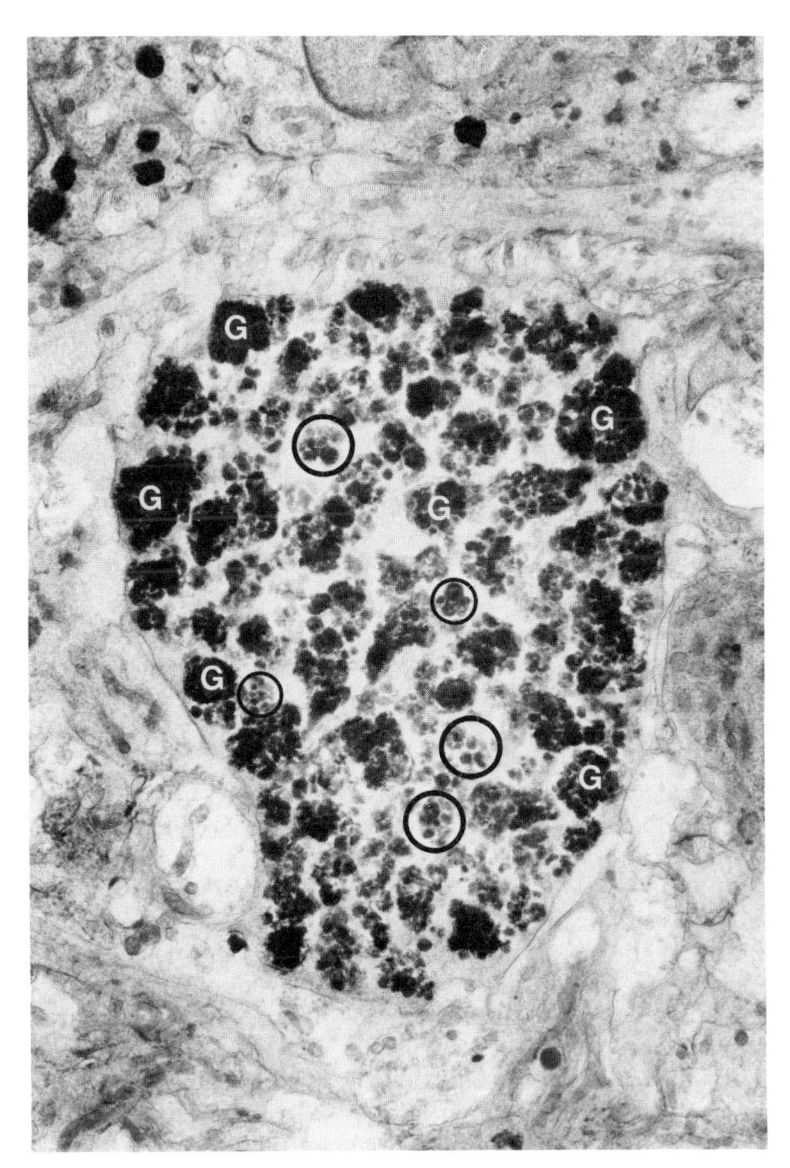

Figure 11. High-voltage electron micrograph of a 0.5 μm thick section. The most concentrated aggregates of PAP and antibody complexes surround neurosecretory granules (G) as found in this cross-section of a fiber stained for LHRH. Smaller clusters of complexes (circles) exist freely within the cytoplasm but their significance is unknown. Pre-embedding technique. From G. P. Kozlowski and G. Hostetter (46).

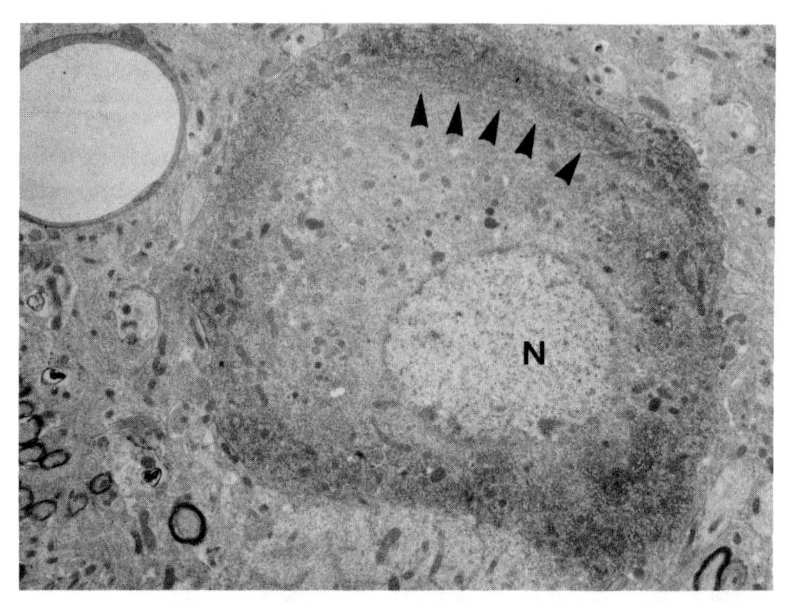

Figure 12. Low-power magnification of a neuron of the rat paraventricular nucleus stained with rabbit anti-rat neurophysin IV using the pre-embedding technique. This cell was located deep within the 50 μm section and the immunoreagents only penetrated the border of the cell, as suggested by the rim of final reaction product (arrowheads) along the cell periphery.

seem to provide adequate penetrability of reagents, preservation of cellular morphology, and retention of antigenicity (Figures 7–10, 14, 15). In addition to the type and concentration of fixatives, other factors such as the choice of buffers (12, 40, 76, 95), length of time of fixation, and pH are important. The rate and degree of fixation has been recently shown to be highly pH dependent. Berod et al. (4) first perfused with 4% paraformaldehyde at pH 6.5 followed by a quick change to 4% paraformaldehyde at a higher pH, which rapidly increases the cross-linking reaction. They found that complete fixation of tyrosine hydroxylase in the dopaminergic system with paraformaldehyde may be obtained with a very basic formaldehyde solution (pH 11) while still retaining immunoreactivity of the enzyme.

For ICC, the improper choice of fixation can not only lead to false positives due to translocation of antigen, but can also yield false negatives due to alteration or dissolution of the antigen. False negativity in ICC tends to be a major problem when one employs single antibody populations directed to restricted amino acid sequences in a peptide molecule. This is particularly seen with monoclonal antibodies and is well exemplified by recent ICC studies with monoclonal IIID-7 to vasopressin (33). The

IIID-7 clone was directed to the ring sequence of vasopressin and did not cross-react with arginine vasotocin or oxytocin as demonstrated by RIA. When brain sections were fixed with 4% buffered paraformaldehyde, periodate-lysine-paraformaldehyde (PLP) (57), and 2% glutaraldehyde and then immunoreacted with IIID-7, different results were obtained with the three fixatives. In formaldehyde-fixed tissue, application of IIID-7 resulted in reactivity mainly in perinuclear portions of perikarya of the supraoptic and paraventricular nuclei, but showed very little reactivity in cells of the suprachiasmatic nucleus. Reactive fibers were scarce in the brain, including the median eminence and posterior pituitary, whereas a rabbit antiserum to vasopressin stained cell bodies as well as fibers in these same sections. PLP-fixed sections immunoreacted with IIID-7 demonstrated better vis-

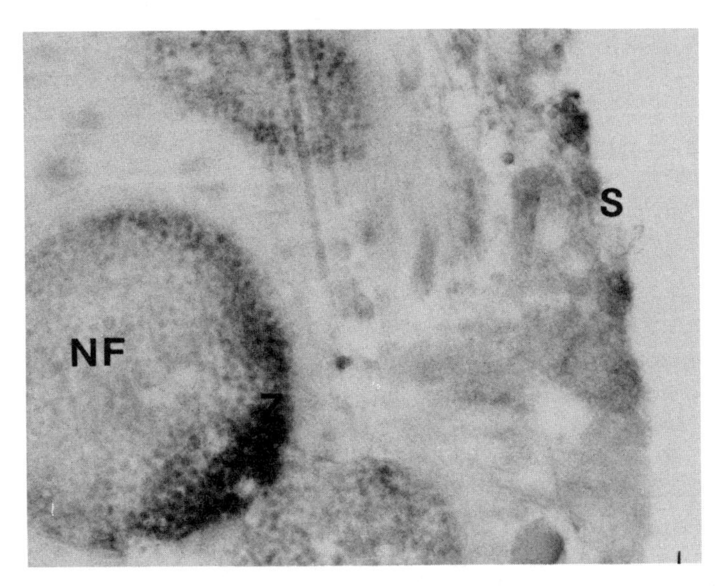

Figure 13. An example of poor penetration of immunoreagents throughout the thickness of the section. A 50 μm thick vibratome section was reacted with rabbit anti-bovine neurophysin II and embedded on edge. Ultrathin sections were then taken through the width of the vibratome section. Beneath the vibratome section surface (or surfaces) there is a cross-section of a neurosecretory fiber (*NF*) with a zone (*Z*) of reactivity facing the section surface. Studies of this type suggest that the immunoreagents pass as a moving front through the width of the section. When nerve fibers were found to communicate with the section surface, the immunoreagents enter the cut surface of the nerve fiber and penetrate at a greater rate so that most of such a fiber becomes stained. The fiber shown here probably coursed within the section without making contact with its surface. Because of the difficult sampling problem encountered in attempting serial-sections through an area of interest, this study could only show that some fibers, even though close to the section surface, may be stained for just a small portion of their length. Various degrees of reagent penetration occur within the same reaction.

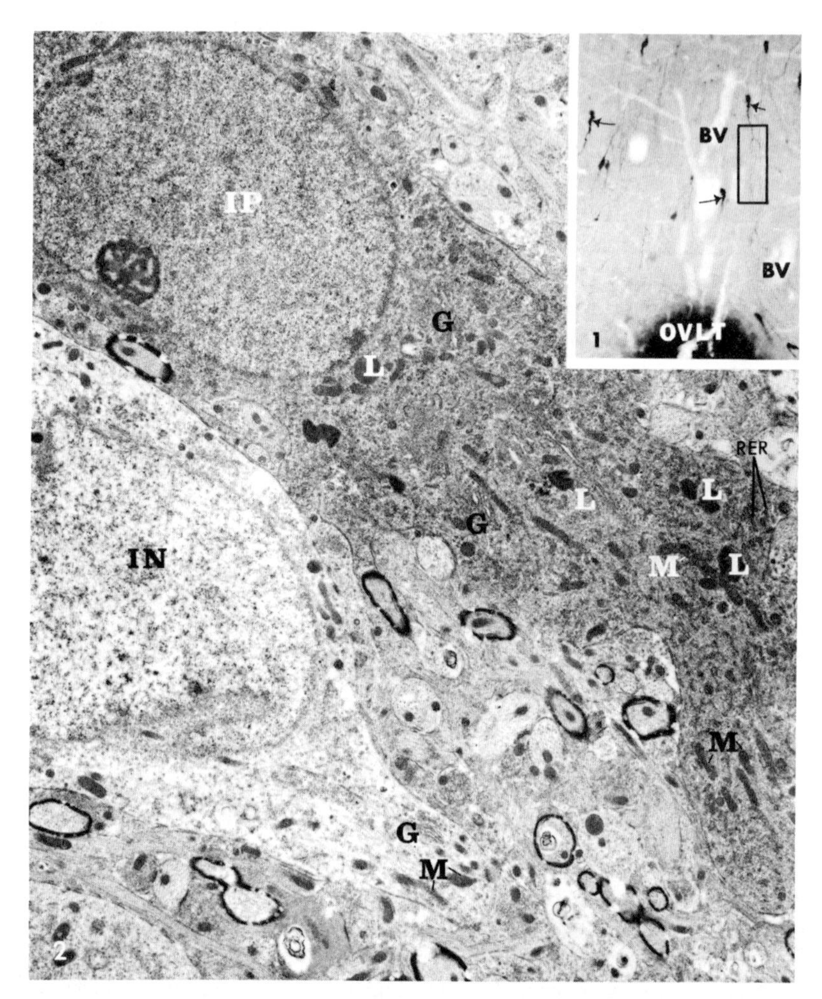

Figure 14. (*1*) Light micrograph of hypothalamic LHRH perikarya in vertical limb of diagonal band of Broca near organum vasculosum of the lamina terminalis (*OVLT*). The processes (arrows) are dendrites, which become beaded (boxed area) distal to the cell body of origin. Cell bodies are in close proximity to the numerous blood vessels (*BV*) found here. Kozlowski et al. (44). (*2*) Two perikarya in the diagonal band of Broca showing comparison of an immunonegative neuron (*IN*) with an immunopositive one (*IP*). The proximal portion of the dendrite is broad and contains elements common to the perikaryon such as rough endoplasmic reticulum (*RER*), mitochondria (*M*), free ribosomes, and lysosome-like bodies (*L*). The Golgi complex (*G*) is quite extensive. The final reaction product imparts a generalized electron density throughout the cell. Kozlowski et al. (44).

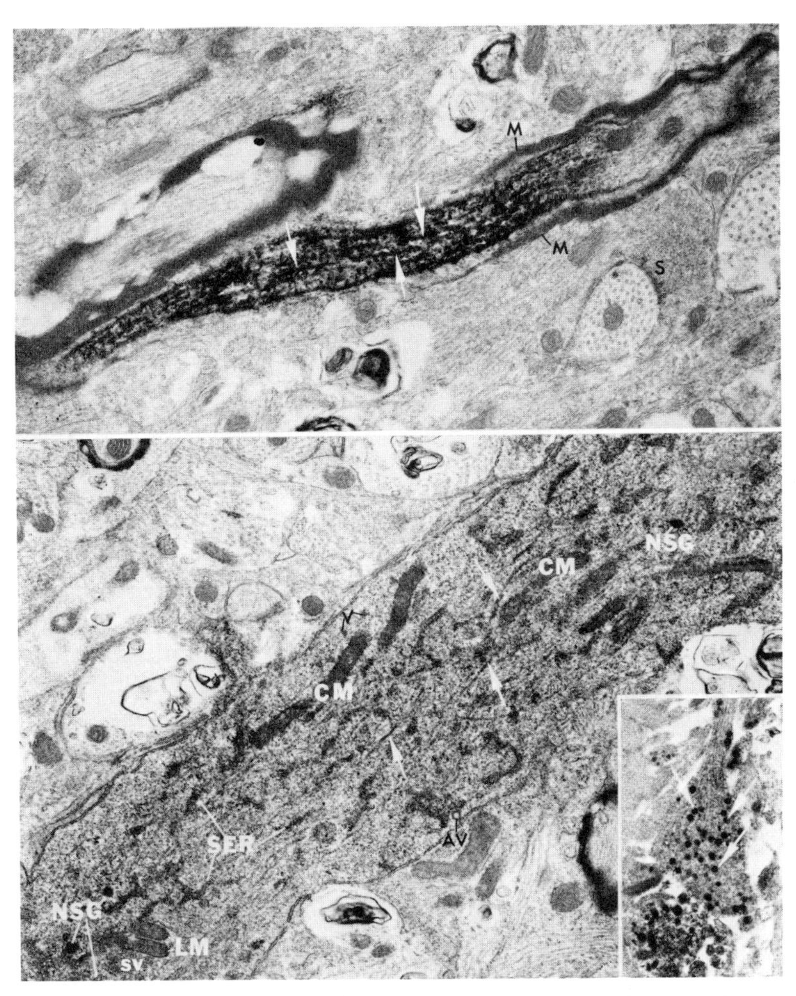

Figure 15. (*1*) The upper panel shows an LHRH process in the preoptic area which is partially myelinated (*M*). The dense reaction product appears to be associated with microtubules (arrows). S, synapse. Kozlowski et al. (44). (2) The lower panel shows a longitudinal section through a fiber of the preoptic area having neurosecretory granules (*NSG*), vesicles (*V*), synaptoid vesicles (*sv*), smooth endoplasmic reticulum (*SER*), microtubules with electron-dense areas (arrows), lamellar (*LM*) and cristal mitochondria (*CM*), and an alveolate vesicle (*AV*). Kozlowski et al. (44). (3) The insert shows a dilated portion of a fiber in the median eminence containing large numbers of immunopositive neurosecretory granules (arrows). Kozlowski et al. (44).

ualization of both cell bodies and fibers. Glutaraldehyde fixation was found to yield the best immunoreactivity, permitting visualization not only of all hypothalamic vasopressin-containing perikarya, but also extrahypothalamic fibers in amygdala, midbrain (central gray), and brainstem (nucleus tractus solitarius). The effects of fixatives, particularly formaldehyde, on aromatic amino acids in tissue antigens may explain these different results of staining with the IIID-7 clone on tissue prepared with these three fixatives. A major problem with formaldehyde fixation vasopressin for IIID-7 was the irreversible reaction of formaldehyde with hydrogens in aromatic amino acids, such as tyrosine or phenylanine (69), which were important determinants for antibody IIID-7 as shown by absorption tests. Glutaraldehyde, which reacts very poorly, if at all, with aromatic rings, spared the antigenic determinants of vasopressin as seen by clone IIID-7, and provided fixation by reacting with alpha-amino groups (70). PLP probably retained more antigenicity because of its lower concentration of formaldehyde (less than 2%), while fixation was maintained by cross-linking of lysine groups to carbohydrates oxidized by periodate (57). This experience with monoclonal antibodies thus also serves to emphasize the effects of fixatives in altering tissue antigenicity, which could account for the false negative results sometimes obtained in immunostaining.

Fixatives may also result in the so-called masking of antigens, which can be reversed by pretreatment of sections with proteolytic enzymes prior to the ICC procedure. Finley et al. (24) showed that somatostatinlike immunoreactivity improved when pronase was used to reveal (unmask) more antigenic determinants, thereby increasing the staining intensity of fibers.

Preembedding ICC can yield both discrete staining of organelles (Figures 7–9) and a generalized cytoplasmic staining in which the final reaction product is distributed in presumably sol portions of the cell or its processes (Figure 11). In addition, dense final reaction product can be associated with microtubules and other organelles as reported for LHRH (44), tyrosine hydroxylase (77, 80), tryptophan hydroxylase (78), substance P (79), the enkephalins (19, 79), glutamic acid decarboxylase (82), and dopamine-β-hydroxylase (67). The functional significance of final reaction product in association with cytoskeletal structures such as microtubules cannot, at present, be determined. These molecules may just indiscriminately bind to nearby structural elements during the process of fixation. On the other hand, one cannot completely dismiss the putative role microtubules may have for the transport of substances along the cellular processes (50); perhaps pre-embedding ICC is capable of visualizing such a process.

Nevertheless, until convincing evidence comes along in the form of improved preservation of enzymes, neuropeptides, and ultrastructure, we are forced to consider such localizations as an artifact of fixation.

At the level of the perikarya, the presence of final reaction product generally distributed throughout the cytoplasm is probably not only due to lack of fixation but can also result from the use of detergents in pre-embedding ICC. Olschowka et al. (67), and we ourselves (Figures 12 and 13) have examined immunoreacted sections cut perpendicular to their surfaces in order to determine the depth of penetration of reagents. Olschowka et al. achieved 20–25 μm penetration of reactants on both sides of 100 μm vibratome sections in the case of cellular areas like the locus coeruleus; whereas in the case of catecholaminergic terminal field areas in the paraventricular nucleus, nearly complete penetration (45 μm from the surface) was achieved. In both their light microscopic and our electron microscopic studies, it was suggested that if fibers or cell bodies communicated with the surface of the section, good penetrability was achieved; however, cells lying in the central zone of thick sections either showed no distribution (67) or partial distribution of reaction product (Figure 12) along their periphery. These kinds of studies should be expanded in order to investigate more thoroughly the effects of different fixatives and detergents on penetrability of reagents.

Most pre-embedding ICC studies include the use of Triton X-100 or some similar detergent to improve visualization of the antigen to be studied. Grzanna et al. (31) suggested a dual function for Triton X-100; it exposes antigenic sites and it facilitates penetration of antibodies into tissue sections. Others (8) avoid its use altogether, preferring to use only the first ultrathin sections cut from the block for electron microscopic study.

Many investigators utilizing Triton X-100 for pre-embedding ICC have first adjusted the concentration of the detergent to suit their particular application (40, 44, 67, 76). A reasonable compromise can be obtained between the degree of penetration desired and avoidance of the destruction of membranes which results when the detergent is used in high concentrations. At the subcellular level, if detergents are used to aid in penetrability of antibodies, an undesirable side effect is that the antigen intended to be localized may also diffuse out of membrane-delimited compartments. This, of course, greatly hinders the task of proving the existence of extragranular, that is, noncompartmentalized, hormone using pre-embedding ICC. Compartmentalized hormone has been demonstrated in pituitary cells processed for post-embedding ICC by Tougard et al. (101) for prolactin and by Moriarity and Tobin (60) for thyroidectomy cells. With new improvements in mechanical fixation of tissue, post-embedding ICC may yet prove to be the method of choice for detecting peptidic antigens that exist in soluble pools.

Another possible source of artifact when using immunoenzymatic techniques could occur during development of the colored final reaction product. An interesting dialogue concerning possible artifacts associated with the use of DAB has appeared as "Letters to the Editor" in the *Journal of Histochemistry and Cytochemistry* (28, 65, 66, 89). That these techniques may be especially suitable for detecting noncompartmentalized antigens is a subject of debate. Goldfischer (28) concludes that DAB artifacts may be caused by diffusion of reaction product, enzyme or antigen adsorption to nonreactive sites, and poor preservation of tissue. Artifacts that may occur because of enzyme translocation could be determined by comparing staining results using the PAP technique with those obtained with nonenzyme electron-dense markers such as protein A-gold (32, 83, 84, 92), even in combination with autoradiography for the detection of labeled antigen (51, 52; Figures 2–4) or antibody (23). One means of avoiding noncompartmentalized staining is to monitor the developing final reaction product in sections with a stereomicroscope during the DAB/H_2O_2 step. Broadwell et al. (9) suggested that incubation of the tissue in DAB/H_2O_2 medium should be interrupted when brown granules begin to appear within the cells. This practice allowed them to select that point which delineated specific from nonspecific staining of structural components of the cytoplasm.

Despite the problems with pre-embedding ICC in resolving with certainty the meaning of some subcellular localizations, advances have been made in the visualization of specific neuropeptide-containing cells and the systems they form with other neurons of the CNS. The synaptic architecture of several peptidergic circuits has confirmed the usefulness of pre-embedding ICC for these types of studies (10, 19, 35, 40, 63, 81, 85).

9 PRE- AND POST-EMBEDDING ICC TECHNIQUES: ADVANTAGES AND DISADVANTAGES

The intended objective of any study to be undertaken will determine the choice of either the pre-embedding or post-embedding ICC technique. Figure 16 outlines the major steps used in both techniques.

There are several advantages and disadvantages associated with each procedure, which can determine the choice of method to be used. A

POST-EMBEDDING	PRE-EMBEDDING
FIX	FIX
EMBED	SECTION (50 μm)
SECTION (80 nm)	TRITON X-100
ETCH	ICC RX
ICC RX	EMBED
E.M.	L .M. AND E.M.

Figure 16. Major steps in the post- and pre-embedding techniques. L. M., light microscopy; E. M., electron microscopy.

major advantage of the post-embedding technique is that one can localize several antigens through the same serial-sectioned cell or granule and even test specificity of staining on different sections through the same cell (12, 13). A major disadvantage is that globulins can nonspecifically adhere to the embedding medium. In some cases, the final reaction product can indeed appear structure-specific (10). Another possible disadvantage could be the lack of penetrability of reagents due to the incomplete effects of the etching reagent. A major advantage of the pre-embedding method arises from the fact that one can examine sections at the light microscopic level for intensity and specificity of staining prior to further processing of the tissue for the electron microscope (76, 95). Also, with thick vibratome sections one obtains a "Golgi-like" three-dimensional image of cells and their processes (31). With thick, serial sections of brains, one can map the pathways taken by peptide-containing neuronal systems from their sites of origin to their terminal field areas, with the added capability of being able to examine those same areas with the electron microscope. The pre-embedding technique preserves antigenic sites more completely, since the ICC procedure occurs prior to the use of solvents, reagents, and embedding media which could denature or eliminate the antigen. With post-embedding ICC, these treatments could lead to false negative results. However, the use of detergents such as Triton X-100, DMSO, or Saponin in pre-embedding ICC could result in false-positive localizations due to diffusion of the antigen. This can be a particularly difficult problem to resolve when interpreting the presence of final reaction product on subcellular structures. On the other hand, false negative results can occur with pre-embedding ICC due to the lack of penetrability of reagents through the entire thickness of the section or even throughout the cell within the section.

10 CONTROLS

In general, the purpose of ICC is to localize, using antibodies, a specific substance in relation to tissues, cells or organelles containing that substance. The use of method and serum controls serve to determine whether or not such a definition has been fulfilled during conduct of the study. Judging from the explosion of ICC publications during the past few years, the union of antibodies and antigens and their subsequent visualization in tissues has been a most prosperous one. Although it is exciting to appreciate both the real and potential value of ICC, at the same time it is important to realize its limitations. The subject of controls emphasizes the limitations by which ICC operates; namely, there are problems with tissue preparation and the cross-reactivity of antisera. For the former case, the antigen pursued should exist in a state as near as possible to its normal state in the living cell. For the latter case, the antibody used should be monospecific for the antigen, and only that antigen, intended to be localized. It remains to be proven that these theoretical circumstances have yet to be achieved; in practice, new and more refinements in these areas have been encouraging in attempting to conquer some of the pitfalls that beset ICC. The proper use of controls imparts standards to the procedure and is important in preventing reports of false positives or negatives. As with most experimental studies, although true positive results are achievable, true negative results are far more difficult to prove in most circumstances. The interpretation of results and their controls must be objective and carefully considered in order to draw valid conclusions. A number of papers have enumerated various considerations that should be taken into account in determining specificity of staining (10, 12–14, 22, 23, 36, 38, 72–74, 76, 95, 98–100, 102–104, 108).

10.1 Method Controls

These largely consist of substituting buffer for each of the immunoreagents, or of substituting preimmune serum from the same rabbit whose immune antisera is used in the staining procedure. No reaction product should form unless either pseudoperoxidase is present or the blood collected from the rabbit already contains interfering antibodies (14). The presence of pseudoperoxidase in glial and neuronal cells (49) can become an especially acute problem when using tissues from older animals (111). In some cases, preimmune serum may react nonspecifically (76, 111).

10.2 Serum Controls

10.2.1 Liquid Phase Immunoabsorption

Small amounts of antigen are added to aliquots of antisera previously shown to provide staining in the ICC procedure used. Staining with supernatant should disappear unless soluble complexes form that can react with antigenic tissue sites. Increasing amounts of 1 to 10 μg peptide are added to 0.1 mL of a 1:100 dilution of antisera, incubated overnight at 4° with shaking, centrifuged, and the supernatant used at final dilution as a substitute for the first immunoreactant. Increasing amounts of peptide added to the antisera should cause a corresponding reduction in staining intensity. In some cases, one can actually achieve an augmentation of staining when increasing amounts of antigen are added, either due to the presence of antigen–antibody complexes that can react with tissue antigen (110) or to free antigen which reacts with binding or receptor sites in tissue. Quantitation of the augmented staining can be linear, as described by Sternberger et al. (97) for LHRH receptors in gonadotrope cells, or biphasic, as described by May et al. (12, 55) for thyrotropin-releasing hormone (TRH) receptors in thyrotrope cells. There is debate as to whether the liquid-phase immunoabsorption test determines method (100) or serum specificity (73, 74); whether it should be included under both of these categories depends upon the circumstances of the reaction. If an impure antigen is used both as an immunogen and as an absorbent, then this test will only prove method specificity because even the undesirable antibodies would react during an incubation. Serum specificity could be proved if it were determined that the antigen–antibody complexes formed were precipitable and that both these complexes and any free antibody were effectively eliminated as a possible source of the specific staining.

10.2.2 Solid Phase Immunoabsorption

Since the antigen–antibody equilibrium reaction previously mentioned allows for the presence of free antibody, and since soluble antigen–antibody complexes can still form final reaction product with tissue sections, the liquid-phase type of immunoabsorption test can lead to false conclusions. It is preferable, therefore, to couple free antibodies and soluble antigen–antibody complexes to some form of substrate that can be physically eliminated from the incubation medium. Briefly, cyanogen-bromide (CNBr)-activated Sepharose 4B agarose gel beads (Pharmacia

Fine Chemicals, Piscataway, NJ) are swollen and reacted with the dissolved antigen in a coupling buffer. Any remaining active groups on the beads are blocked, using ethanolamine or glycine, and the beads are then filtered and washed. They are resuspended in buffer and incubated with the antiserum to be absorbed. The beads are filtered and the dilution of the absorbed antiserum is adjusted to the same concentration as used in the technique as unabsorbed antiserum. The absorbed antisera is then substituted as the first immunoreagent in the staining procedure. Detailed instructions can be obtained from the manufacturer (75).

Affinity chromatography is useful for a wide variety of applications. Swaab and Pool (99) purified anti-vasopressin and anti-oxytocin of cross-reacting antibodies by incubating the antisera in the presence of beads coupled with the heterologous antigens which were then eliminated from the antiseras by centrifugation. They also used an immunofluorescent technique employing hormone-coupled beads on glass slides to demonstrate the cross-reactive nature of their antisera to vasopressin and oxytocin. Incubation of anti-vasopressin and anti-oxytocin with beads coupled to the respective homologous hormone (vasopressin or oxytocin) abolished their immunofluorescent staining capacity. Nilaver et al. (64) used affinity chromatography to purify antisera to oxytocin-neurophysin and vasopressin-neurophysin for characterizing magnocellular hypothalamic neurons, containing a particular neurophysin, which project to a variety of CNS sites. Leu- and Met-enkephalin bound to beads can be useful for testing the effects of fixation on hormone antigenicity when stained and examined with the light microscope and electron microscope (6, 7, 12). Affinity chromatography is also useful for eliminating interfering antibodies to carrier proteins of the immunogen such as the anti-BSA present in anti-LHRH No. 42 (Kozlowski, unpublished observations).

In some cases, *biologic controls* are available for determining serum specificity. A fortunate test is available for anti-vasopressin when reacted with brains of the homozygous Brattleboro rat having complete diabetes insipidus (DI). The DI rat has oxytocin and its associated neurophysin but is deficient in vasopressin and vasopressin-neurophysin. If a particular anti-vasopressin stains in sections of DI rat brain, it can be concluded that there are cross-reacting antibodies to oxytocin present. Similarly, completeness of purification of the anti-vasopressin intended to eliminate anti-oxytocin antibodies can be tested (99, 111). Also, since it is known that the suprachiasmatic nucleus has vasopressin but not oxytocin, one can test for the presence of antibodies that cross-react with vasopressin in anti-oxytocin. If a particular anti-oxytocin stains sections of the suprachiasmatic nucleus, it can be concluded that there are cross-reacting antibodies to vasopressin present (104).

11 CONCLUSIONS AND FUTURE PROSPECTS

Many excellent studies using immunoelectron microscopy for localizing neuropeptides have appeared in recent years. In this chapter we reviewed rather basic issues of methodology common to most of these studies, and their theoretical mechanisms of action as we currently understand them. The field of study using immunoelectron microscopic techniques for the localization of neuropeptides continually strives to keep pace with the rapid developments now occurring in the general area of ICC. Advances in light microscopic ICC using new bridging techniques (68) and stronger coupling agents, such as biotin-avidin (34), will soon be utilized for demonstrating neuropeptides at the ultrastructural level. Combination techniques that simultaneously demonstrate the presence of two or more antigens will appear as variants of already established techniques or may develop along with the discovery of new and more precise electron-dense markers. Improvements in mechanical fixation techniques may prove helpful in the quest for localizing CNS receptor sites of neuropeptides, an area that has been problematic when chemical fixatives and current techniques are used.

We are witnessing the development of monoclonal antibodies as potentially powerful probes for identifying neural antigens in a remarkably specific manner (56). The complexities of neural systems can best be unravelled with the use of substance-specific marker methods, and these tools may add immense capabilities to the already rich armamentarium of available techniques.

ACKNOWLEDGMENTS

The ultrastructural immunocytochemistry studies performed by G. P. Kozlowski at Colorado State University were assisted by Drs. Samy Frenk (Boulder, Colorado), William Todd (Bozeman, Montana), Mark Brownfield (Madison, Wisconsin), and Gayle Hostetter (Portland, Oregon). The HVEM used in this study was under the direction of Prof. Keith Porter and Miricea Fotino of the Department of Molecular, Cellular, and Developmental Biology, University of Colorado, Boulder. This instrument facility is supported by Grant No. 1-PO7-RR00592 from the Division of Research Facilities and Resources-NIH. Gifts of antisera were by: Drs. Nett and Niswender (anti-LHRH No. 42), Bernard Kerdelhué (anti-LHRH No. 4150-11), Zimmerman and Robinson (anti-bovine NPII, anti-rat NPIV). Through the years, many valuable suggestions have been received from Drs. Ludwig Sternberger (Rochester, NY), Virginia Pickel (New York,

NY), Gwenn Childs (Galveston, TX), and Earl Zimmerman (New York, NY). We thank Mrs. Lena Chu (Houston, TX) for her excellent technical assistance and Mrs. Diane Doach for typing the manuscript. Studies by G. P. K. were supported by grants from NSF (BNS 77-00176) and NIH (NS 15337, NS 17325, HD 15040, HD 12781).

REFERENCES

1. S. Avremeas, *Int. Rev. Cytol.* **27**, 349 (1970).
2. J. C. Beauvillain and G. Tramu, *J. Histochem. Cytochem.* **28**, 1014 (1980).
3. J. C. Beauvillain, G. Tramu, and M. P. Dubois, *Cell Tiss. Res.* **218**, 1 (1981).
4. A. Berod, B. K. Hartman, and J. F. Pujol, *J. Histochem. Cytochem.* **29**, 844 (1981).
5. J. W. Bigbee, J. C. Kosek, and L. F. Eng, *J. Histochem. Cytochem.* **25**, 443 (1977).
6. K. M. Braas, G. V. Childs, M. Kubek and J. F. Wilber, Program of the 31st Annual Meeting of the Histochemical Society, New Orleans 1980, Abstract No. 41.
7. K. M. Braas, J. F. Wilber, and G. V. Childs, *J. Cell Biol.* **87**, 173a (1980).
8. R. D. Broadwell, "Enzyme Cytochemical and Immunocytochemical Investigations of the Secretory Process in Peptide-Secreting Cells," Chapter 5 in this volume.
9. R. D. Broadwell, C. Oliver and M. W. Brightman, *Proc. Natl. Acad. Sci. USA* **76**, 5999 (1979).
10. R. M. Buijs and D. F. Swaab, *Cell Tiss. Res.* **204**, 355 (1979).
11. V. Chan-Palay, *Anat. Embryol.* Berlin **156**, 225 (1979).
12. G. V. Childs (Moriarity), "The Use of Immunocytochemical Techniques in Cellular Endocrinology," in L. Sternberger and J. Griffith, Eds., *Electron Microscopy in Biology*, Vol. 2, Wiley, New York, in press.
13. G. V. Childs and D. G. Ellison, *Histochem. J.* **12**, 405 (1980).
14. C. J. Clayton and G. E. Hoffman, *Amer. J. Anat.* **155**, 139 (1979).
15. A. H. Coons, H. J. Creech, R. N. Jones, and E. Berliner, *J. Immunol.* **45**, 159 (1942).
16. H. D. Coulter and R. P. Elde, *Anat. Rec.* **190**, 369 (1978).
17. H. D. Coulter and R. P. Elde, *J. Histochem. Cytochem.* **27**, 1293 (1979).
18. F. Dacheux and M. P. Dubois, *Cell Tiss. Res.* **174**, 245 (1976).
19. M. DiFiglia, N. Aronin, and J. B. Martin, *J. Neurosci.* **2**, 303 (1982).
20. B. Droz, "Synthetic Machinery and Axoplasmic Transport: Maintenance of Neuronal Connectivity," in D. B. Tower, Ed., *The Basic Neurosciences, Vol. 1: The Nervous System*, Raven Press, New York, 1975, pp. 111–127.
21. R. W. Dudek, G. V. Childs, and A. F. Boyne, *J. Histochem. Cytochem.* **30**, 129 (1982).
22. G. Feldmann, P. Druet, J. Bignon and S. Avrameas, Eds., *Immunoenzymatic Techniques*, INSERM Symposium No. 2, American Elsevier, New York, 1976.
23. A. Fețeanu, *Labeled Antibodies in Biology and Medicine*, 2nd ed., Abacus Press, Trumbridge Wells, Kent, and McGraw-Hill, New York, 1978.
24. J. C. W. Finley, G. H. Grossman, P. Dimes, and P. Petrusz, *Amer. J. Anat.* **153**, 483 (1978).
25. R. M. Franklin and M.-T. Martin, *Eur. J. Cell Biol.* **21**, 134 (1980).

26. V. L. Fredrick, Jr., and E. Mugnaini, "Electron Microscopy: Preparation of Neural Tissues for Electron Microscopy," in L. Heimer and M. J. Robards, Eds., *Neuroanatomical Tract-Tracing Methods*, Plenum Press, New York, 1981, p. 345.

27. H. Gainer, J. T. Russell, and D. J. Fink, "The Hypothalamo-Neurohypophysial System: A Cell Biological Model for Peptidergic Neurons," in A. M. Gotto, Jr., E. J. Peck, Jr., and A. E. Boyd III, Eds., *Brain Peptides: A New Endocrinology*, Elsevier/North-Holland Biomedical Press, 1979, pp. 139–159.

28. S. Goldfischer, *J. Histochem. Cytochem.* **28**, 1360 (1980).

29. P. C. Goldsmith and W. F. Ganong, *Brain Res.* **97**, 181 (1975).

30. N. K. Gonatas, A. Stieber, J. O. Gonatas, J.-U. Antoine, and S. Avremeas, "I. Demonstration of Intracellular Immunoglobulin with ^{125}I Fab Antibody Fragment. II. Segregation and Internalization of Immunoglobulins and of Lactoperoxidase-Iodinated Proteins of Lymphoid Cell Plasma Membranes," in G. Feldmann, P. Druet, J. Bignon, and S. Avrameas, Eds., *Immunoenzymatic Techniques*, INSERM Symposium No. 2, American Elsevier, New York, 1976, p. 231.

31. R. Grzanna, M. E. Molliver, and J. T. Coyle, *Proc. Natl. Acad. Sci. USA* **75**, 2702 (1978).

32. M. Horisberger, *Biol. Cellulaire* **36**, 253 (1979).

33. A. Hou-Yu, P. Ehrlich, G. Valiquette, D. L. Engelhardt, W. H. Sawyer, G. Nilaver, and E. A. Zimmerman, *J. Histochem. Cytochem.* **30**, 1249 (1982).

34. S. M. Hsu, L. Raine, and H. Fanger, *J. Histochem. Cytochem.* **29**, 577 (1981).

35. S. P. Hunt, J. S. Kelly, and P. C. Emson, *Neuroscience* **5**, 1871 (1980).

36. J. C. Hutson, G. V. Childs, and P. J. Gardner, *J. Histochem. Cytochem.* **27**, 1201 (1979).

37. O. Johansson, T. Hökfelt, B. Pernow, S. L. Jeffcoate, N. White, H. W. M. Steinbusch, A. A. J. Verhofstad, P. C. Emson, and E. Spindel, *Neurosceince* **6**, 1857 (1981).

38. E. G. Jones and B. K. Hartman, *Ann. Rev. Neurosci.* **1**, 215 (1978).

39. S. A. Joseph, D. T. Piekut, and K. M. Knigge, *J. Histochem. Cytochem.* **29**, 247 (1981).

40. K. Kakudo, W. K. Paull, Jr., and L. L. Vacca, *Histochemistry* **71**, 17 (1982).

41. G. P. Kozlowski, "Fine Structural Features of Immunolabeled Neurons Containing Luteinizing Hormone-releasing Hormone, Vasopressin, and Neurophysin," in D. S. Farner and K. Lederis, Eds., *Neurosecretion: Molecules, Cells, Systems*, Plenum Press, New York, 1982, pp. 61–70.

42. G. P. Kozlowski, "Ventricular Route Hypothesis and Peptide-Containing Structures of the Cerebroventricular System," in Tj. B. Van Wimersma Greidanus, Ed., *Frontiers of Hormone Research*, Vol. 9, Karger, Basel, 1982, pp. 105-118.

43. G. P. Kozlowski, "Comparative Ultrastructure of Neuropeptide-Containing Cells of the Parvo- and Magnocellular Neurosecretory System," in H. Sano and E. A. Zimmerman, Eds., *Structure and Function of Peptidergic and Aminergic Neurons*, 5th International Symposium of the Taniguchi Foundation, Wiley, New York, in press.

44. G. P. Kozlowski, L. Chu, G. Hostetter, and B. Kerdelhué, *Peptides* **1**, 37 (1980).

45. G. P. Kozlowski, S. Frenk, and M. S. Brownfield, *Cell Tiss. Res.* **179**, 467 (1977).

46. G. P. Kozlowski and G. Hostetter, "Cellular and Subcellular Localization and Behavioral Effects of Gonadotropin-Releasing Hormone (Gn-RH) in the Rat," in D. E. Scott, G. P. Kozlowski, and A. Weindl, Eds., *Brain-Endocrine Interaction, Vol. 3: Neural Hormones and Reproduction*, Karger, Basel, 1977, p. 138.

47. G. P. Kozlowski, T. M. Nett, and E. A. Zimmerman, "Immunocytochemical Localization

of Gonadotropin-Releasing Hormone (Gn-RH) and Neurophysin in the Brain," in W. E. Stumpf and L. D. Grant, Eds., *Anatomical Neuroendocrinology*, Karger, Basel, 1975, p. 185.

48. B. Krisch, "Immunocytochemistry of Neuroendocrine Systems," Vol. 13, No. 2, *Prog. Histochem. Cytochem.*, Gustav Fischer Verlag, Stuttgart, New York, 1981, pp. 1–163.

49. T. Kumamoto, *Acta Histochem. Cytochem.* **14**, 173 (1981).

50. R. J. Lasek and M. L. Shelanski, "Cytoskeletons and the Architecture of Nervous Systems," in *Neurosci. Res. Prog. Bull.* (MIT Press, Cambridge, Mass.) **19**, 1–153 (1981).

51. L.-I. Larsson, *Peptide Immunocytochemistry*, Vol. 13, No. 4, *Prog. Histochem. Cytochem.*, Gustav Fischer Verlag, Stuttgart, New York, 1981, pp. 1–85.

52. L. I. Larsson and T. W. Schwartz, *J. Histochem. Cytochem.* **25**, 1140 (1977).

53. B. G. Livett, *Int. Rev. Cytol.* Suppl. 7, 53–237 (1978).

54. T. C. Mason, R. F. Phifer, S. S. Spicer, R. A. Swallow, and R. B. Dresken, *J. Histochem. Cytochem.* **17**, 563 (1969).

55. V. May, J. F. Wilber, and G. V. Childs, 32nd Annual Meeting of the Histochemical Society, Minneapolis, 1981.

56. R. McKay, M. C. Raff, and L. F. Reichardt, Eds., *Monoclonal Antibodies to Neural Antigens*, Vol. 2, Cold Spring Harbor Reports in the Neurosciences, Cold Spring Harbor, N.Y., 1981.

57. I. W. McLean and P. K. Nakane, *J. Histochem. Cytochem.* **27**, 1077 (1974).

58. T. H. McNeill and C. D. Sladek, 31st Annual Meeting of the Histochemical Society, New Orleans, 1980, Abstract No. 20.

59. G. C. Moriarity, C. M. Moriarity, and L. A. Sternberger, *J. Histochem. Cytochem.* **21**, 825 (1973).

60. G. C. Moriarity and R. B. Tobin, *J. Histochem. Cytochem.* **24**, 1140 (1976).

61. D. V. Naik, *Anat. Rec.* **178**, 424 (1974).

62. P. K. Nakane and G. B. Pierce, *J. Histochem. Cytochem.* **14**, 291 (1966).

63. G. Nilaver, E. A. Zimmerman, J. G. Linner, L. Chu, and G. P. Kozlowski, 10th Annual Meeting of the Society for Neuroscience, Cincinnati, 1980, Abstract No. 62.4.

64. G. Nilaver, E. A. Zimmerman, J. Wilkins, J. Michales, D. Hoffman, and A.-J. Silverman, *Neuroendocrinology* **30**, 150 (1980).

65. A. B. Novikoff, *J. Histochem. Cytochem.* **28**, 1036 (1980).

66. A. B. Novikoff, P. M. Novikoff, N. Quintana, and C. Davis, *J. Histochem. Cytochem.* **20**, 745 (1972).

67. J. A. Olschowka, M. E. Molliver, R. Grzanna, F. L. Rice, and J. T. Coyle, *J. Histochem. Cytochem.* **29**, 271 (1981).

68. P. Ordronneau, P. B.-M. Lindström, and P. Petrusz, *J. Histochem. Cytochem.* **29**, 1397 (1981).

69. A. G. E. Pearse, *Histochemistry, Theoretical and Applied*, 3rd ed., Vol. 1, Little, Brown, Boston, 1968, pp. 70–105.

70. A. G. E. Pearse and P. J. Steward, Eds., *Fixation in Histochemistry*, Chapman and Hall, London, 1973, pp. 47–84.

71. G. Pelletier, R. Ruviani, O. Bosler, and L. Descarries, *J. Histochem. Cytochem.* **29**, 759 (1981).

72. P. Petrusz, P. Ordronneau, and J. C. W. Finley, *Histochem. J.* **12**, 333 (1980).
73. P. Petrusz, M. Sar, P. Ordronneau, and P. Dimeo, *J. Histochem. Cytochem.* **24**, 1110 (1976).
74. P. Petrusz, M. Sar, P. Ordronneau, P. Dimeo, *J. Histochem. Cytochem.* **25**, 390 (1977).
75. Pharmacia Fine Chemicals, *Affinity Chromatography Principles and Methods*, Piscataway, N.J., 1979.
76. V. M. Pickel, "Immunocytochemical Methods," in L. Heimer and M. J. Robards, Eds., *Neuroanatomical Tract-Tracing Methods*, Plenum Press, New York, 1981, pp 483–509.
77. V. M. Pickel, T. H. Joh, and D. J. Reis, *Brain Res.* **85**, 295 (1975).
78. V. M. Pickel, T. H. Joh, and D. J. Reis, *Brain Res.* **131**, 197 (1977).
79. V. M. Pickel, T. H. Joh, D. J. Reis, S. E. Leeman, and R. J. Miller, *Brain Res.* **160**, 387 (1979).
80. V. M. Pickel, T. H. Joh, T. Shikimi, and D. J. Reis, "Immunocytochemical Localization of Tyrosine Hydroxylase and Tryptophan Hydroxylase in Relation to Microtubules in Rat Brain," in L. Ahtea, Ed., Proceedings of the 6th International Congress of Pharmacology, *Vol. 2: Neurotransmission*, Pergamon Press, Oxford, 1976, p. 195.
81. V. M. Pickel, D. J. Reis, and S. E. Leeman, *Brain Res.* **122**, 534 (1977).
82. C. E. Ribak, J. E. Vaughn, and E. Roberts, *J. Comp. Neurol.* **187**, 261 (1979).
83. J. Roth, M. Bendayan, and L. Orci, *J. Histochem. Cytochem.* **26**, 1974 (1978).
84. J. Roth, M. Ravazzola, M. Bendayan, and L. Orci, *Endocrinology* **108**, 247 (1981).
85. M. A. Ruda, *Science* **215**, 1523 (1982).
86. H. Salih, G. S. Murthy, and H. G. Friesen, *Endocrinology* **105**, 21 (1979).
87. M. Schultzberg, G. J. Dockray, and R. G. Williams, *Brain Res.* **235**, 198 (1982).
88. A. M. Seligman, M. J. Karnovsky, H. L. Wasserkrug, and J. S. Hanker, *J. Cell Biol.* **38**, 1 (1968).
89. A. M. Seligman, W. A. Shannon, Y. Hochino, and R. E. Plapinger, *J. Histochem. Cytochem.* **21**, 756 (1973).
90. A. J. Silverman and E. A. Zimmerman, *Cell Tiss. Res.* **159**, 291 (1975).
91. W. R. Skowsky and D. A. Fisher, *J. Lab. Clin. Med.* **80**, 134 (1972).
92. J. W. Slot and H. J. Geuze, *J. Cell Biol.* **90**, 533 (1981).
93. H. W. Sokol, E. A. Zimmerman, W. H. Sawyer, and A. G. Robinson, *Endocrinology* **98**, 1176 (1976).
94. M. Stefanini, C. DeMartino, and L. Zamboni, *Nature* **216**, 173 (1967).
95. L. A. Sternberger, *Immunocytochemistry*, 2nd ed., Wiley, New York, 1979.
96. L. A. Sternberger, P. H. Hardy, Jr., J. J. Cuculis, and H. G. Meyer, *J. Histochem. Cytochem.* **18**, 315 (1970).
97. L. A. Sternberger, J. P. Petrali, S. A. Joseph, H. C. Meyer, and K. R. Mills, *Endocrinology* **102**, 63 (1978).
98. D. F. Swaab and G. J. Boer, Immunocytochemistry and Its Application in Brain Research, EMBO practical course, Amsterdam, May 27–30, 1980, Netherlands Institute for Brain Research.
99. D. F. Swaab and C. W. Pool, *J. Endocrinol.* **66**, 263 (1975).
100. D. F. Swaab, C. W. Pool, and F. W. Van Leeuwen, *J. Histochem. Cytochem.* **25**, 388 (1977).
101. C. Tougard, R. Picart, and A. Tixier-Vidal, *Amer. J. Anat.* **158**, 471 (1980).

102. F. Vandesande, *J. Neurosci. Methods* **1**, 3 (1979).

103. F. W. Van Leeuwen, *J. Histochem. Cytochem* **25**, 1213 (1977).

104. F. W. Van Leeuwen, *An Introduction to the Immunocytochemical Localization of Neuropeptides and Neurotransmitters*, Acta Histochemica, Suppl. Vol. 24 (1980).

105. R. L. Wasserman and J. D. Capra, "Immunoglobulins," in W. Pigman and M. I. Horowitz, Eds., *The Glycoconjugates*, Academic Press, New York, 1977, p. 323.

106. E. Weber, C. J. Evans, and J. D. Barchas, Program of 32nd Annual Meeting of the Histochemical Society, Minneapolis, 1981, Abstract No. 40.

107. K. Weber, P. C. Rathke, and N. Osburne, *Proc. Natl. Acad. Sci. USA* **75**, 1820 (1978).

108. M. A. Williams, *Autoradiography and Immunocytochemistry*, Vol. 6, Part 1, *Practical Methods in Electron Microscopy*, A. M. Glauert, Ed., North-Holland, Amsterdam, 1977.

109. E. A. Zimmerman, "Localization of Hypothalamic Hormones by Immunocytochemical Techniques," in L. Martini and W. F. Ganong, Eds., *Frontiers in Neuroendocrinology*, Vol. 4, Raven Press, New York, 1976, p. 25.

110. E. A. Zimmerman, K. C. Hsu, M. Ferin, and G. P. Kozlowski, *Endocrinology* **95**, 1 (1974).

111. E. A. Zimmerman, L. Krupp, D. L. Hoffman, E. Matthew, and G. Nilaver, "Exploration of Peptidergic Pathways in Brain by Immunocytochemistry: A Ten-Year Perspective," in D. E. Scott and J. R. Sladek, Jr., Eds., *Brain-Endocrine Interaction Symposium 4: Neuropeptides, Development and Aging*, Peptides, 1, Ankho Int. Inc., Fayetteville, NY, 1980, p. 3.

Chapter 5

Enzyme Cytochemical and Immunocytochemical Investigations of the Secretory Process in Peptide-Secreting Cells

Richard D. Broadwell

Division of Neuropathology
Department of Pathology
University of Maryland School of Medicine, Baltimore

1 INTRODUCTION

The term *secretory process* refers to the elaboration of secretory protein (68). It includes consideration of the subcellular components associated with the synthesis and refinement of secretory protein as well as release of this protein from the cell. The mammalian cell type used extensively to define the secretory process has been the exocrine acinar cell of the guinea pig pancreas. Combining techniques of cell fractionation, autoradiography, and electron microscopy, Palade and coworkers (16, 39–42, 67, 69, 74) have demonstrated in convincing fashion that the secretion of pancreatic digestive enzymes is a sequential multistep process proceeding as follows: the secretory product (polymeric chains of amino acids) is synthesized on polyribosomes attached to the endoplasmic reticulum (ER); the forming polypeptide chain is transferred across the ER membrane to become segregated within the cisternal space of the rough ER. Soon thereafter the protein is conveyed in transfer vesicles of ER origin to condensing vacuoles located in the vicinity of the Golgi apparatus; here, the concentration of the secretory product and the formation of mature secretion granules occur. In response to the appropriate stimulus, the secretion granules storing the secretory product fuse with the plasmalemma for discharge of their contents outside the cell. Hence, within the secretory cell six successive steps, each occurring in a separate intracellular compartment, are involved in the production of secretory protein: (i) synthesis, (ii) segregation, (iii) intracellular transport, (iv) concentration, (v) storage, and (vi) discharge. Collectively these successive steps constitute the secretory process.

Given the high level of interest in and enthusiasm for peptidergic neurons currently pervading the field of neurobiology, consideration must necessarily be directed to the applicability of the secretory process to this particular class of cells. Neuropeptides represent chains of amino acids of varying length. For example, carnosine is only two amino acids in length, while corticotropin is 39 amino acids long. None of the neuropeptides is believed to contain carbohydrate moieties. The question most germane to this discussion is, to what extent is the secretory process followed in peptidergic neurons? The answer need not necessarily be the same for each neuron producing a different peptide. The answer may depend heavily upon whether the neuropeptide is derived from a large precursor molecule, synthesized, most likely, by ribosomes restricted to the cell body. Di- and tripeptides (e.g., carnosine, thyrotropin-releasing hormone) may by synthesized, perhaps in the axon terminal, by enzymatic processes similar to those for neurotransmitters such as acetylcholine, GABA, and the biogenic amines.

Because the oxytocin- and vasopressin-producing neurons of the hypothalamo-neurohypophysial system are the best understood anatomically, physiologically, and biochemically of all the peptidergic neurons, these neurosecretory cells have been proposed as a cell biological model for peptidergic neurons (26). In this context, the hypothalamo-neurohypophysial system serves as an ideal model with which to study the secretory process for two reasons: (i) for sampling purposes its cell bodies are grouped into the supraoptic and paraventricular nuclei (5, 9), and (ii) vasopressin-producing neurons can be stimulated hyperosmotically to increase the production and release of their secretory product *in vivo* (22, 28, 43). Vasopressin acts on kidney tubules for retention of water to preserve the normal osmolarity of body fluids.

This presentation focuses on our investigations of the morphological aspects of the secretory process in the hypothalamo-neurohypophysial system of normal mice and mice hyperosmotically stressed by imbibing a 2% solution of sodium chloride (7–9, 11–13). Enzyme cytochemical and immunocytochemical techniques have been applied at the ultrastructural level to determine the intracellular pathways involved in the processing of vasopressin (molecular weight 1000) and the carrier protein to which it is complexed, neurophysin (molecular weight 10,000). Particular emphasis is directed to the role of the Golgi apparatus and GERL[1] in the packaging of secretory protein. Horseradish peroxidase cytochemistry has been employed to trace the fate of internalized axon terminal membrane in the neurosecretory neuron following discharge of the secretory product. Wherever possible, comparisons are drawn between aspects of the secretory process involving the neurosecretory cell and similar events we have been studying in the different types of peptide-secreting cells of the anterior pituitary gland (10).

2 CYTOCHEMICAL METHODS

In our hands (13) the pre-embedding staining approach to Sternberger's (83) unlabeled antibody–enzyme or peroxidase-antiperoxidase (PAP) method has offered far better immunocytochemical localization of tissue antigen ultrastructurally than has the alternative post-embedding staining approach.[2] The incubation sequence for immunocytochemical staining with the PAP method is given in Table 1 and is the same regardless of which of the two staining approaches is employed.

With pre-embedding staining, tissue chopper or vibratome cut sections prepared from fixed material are incubated free-floating in plastic culture dishes. The sections are trimmed down to only the areas of interest and then are osmicated, dehydrated through a graded alcohol series, and finally embedded in plastic for ultrathin sectioning. The concentration of the primary antiserum for the initial incubation and the length of time the tissue is incubated in the diaminobenzidine-H_2O_2 medium are two interrelated critical variables in the pre-embedding staining regimen. A primary antiserum dilution curve (i.e., dilutions ranging from 1:100–

[1] Novikoff (52–55) has proposed the acronym GERL for an acid phosphatase reactive, smooth membrane system that lies adjacent to the trans face of the Golgi apparatus, is connected to the Endoplasmic Reticulum, and gives rise to Lysosomes.

[2] For a detailed discussion regarding the preservation of antigenicity and cellular morphology in ultrastructural immunocytochemistry, see Broadwell in "Strategies for Studying the Roles of Peptides in Neuronal Function," Society for Neuroscience Short Course Syllabus, p. 27 (1982).

Table 1. Four-Step Incubation Sequence for Immunocytochemical Staining Using the Unlabeled Antibody–Enzyme Method (83)

1. *1° Antibody*
   Rabbit anti-X (IgG)
2. *2° Antibody Bridge*
   Sheep anti-rabbit (IgG)
3. *3° Antibody Enzyme Complex*
   Rabbit peroxidase-antiperoxidase complex (PAP)
4. *Detection of Peroxidase Reaction Product*[a] *(30)*
   Tris buffer, pH 7.6, diaminobenzidine, H_2O_2

[a] For ultrastructural inspection, the oxidized diaminobenzidine is converted to osmium black by rinsing the tissue sections in osmium tetroxide.

1:10,000 or higher) must be conducted for each primary antiserum. This curve is instrumental in establishing the concentration of the primary antiserum that will yield specific staining and low background. Because of their large molecular weight and physical size, antibodies do not penetrate deeply into the tissue during incubation. For this reason the PAP reaction product is restricted, for the most part, to the surfaces of the tissue block; therefore, when cutting ultrathin sections for electron microscopy care must be taken not to discard those sections in which the tissue first appears. Exposure of the tissue sections to detergents (e.g., Triton X-100) prior to incubation in the primary antiserum is believed to open cellular membranes, thereby improving antibody penetration into the cells. Our experience suggests that detergents are of questionable value in this regard at the ultrastructural level and only succeed in adversely affecting morphology. Incubation of the tissue in the diaminobenzidine-H_2O_2 medium should be stopped when brown granules begin to appear within the cells. The granules are indicative of immunoreactive organelles; the brown color signifies oxidized diaminobenzidine. The color change may occur rapidly or gradually over a period of several minutes, depending upon the concentration of the primary antiserum employed. We have found that proper dilution of the primary antiserum results in a diaminobenzidine-H_2O_2 incubation time of 7–10 min. If the primary antiserum is too concentrated, even brief incubation in the diaminobenzidine-H_2O_2 medium may produce a nonspecific, homogeneous brown staining of the cytoplasm of virtually all cells in the tissue.

Post-embedding staining entails incubating nonosmicated, plastic-embedded, ultrathin sections that have been placed on nickel or gold grids. The sections must be etched with H_2O_2 before the four-step incubation process. Following incubation the sections are osmicated. Histological processing of the tissue for ultrastructural examination prior to immunocytochemical staining may well be a major contributing factor to the apparent loss of antigenicity compared with that obtained by pre-embedding staining (13).

The immunocytochemical localization of antigen is dependent upon a number of factors, all of which interrelate with each other and are important in yielding reliable data (83). Two major considerations are preserving morphology and as muc antigenic activity as possible. If the tissue is not fixed adequately, the antigenic determinants may be lost during histological processing, or the antigen may diffuse from its original location. Fixing the tissue may alter the antigen in such a fashion that the alteration may be an asset or perhaps a liability for immunocytochemical localization. For example, antigens have very great differences in stability to fixation methods (e.g., freezing, freeze substitution, perfusion, immersion) and in interaction with the type of fixative (e.g., glutaraldehyde, formaldehyde). Antigenic sites on the molecule are altered when they are cross-linked with fixatives. For this reason proper fixative components and concentrations are of major importance in preserving antigenicity. The high concentration of glutaraldehyde (4%) used routinely in preparing tissue for electron microscopy may destroy the most labile antigens in tissues. Conversely, antigenicity of the most stable antigens may be enhanced by strong fixation. If the antigenic site is sterically blocked by the way the antigen is bound in the tissue, then the fixation process, by disrupting this binding, may cause more antigen to become visible to the antibody against that antigen. Our best results in localizing neurophysin ultrastructurally in brain have been in mice fixed by perfusion with 100 mL of 1.25% glutaraldehyde and 1% paraformaldehyde (13).

Excellent enzyme cytochemistry, like immunocytochemistry, is dependent upon preservation of cell morphology and activity of the enzymes present in the tissue under investigation. Enzymatic activity is destroyed by high concentrations of glutaraldehyde. In our studies good morphology as well as acid hydrolase and thiamine pyrophosphatase activities are obtained in tissues fixed by perfusion with a mixture of 1.25% glutaraldehyde and 1% paraformaldehyde (11, 12); the morphology of cells incubated for acid hydrolase activity may be compromised somewhat by the acidic pH of the incubation medium (pH 3.9–5.0). Glucose-6-phosphatase is extremely sensitive to fixation and is best preserved with a 5 min, 2% glutaraldehyde perfusion (8).

Acid phosphatase activity is a marker for lysosomes and is demonstrated cytochemically using the Gomori lead capture method (23) with either beta-glycerophosphate (2) or cytidine 5′-monophosphate (51) as the substrate. Trimetaphosphatase activity, also a lysosomal marker, hydrolyzes the substrate trimetaphosphate (20). Although both cytochemical methods yield reaction products confined largely to lysosomes, significant differences in the localizations of the two enzyme activities are becoming well documented. For instance, acid hydrolase activity is localized in GERL with the substrates for demonstrating acid phosphatase activity (Figures 4 and 5) but not with the substrate for trimetaphosphatase activity. The differences in localizations of acid phosphatase and trimetaphosphatase activities in neurons and other cell types along with the chemical reactions involved have been discussed in detail by us elsewhere (4, 63).

Thiamine pyrophosphatase activity is demonstrated cytochemically by the lead capture method with thiamine pyrophosphate as substrate (57). The enzyme is thought to be involved in distal glycosylation of glycoproteins (19, 31, 54) and is localized predominantly in the inner saccules of the Golgi apparatus (Figure 6).

Glucose-6-phosphatase activity is localized cytochemically in the rough and smooth endoplasmic reticulum (8; see Addendum and Figure 13). This enzyme may be involved in cerebral gluconeogenesis from potential energy sources like glycogen. Neuronal glucose-6-phosphatase is known to hydrolyze glucose-6-phosphate to glucose (44). In addition, it may have a phosphotransferase activity, as does the similar enzyme in hepatocytes; however, this enzymatic function in the neuron has yet to be demonstrated biochemically.

3 ORGANELLES INVOLVED IN THE SECRETORY PROCESS

Biochemical (14, 27, 76, 77), autoradiographic (45, 50), and immunocytochemical (13) data have combined to demonstrate conclusively that the secretory process in the hypothalamo-neurohypophysial system utilizes the same macromolecular structures as those in other protein-secreting cells. The secretory product is synthesized by ribosome-studded ER and then delivered to the Golgi apparatus for packaging into secretory granules. Our extensive enzyme cytochemical and immunocytochemical studies of supraoptic somata (9, 11, 13) indicate that although Golgi saccules are involved in the secretory process, the formation of secretory granules under normal ("resting") conditions occurs predominantly from GERL and not from Golgi saccules (Figure 1). With hyperosmotic stress, secretory granule production is demonstrably increased off all Golgi saccules and GERL (Figure 7).

Perikarya of the supraoptic nucleus contain a well defined network of ribosomal ER. This rough ER frequently appears as parallel lamellae concentrated to one side of the cell body (Figure 1). Single strands of rough ER course though the cytoplasm encircling the cell nucleus. Polyribosomes, in addition to being associated with the ER, are attached to the nuclear envelope. The ER and nuclear envelope are in continuity with each other (Figure 2, inset). Biochemical data obtained from a variety of secretory cells have shown that once the secretory product enters the rough ER cisternae, enzymes within the membrane or cisternal space modify the protein. These enzymes are responsible for disulfide bridge formation, hydroxylation of proline and lysine residues, proximal glycosylation of the polypeptide chain, and perhaps partial proteolysis (68, 73, 88).

The Golgi apparatus of the supraoptic neuron (Figure 1) is located predominantly in the perinuclear region. It consists of three to seven stacks of saccules of irregular width and length. The trans saccules, often referred to as the inner saccules or maturing face of the Golgi apparatus, are narrow in width, regular in contour, and rarely appear dilated. The cis Golgi saccules, also called the outer saccules or forming face, consist of short, dilated, and discontinuous segments. Ribosome-studded or ribosome-free portions of the ER and smooth-surfaced vesicles lie in proximity to the cis saccules more than to the trans saccules (Figure 1). The vesicles may be those of rough ER origin hypothesized to transport membranes and the forming secretory product to the cis Golgi saccules (8, 39, 40, 48, 73). The Golgi apparatus of secretory cells is known to contain enzymes active in the synthesis of sulfate-containing macromolecules, terminal glycosylation of secretory protein, and partial proteolysis of the polypeptide chain (49, 68, 73, 88).

Granules 100–200 nm in diameter and containing an electron-dense core are scattered throughout the supraoptic perikaryal cytoplasm; many are located close to the Golgi apparatus. These granules are most likely an immature form of the neurosecretory granule. Granules of a similar size but exhibiting complete electron density are prevalent in pituitary stalk axons and posterior lobe Herring bodies. This form of granule may represent the mature form of the neurosecretory granule. The 100–200 nm granules distributed throughout the hypothalamo-neurohypophysial system are immunoreactive against antisera for neurophysin, vasopressin, and oxytocin (13, 46, 80, 89). As is the case with the rough ER and Golgi apparatus, the secretory granule is believed to contribute to the multiplicity of functions in the processing of secretory product. Gainer and coworkers (14, 27) have demonstrated that within neurosecretory granules vasopressin, oxytocin, and their respective neurophysins are derived from the posttranslational cleavage of a precursor molecule.

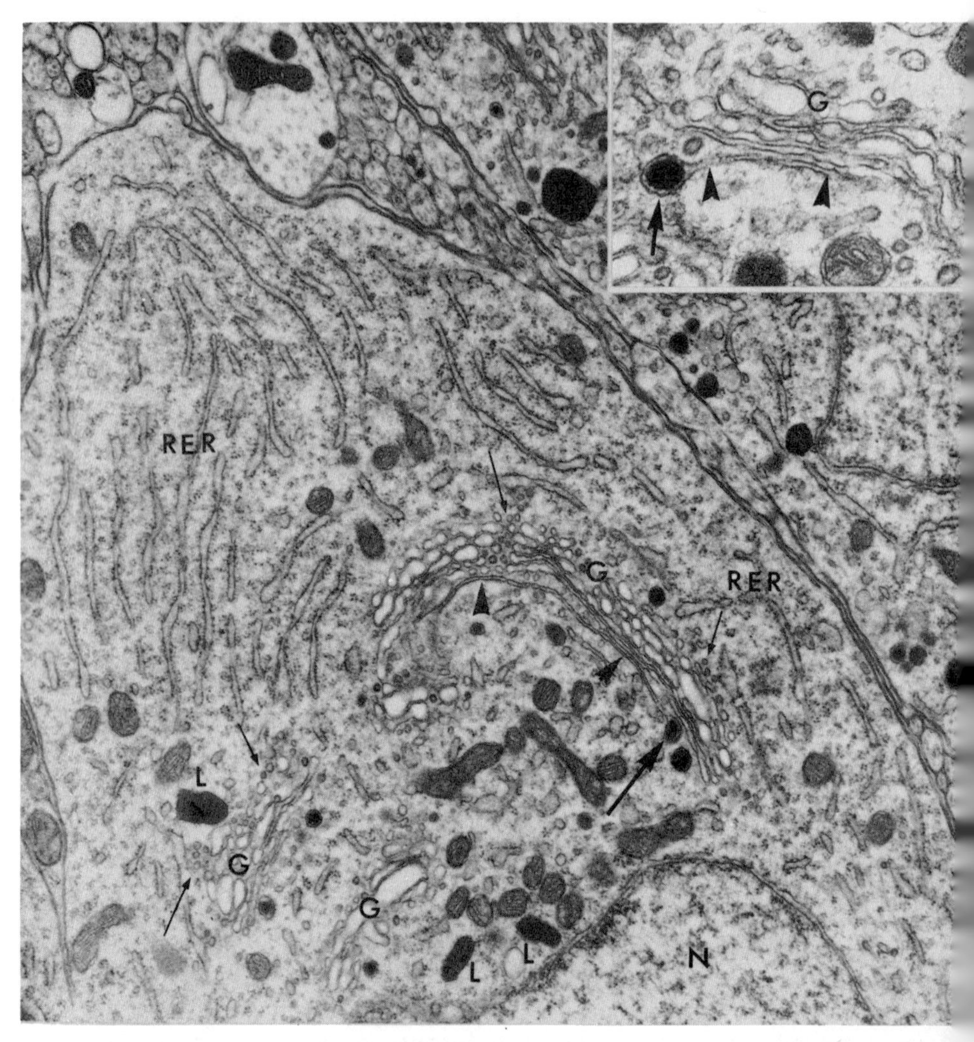

Figure 1. Perikarya of neurosecretory neurons in the mouse supraoptic nucleus contain parallel stacks of rough endoplasmic reticulum (*RER*). The Golgi apparatus (*G*) is located in proximity to the cell nucleus (*N*). Secondary lysosomes (*L*) appear throughout the cell body and are frequently in the vicinity of the Golgi apparatus. Under resting conditions the formation of secretory granules (large arrows) occurs predominantly from GERL (arrowheads) and less so from Golgi saccules. Transfer vesicles (small arrows), presumably of RER origin, may serve to interconnect the RER and the cis face of the Golgi apparatus. Magnification ×34,000, inset ×45,000. **Unless stated otherwise, all subsequent electron micrographs are of supraoptic perikarya from control or hyperosmotically stressed mice.**

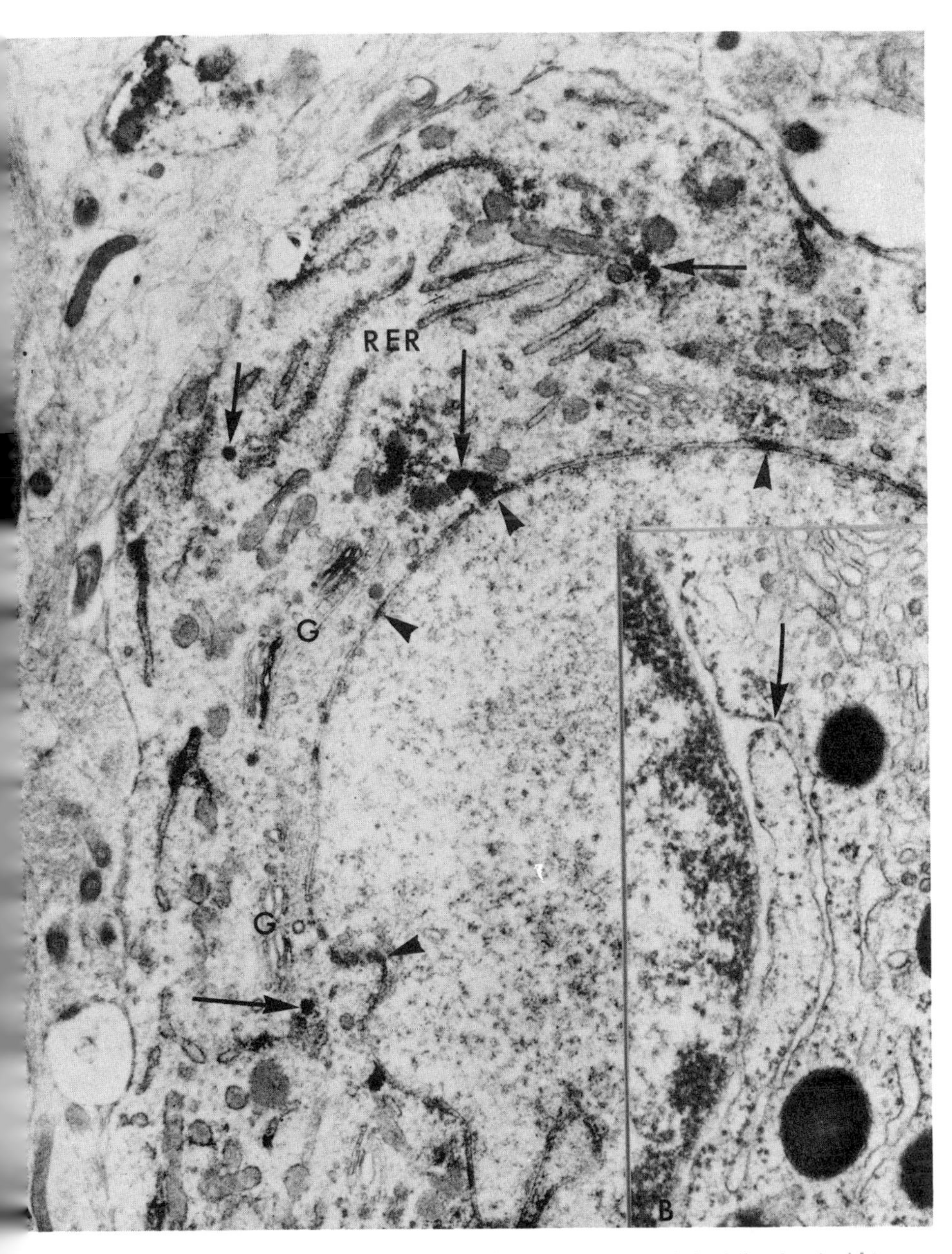

Figure 2. (A) PAP reaction product for neurophysin immunoreactivity is localized within the nuclear envelope (arrowheads), rough endoplasmic reticulum (*RER*), Golgi apparatus (*G*), and secretory granules (arrows). Magnification ×25,500. (B) The nuclear envelope and rough endoplasmic reticulum are confluent with each other (arrow); anterior pituitary somatotroph. Magnification ×45,000.

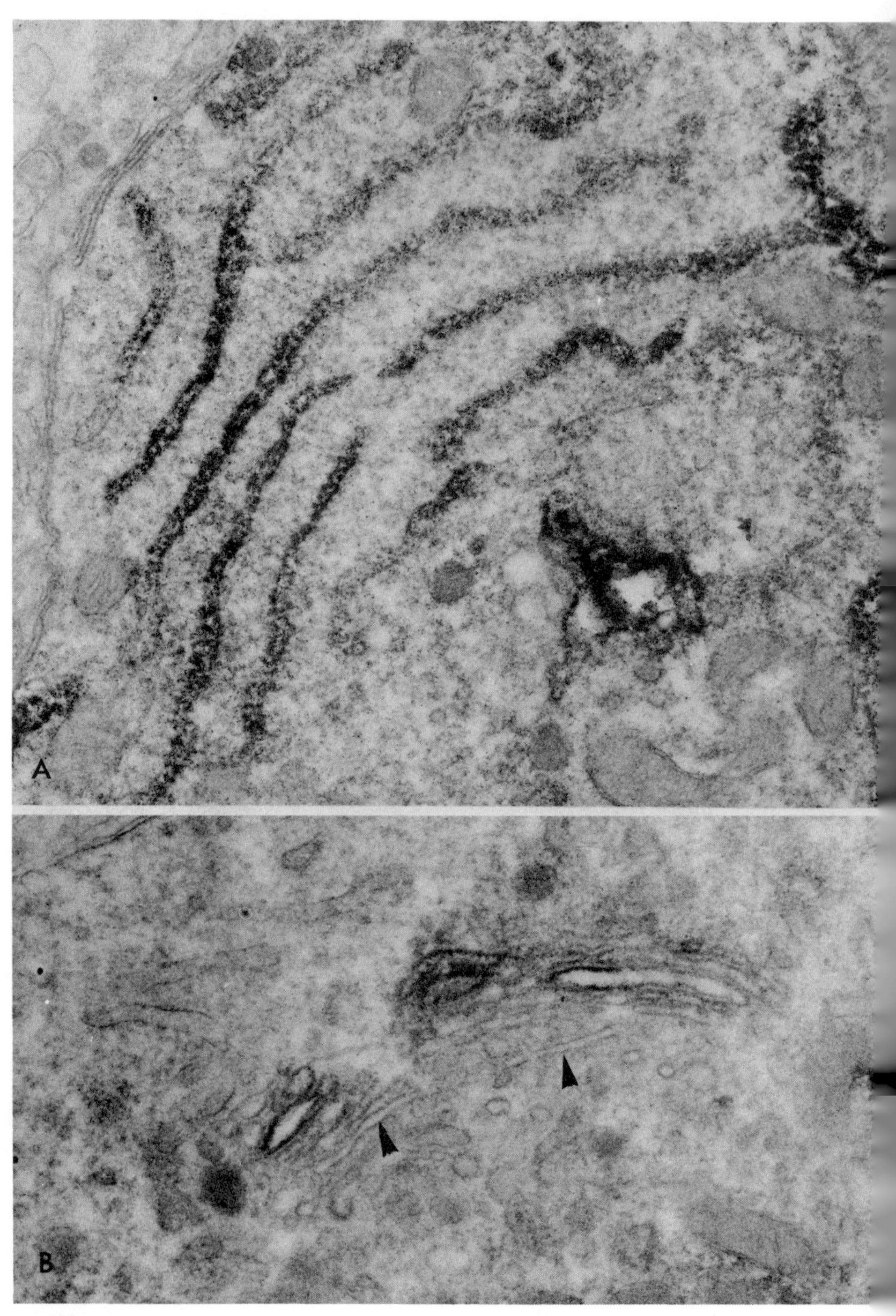
A
B

In "resting" supraoptic perikarya and in the five different types of secretory cells in the anterior pituitary gland, dense core granules are observed rarely in continuity with Golgi saccules but are seen frequently to be confluent with a narrow (30–40 nm wide), smooth-membrane cisterna located at the trans face of the Golgi apparatus (Figure 1). This smooth-membrane cisterna is equivalent morphologically and cytochemically to GERL (see Figures 1, 4, 5). The GERL membrane is 7–8 nm wide and is slightly thicker than that of the trans Golgi saccule (5 nm wide). Portions of the GERL membrane appear coated. Primary lysosomes 40–70 nm wide may arise from GERL (Figure 4). Occasionally, cisternae of rough ER lie in proximity to GERL; however, we (8–11) and others (32, 33, 38) have yet to observe direct continuities between these two structures as Novikoff and his associates (54, 55, 58–60, 62) reportedly have seen in a variety of cell types.

By applying the pre-embedding staining approach to the PAP method of Sternberger (83), with primary antibody directed against the neurophysins for both vasopressin and oxytocin, we have been able to obtain precise localization of immunoreactivity for neurophysin within organelles associated with protein synthesis and packaging in supraoptic perikarya from normal and hyperosmotically stressed mice (13). The immunoreactive organelles include the nuclear envelope, rough ER, Golgi apparatus, and secretory granules (Figures 2, 3). Lysosomes were likewise immunoreactive. Staining of the cell body cytoplasm, axoplasm, microtubules, axonal smooth ER, and 40–70 nm wide vesicles in the axon terminals was not observed. Neurophysin immunoreactivity in rough ER and Golgi saccules is most likely not indicative of neurophysin per se but of neurophysin as part of a precursor molecule.

A major disappointment in our immunocytochemical study was the inability to obtain immunoreactivity in GERL (Figure 3B). The explanation offered for this negative result was inadequate penetration of antibody into the tissue, an inherent shortcoming with the pre-embedding staining approach to the PAP method (13, 83). This explanation seemed a plausible one a priori, since many nonimmunoreactive secretory granules were scattered among those that stained positively in the cell body and neurohypophysis.

Figure 3. PAP reaction product for neurophysin immunoreactivity appears within cisternae of the rough endoplasmic reticulum (A) and the Golgi apparatus (B). GERL (B, arrowheads) is unlabeled. Magnification ×60,000.

4 PACKAGING OF SECRETORY PRODUCT: GERL VS. GOLGI APPARATUS

The relationship between GERL and the Golgi apparatus and the role these organelles play in the packaging of secretory proteins into granules have been topics of speculation and controversy since Novikoff (52) introduced the concept of GERL. Novikoff and his associates (54, 55, 58, 61, 62) consider GERL to be a morphologically and functionally distinct organelle from the Golgi apparatus. Novikoff and numerous others (31–33, 35, 38, 61, 64, 66, 71) have advocated the origin of secretory granules from GERL in a host of endocrine and exocrine cell types. Other investigators have thought of GERL as a component of the Golgi apparatus (8, 37, 87); most histology and cell biology textbooks fail to acknowledge GERL at all.

Morphologically, the GERL membrane is distinct from that of the Golgi saccules (see above). Enzyme cytochemical markers for acid phosphatase (AcPase) and thiamine pyrophosphatase (TPPase) activities are particularly useful for distinguishing GERL from the Golgi saccules (51, 55, 57). AcPase activity appears in GERL, in its forming secretory granules, in 40–60 nm wide primary lysosomes which are presumably of GERL origin, and in secondary lysosomes (Figures 4 and 5). The activity of TPPase is normally restricted to the one or two trans Golgi saccules. The localizations of these enzyme activities basically holds true in our preparations of the "resting" neurosecretory cell (Figures 5 and 6) and in the secretory cells of the anterior pituitary gland (8–11). Infrequently, we see some AcPase activity in the trans Golgi saccule and TPPase activity in secretory granules forming off GERL. Such deviations from the normal localizations for AcPase and TPPase activities suggest that the trans Golgi saccule may be a form of transition saccule that loses its TPPase activity, acquires AcPase activity, and undergoes a conversion to GERL. That a structural as well as a functional interrelationship may exist between GERL and the Golgi apparatus is supported further in the hyperosmotically stressed supraoptic neuron.

Demonstrable changes in the morphology of the Golgi apparatus and GERL and alterations in and localizations of AcPase and TPPase activities in these organelles are produced in supraoptic perikarya of mice hyperosmotically stressed by drinking a 2% solution of sodium chloride for five to seven days (9–11). Compared with controls, the Golgi apparatus and GERL have hypertrophied; their saccules are noticeably longer and exhibit numerous fenestrations or anastomosing tubules. Secretory granule production, which normally is from GERL with little involvement of Golgi saccules, has markedly increased off GERL and all Golgi saccules

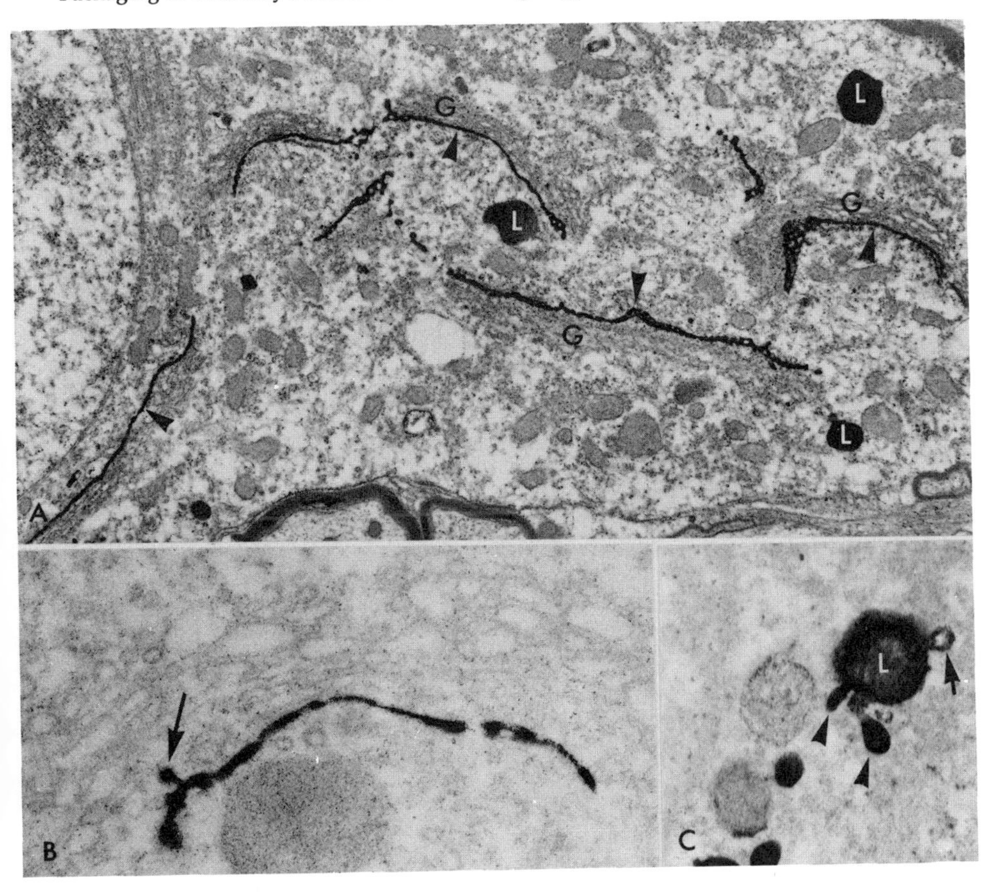

Figure 4. (A) GERL (arrowheads), located at the trans face of the Golgi apparatus (*G*), and lysosomes (*L*) are acid phosphatase-reactive in neuronal cell bodies; VII cranial nerve nucleus. Magnification ×18,000. (B) Primary lysosomes (arrow) 40–70 nm in diameter may be derived from GERL. Magnification ×87,200. (C) A primary lysosome (arrow) is one that is brought to ferry acid hydrolases to secondary lysosomes (*L*). Additional acid hydrolase-reactive structures that may fuse with or bud from spherical secondary lysosomes are tubules (arrowheads). These tubules may represent a segment of the GERL cisterna, in which case they would be a form of primary lysosome or they would be a variant of the typical secondary lysosome (see refs. 4, 7, 10, 12). Magnification ×51,000.

(Figure 7). Secretory granules accumulate in the cell body. More secretory granules appear to be transported anterogradely down pituitary stalk axons than the concentration seen in pituitary stalk axons from hydrated control animals (9). AcPase activity in GERL and its forming secretory granules is nil in comparison to that in controls (9, 11), despite a proliferation of secondary lysosomes throughout the neuron (4, 7, 12). TPPase

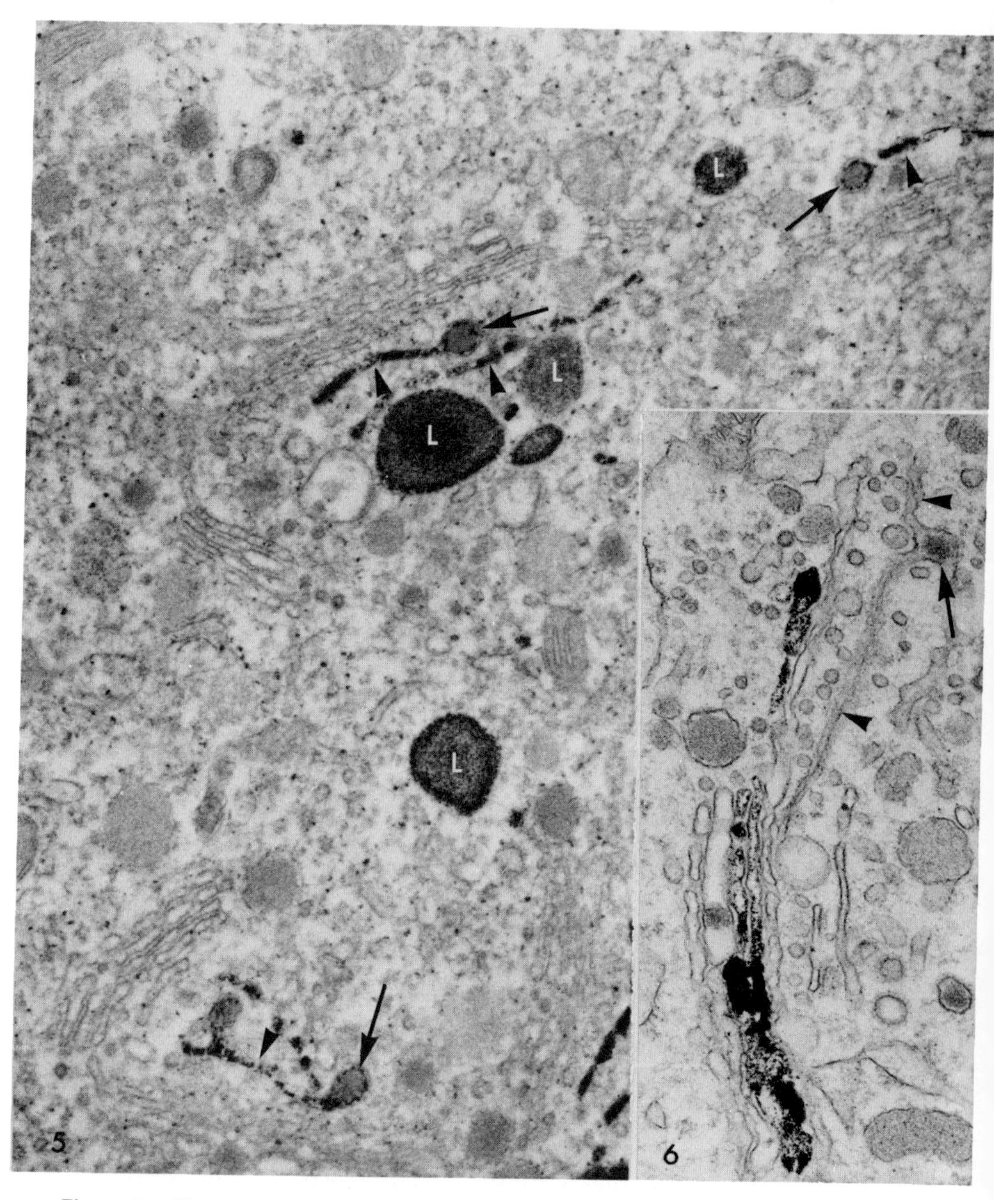

Figure 5. (Top) Acid phosphatase-positive organelles in control supraoptic perikarya include GERL (arrowheads), secretory granules (arrows) forming from GERL, and secondary lysosomes (*L*). Note that in the forming secretory granules, acid phosphatase reaction product surrounds but does not obscure the electron dense core. Magnification ×25,000.

Figure 6. (Bottom) Thiamine pyrophosphatase activity is normally restricted to one or two trans Golgi saccules and does not appear in GERL (arrowheads) or its forming secretory granules (arrow). Magnification ×40,000.

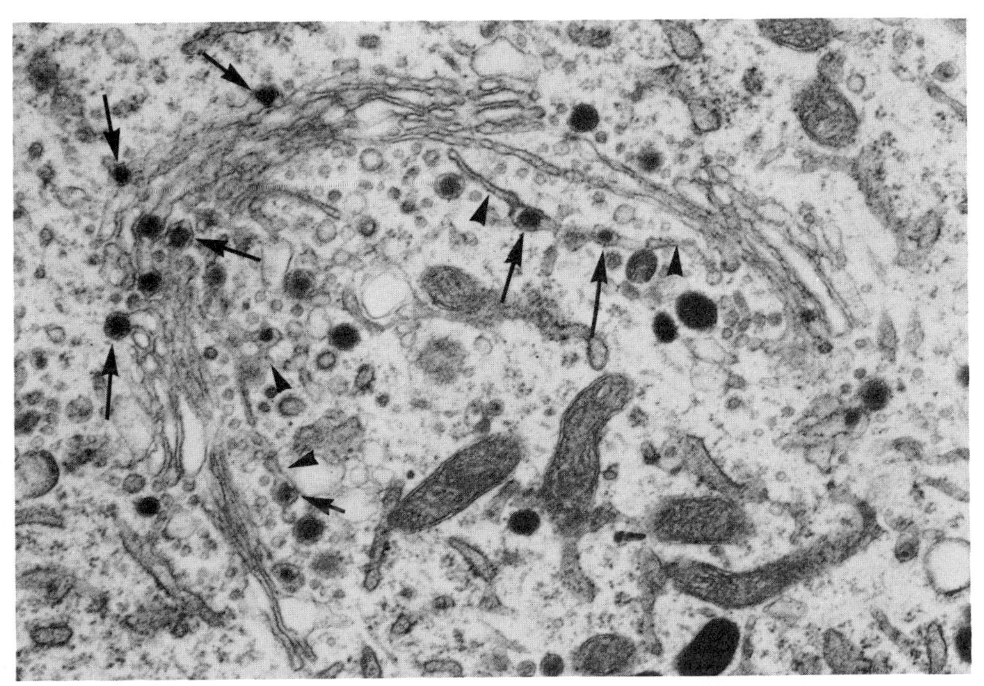

Figure 7. In salt-stimulated cell bodies from the supraoptic nucleus, all Golgi saccules and GERL (arrowheads) are active in producing secretory granules (arrows). Magnification ×18,000.

activity parallels the production of secretory granules and is prominent in all Golgi saccules, GERL, and secretory granules forming from these structures (Figure 8). In some instances TPPase-positive anastomosing tubules may interconnect the trans Golgi saccule and the GERL cisterna (Figure 8).

Although we have not observed neurophysin immunoreactivity or TPPase activity in the smooth ER and 40–70 nm vesicles of neurohypophysial axons and terminals, this membrane bound, cytoskeletal network has been implicated by others in the transport of neurosecretory material and in the formation of neurosecretory granules or vesicles (1, 17). These studies and a recent report suggesting that, at least in adrenergic neurons, the axonal reticulum may be a direct extension of the Golgi apparatus (72) offer the prospect that the packaging and refinement of neuropeptides may be more involved than initially expected.

When hyperosmotically stressed mice are returned to normal drinking water for a period up to ten days, the supraoptic perikarya present morphological and enzyme cytochemical alterations intermediate between

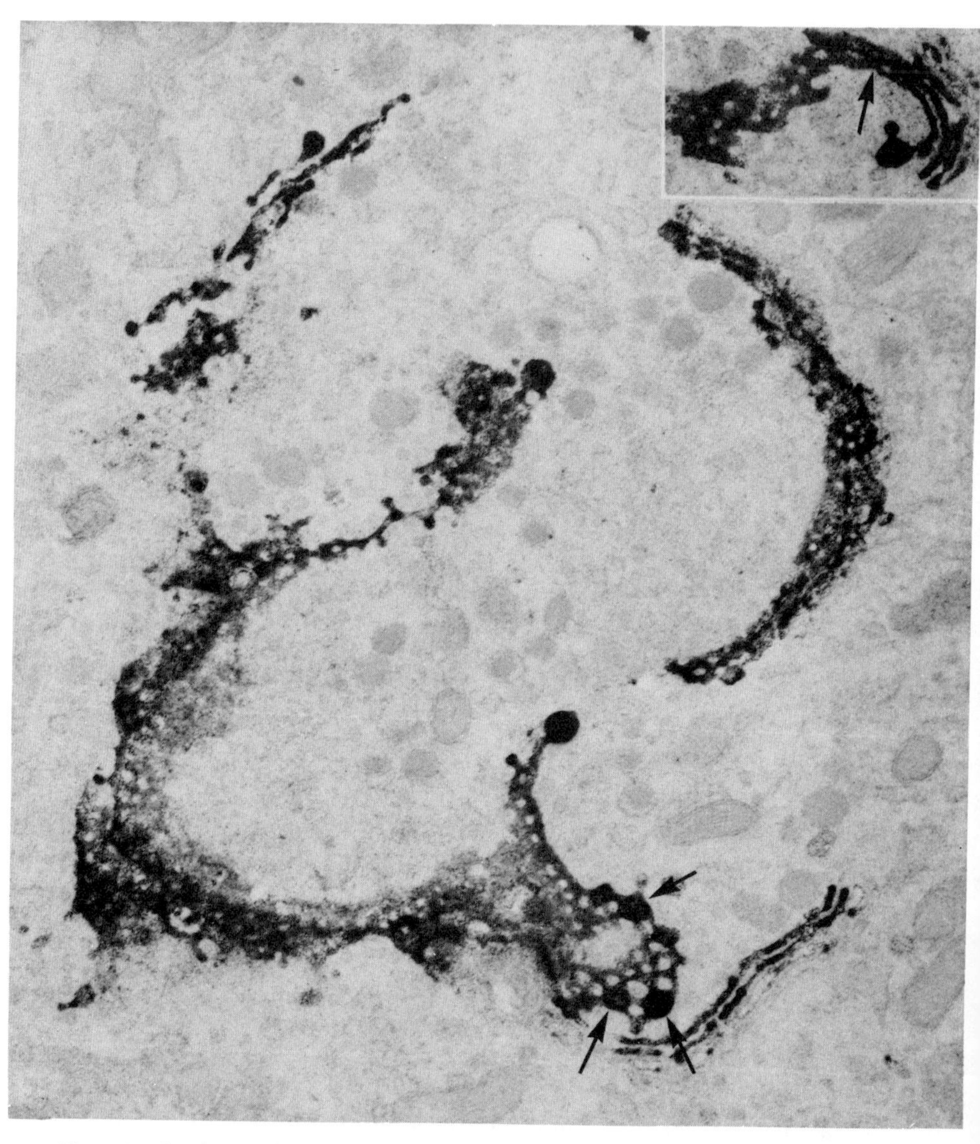

Figure 8. In salt-stimulated supraoptic perikarya, thiamine pyrophosphatase activity appears in most Golgi saccules, in GERL, and in secretory granules (arrows) forming from these structures. In this cytochemical preparation, GERL and the inner Golgi saccule may be interconnected (inset, arrow). Magnification ×48,000; inset, ×36,000.

control and salt-stressed conditions (11). The Golgi apparatus and GERL are not as extensive as in the hyperosmotically stressed state. GERL and most Golgi saccules remain active in secretory granule production. TPPase activity likewise remains elevated in most but not all Golgi saccules, in GERL, and in their forming secretory granules. The most striking change in these cells during the "recovery" phase is the return of AcPase activity in GERL and its forming secretory granules.

Two important conclusions have been drawn from these results. First, a major function of GERL in supraoptic neurons, as well as in anterior pituitary cells, is the production of secretory granules. Under normal conditions the Golgi apparatus appears to function less so in this regard. Nevertheless, the Golgi apparatus actively participates in the processing of secretory products. Concentration and terminal glycosylation of the secretory product certainly represent two important functions of the Golgi apparatus (49, 88). Recent biochemical findings suggest that the terminal portion of the precursor molecule for vasopressin is glycosylated (14). The second important conclusion from our studies is that the modulations in AcPase and TPPase activities between GERL and the Golgi saccules suggest that GERL may be structurally and functionally related to the Golgi apparatus. With the increased production of neurosecretory granules in response salt stress, the conversion of the trans Golgi saccule to GERL may occur at a rate too rapid for normal enzyme modulation to take place. Similar enzyme cytochemical alterations between GERL and the Golgi apparatus have been reported in additional types of secretory cells (64–66). Studies such as these indeed support the notion of a functional relationship between GERL and the Golgi apparatus and strengthen the supposition that GERL may be derived from the trans Golgi saccule (32, 33) (also see the Addendum in this chapter).

The role of GERL in the production of primary lysosomes in the neurosecretory neuron is difficult to evaluate. The 120–180 nm wide, dilated portions of the GERL cisternae, which contain dense cores and are believed to represent forming secretory granules, should not be considered prospective primary lysosomes. Again, similar dense-core dilatations arise from all Golgi saccules during hyperosmotic stimulation. Lysosomes in the supraoptic and anterior pituitary cells have never been seen to contain dense cores. We (4, 9, 11, 13) and others (3, 24, 25, 32–34, 36, 53, 62, 66) have observed AcPase-positive 40–60 nm wide vesicles attached to the ends of the GERL cisterna, but whether or not these vesicular profiles are primary lysosomes is unclear. Because GERL consists of a system of anastomosing tubules at the ends of its cisterna (11, 32, 33), the vesicular profiles may represent sections cut through the GERL tubules. The localization of AcPase activity in forming secretory granules in neurosecretory and anterior pituitary cells suggests that this enzyme may be involved in the processing of secretory product.

Reference has been made to the observation that the GERL membrane possesses certain morphological characteristics that distinguish it from the membrane of Golgi saccules. Membrane of the cis Golgi saccules appears similar morphologically to that of the ER; transfer vesicles are believed to convey membrane from the ER to the cis face of the Golgi apparatus (8, 48, 73). The GERL membrane and that of secretory granules are similar morphologically to the plasmalemma. What differences, if any, distinguish the secretory granules of Golgi saccule origin from those forming from GERL and how such differences may be important to the discharge of secretory product from the cell remain to be elucidated.

The morphological and enzyme cytochemical alterations described above are not restricted to a select population of perikarya in the supraoptic nucleus but appeared in most somata of this cell group. This observation led us to speculate that perhaps the oxytocin-producing neurons in the supraoptic nucleus, as well as those synthesizing vasopressin, respond to hyperosmotic stress (11). Available experimental data indicate that with prolonged osmotic stress the synthesis of vasopressin and oxytocin are increased (27), whereas the neurohypophysis becomes depleted of both hormones (22, 28).

5 STEP 7: RECYCLING OF SECRETORY GRANULE MEMBRANE

If a seventh step were included in the secretory process, this step would consider the internalization and recycling of secretory granule membrane from the plasmalemma after the secretory granule has merged with the plasmalemma for discharge of the secretory product. Were there no mechanism to ensure recovery of this membrane after exocytosis, the secretory cell or, in the case of the neuron specifically, the axon terminal would expand in surface area with each successive exocytotic event. Retrieval of cell surface membrane in the form of endocytotic structures, the intracellular pathways this membrane follows, and the organelles involved in the membrane recycling process can be investigated with horseradish peroxidase (HRP) cytochemistry (30).

In control and hyperosmotically stressed mice injected intravenously with peroxidase, endocytotic vesicles and vacuoles forming in neurohypophysial axon terminals incorporate the protein (Figure 9) that has leaked from the blood and filled the extracellular clefts in the neurohypophysis (6–8, 12). These HRP-positive structures are conveyed by retrograde axoplasmic transport (Figure 9) to cell bodies in the supraoptic

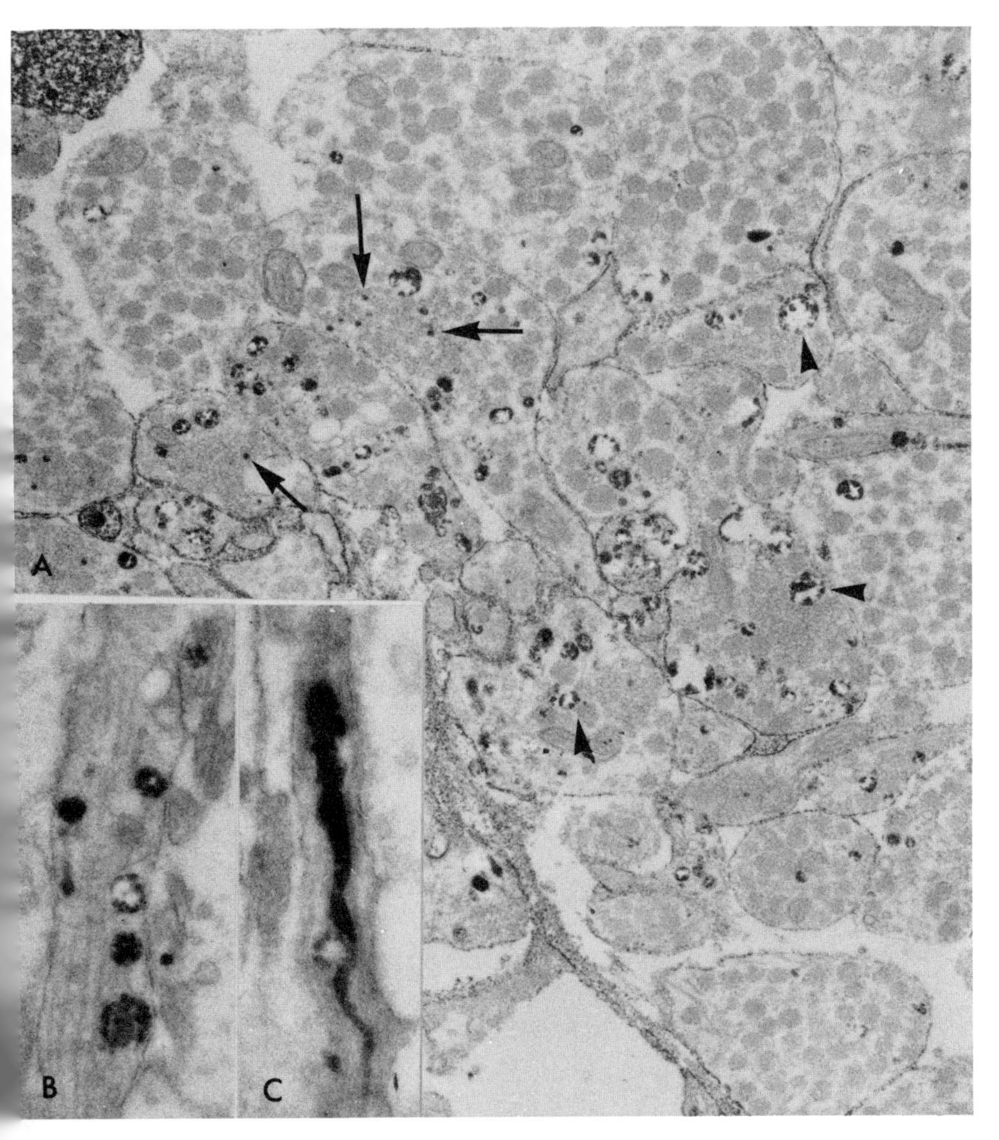

Figure 9. (A) Intravenously administered horseradish peroxidase associated with internalized axon terminal membrane in the mouse neurohypophysis appears in vesicles (arrows) and vacuoles (arrowheads) three hours after injection of the protein. (B) Peroxidase-labeled vacuoles and (C) blunt-ended tubules are transported by retrograde axoplasmic flow in pituitary stalk axons to the neurosecretory cell bodies. Magnification (A) ×28,500; (B) ×48,000; (C) ×51,000.

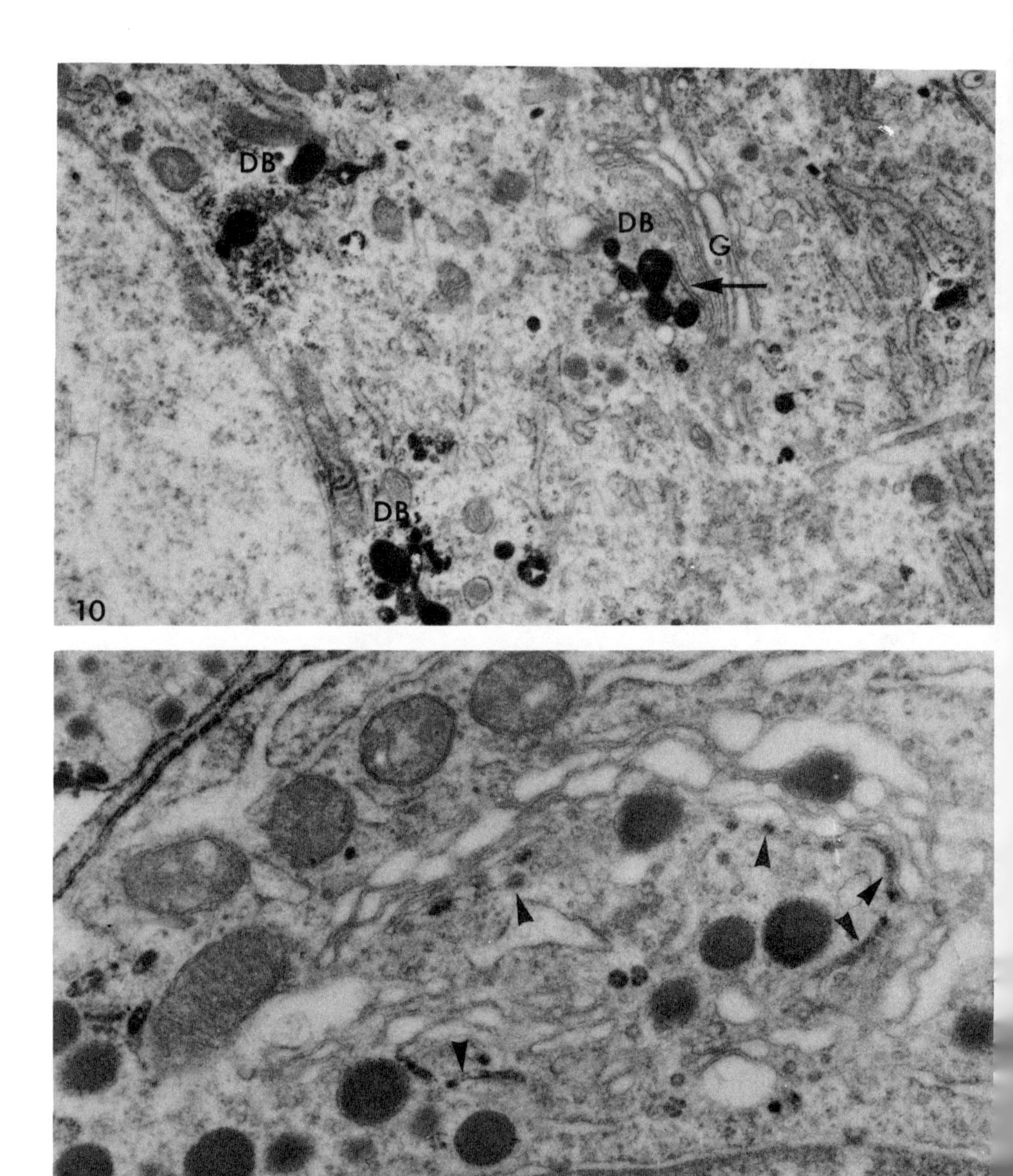

Figure 10. (Top) Peroxidase and internalized axon terminal membrane in neurosecretory neurons are ultimately directed to lysosomal dense bodies (*DB*) in perikarya and apparently not to the Golgi apparatus (*G*) or to GERL (arrow). Magnification ×32,000.

Figure 11. (Bottom) In somatotrophs of the mouse anterior pituitary gland, intravenously injected peroxidase and internalized cell surface membrane are channeled to lysosomal dense bodies and to GERL (arrowheads). Magnification ×45,000.

and paraventricular nuclei. Within the perikaryon the labeled endocytotic vacuoles fuse with lysosomes but apparently not with Golgi saccules or GERL, which produce the secretory granules[3] (Figure 10) (12). The lysosomal membrane is permeable to amino acids, small sugars, and other molecules released through the hydrolysis of macromolecules. These products of digestion can diffuse out of the lysosome and into the cytoplasm where they become available for reutilization within the cell. Because exocytosis, endocytosis, and the concentration of peroxidase-labeled organelles in the neurosecretory neuron are demonstrably increased with hyperosmotic stress (7, 8, 12, 22, 28, 43), the events outlined above may be associated specifically with the retrieval and recycling of secretory granule membrane.

Internalization of cell surface membrane in all anterior pituitary cells exposed to HRP *in vitro* (70) and *in vivo* (10) is likewise channeled to lysosomes. In somatotrophs, however, membrane was also traced to GERL (Figure 11). HRP labeling of GERL in somatotrophs may be indicative of a direct recycling of cell surface membrane for the production of secretory granules. The cell type may very well be a factor in determining the fate of retrieved cell surface membrane. The discrepancy between HRP localization in lysosomes and GERL may also be related to metabolic differences among different cell types.

6 CONCLUSIONS

The morphological aspects of the secretory process in peptidergic neurons, defined above with the use of vasopressin-producing neurons of the hypothalamo-neurohypophysial system as a model, are summarized

[3] The absence of observable activity in GERL of cells exposed extracellularly to native peroxidase addresses considerations about the eventual fate of the tracer versus the fate of the retrieved membrane transporting that tracer. The two fates may be mutually exclusive. Portions of internalized cell surface membrane with tracer may be directed to secondary lysosomes where both would be deposited. In some instances perhaps this membrane, after depositing its contents into the lysosome, would move on to GERL, which then would not exhibit HRP activity. Lectin-conjugated HRP, unlike native HRP, binds to the plasmalemma, including the cell surface receptors. The conjugate with membrane is taken into the cell by adsorptive or receptor-mediated endocytosis. By this process, the conjugate and its associated cell surface membrane (receptors?) have been traced not only to secondary lysosomes but to GERL as well (see Gonatas, *J. Neuropath. Exp. Neurol.* **41**, 6, 1982 for a review). Both may be channeled directly to GERL or perhaps to lysosomes first. For a detailed account of membrane recycling in the neuroendocrine cell, see R. D. Broadwell and M. W. Brightman in *Methods in Enzymology: Neuroendocrine Peptides*, P. Michael Conn, Editor, Academic Press, N.Y., in press.

diagrammatically in Figure 12. Emphasis has centered upon the ribosomal-dependent precursor mode of peptide biosynthesis as opposed to the enzymatic mechanism. Quantitative cell fractionation studies indicate that for the most part proteins destined for export and/or secretion are synthesized on polyribosomes attached to the endoplasmic reticulum (68, 73). That peptides or hormones may be formed from a precursor

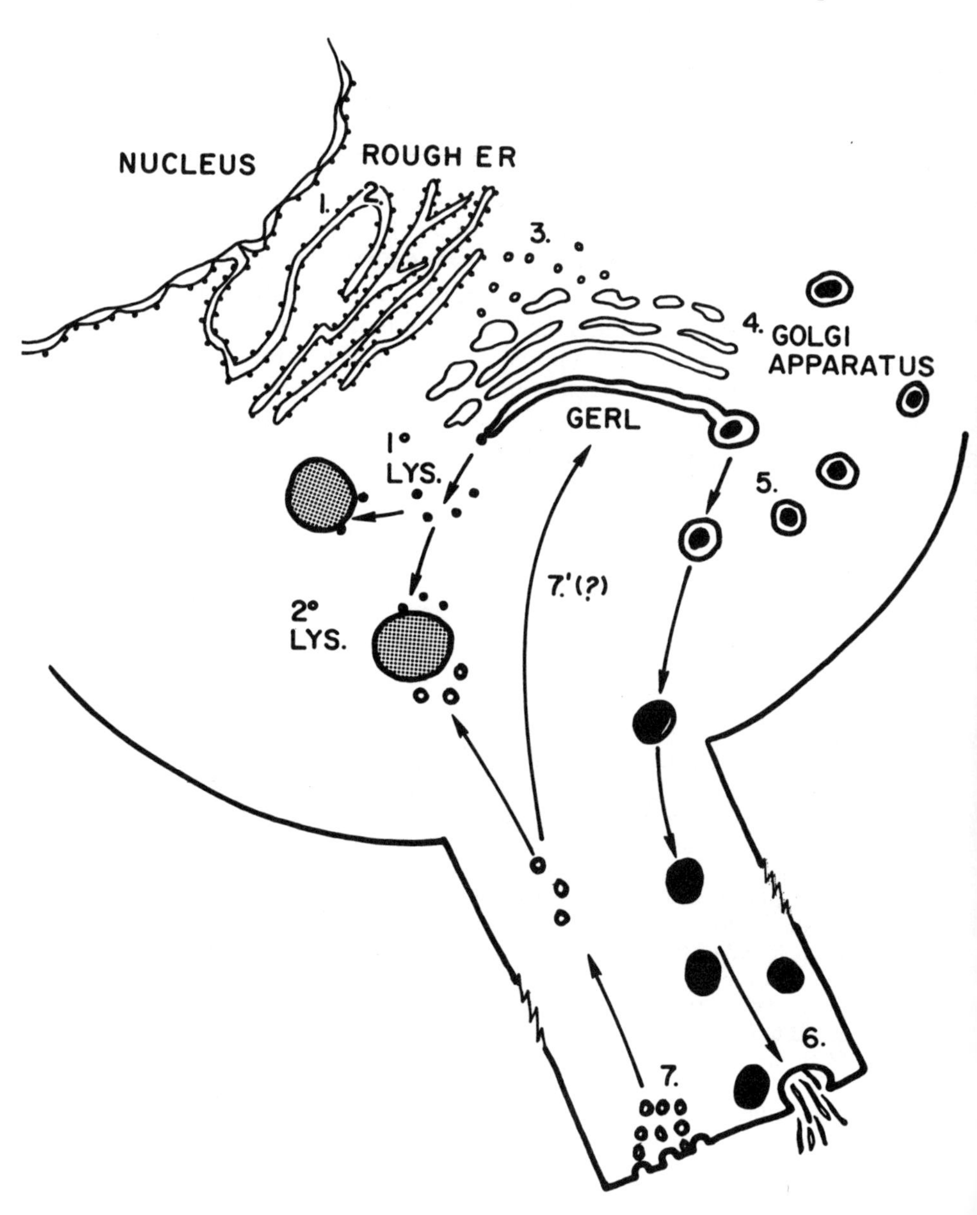

protein synthesized by ribosomes, a view first postulated for vasopressin and neurophysin (76, 77), has now become a familiar theme concerning several hormones and many cells other than the neuron. In addition to the neurophysin-vasopressin complex, precursor molecules are believed to exist for insulin (82), growth hormone (81), glucagon (84), alpha- and beta-MSH (79), ACTH/endorphin (47), and enkephalin (37), to name but a few. Synthesis of peptides and hormones in a precursor form may be necessary to prevent diffusion of the biologically active molecule out of the various intracellular compartments associated with the secretory process. The rough ER membrane, for example, is highly permeable to small molecules of about 10 Å in diameter (68); the binding of peptides to carrier proteins may insure against leakage of low-molecular-weight peptides out of the secretory package in which they are stored prior to exocytosis. Whether or not the secretory process for other neuropeptides parallels that described for the neurophysin-vasopressin complex remains to be determined. In numerous laboratories consideration is directed at determining if more than one peptide is produced by a single neuron, and, if so, whether those peptides are processed and packaged together as part of a common precursor molecule similar to that for ACTH/endorphin/beta-lipotropin in the anterior pituitary (47). In this regard, the hypothalamo-neurohypophysial system may again serve as a valuable model. Recent evidence suggests that leucine-enkephalin (18, 75, 78), dynorphin (29, 86), and somatostatin (15, 18, 21, 85) are localized within this neuronal system in granules similar to those containing oxytocin, vasopressin, and their respective neurophysins. Obvious questions are: Are these peptides together in the same neuron? Are they each in neurons separate from those containing vasopressin or oxytocin? Combined immunocytochemical and biochemical studies will surely answer these questions and others as interest in neuropeptides continues to expand. What Dr. Alex Novikoff (56) stated many years ago applies equally well today and will tomorrow, "Life is wondrous and beautiful, even at the level of cell organelles."

Figure 12. Steps of the secretory process involving the production, packaging, and release of peptides in neurons include: (1) synthesis on ribosomes associated with the endoplasmic reticulum, (2) segregation within the cisternal space of the rough endoplasmic reticulum, (3) transport to the cis face of the Golgi apparatus, (4) concentration and further processing in the Golgi apparatus and GERL, (5) storage in secretory granules, (6) exocytosis of secretory granule content, and (7) internalization of cell surface membrane in the axon terminal following release of the peptide from secretory granules and the delivery of this internalized membrane to perikaryal secondary lysosomes and/or perhaps to GERL (7′). Membranes appearing as either thick lines or thin lines connote morphological similarities in membrane among the intracellular compartments. Note that both types of membrane are present in the Golgi apparatus/GERL, where transition between the two types of membrane may occur. *1° Lys.*, primary lysosomes of GERL origin; *2° Lys.*, secondary lysosomes.

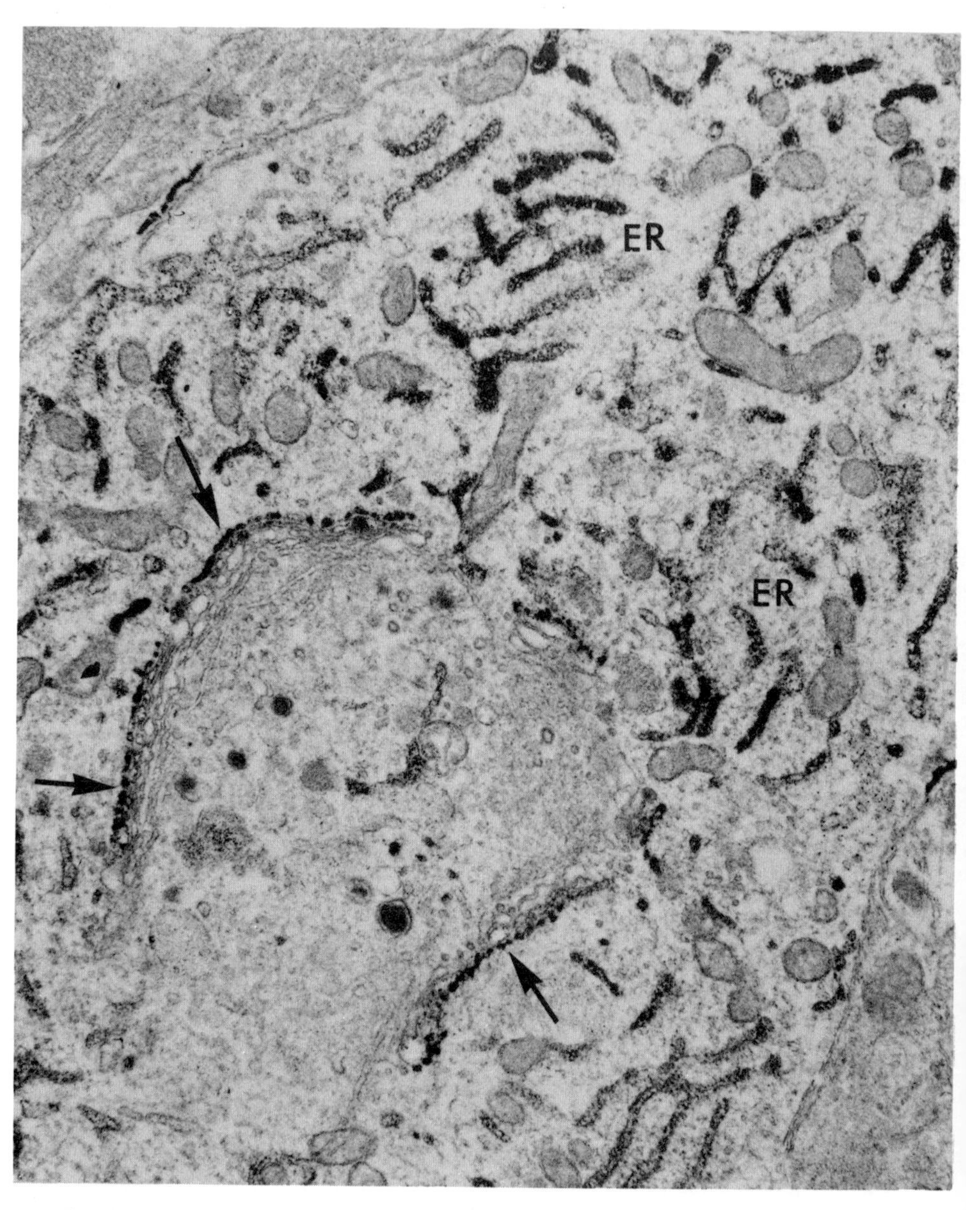

Figure 13. G-6-Pase activity within the supraoptic perikaryon is localized predominantly in the endoplasmic reticulum (*ER*). The outer Golgi saccule (arrows) frequently contains similar enzymatic activity. Magnification ×22,500.

ADDENDUM

We have successfully localized activity for glucose-6-phosphatase (G-6-Pase), a cytochemical marker specific to the endoplasmic reticulum (ER), in perikarya, dendrites, and axons (8). G-6-Pase reaction product was demonstrable predominantly in the rough and smooth ER throughout the cell body and dendrites (Figure 13) but less so in axons under normal conditions. Enzymatic activity was localized frequently in the outer Golgi saccule (Figure 13). Reactive segments of the ER were in continuity with saccules in the cis half of the Golgi apparatus. Despite many G-6-Pase positive profiles of ER in the vicinity of the trans face of the Golgi apparatus, no interconnection between the ER and GERL could be identified; GERL was never G6Pase-reactive. The most striking observation in the G-6-Pase preparations was an increase in the number of reactive segments of the smooth ER in neurohypophysial axons and Herring bodies from salt-stressed mice compared with hydrated controls. The G-6-Pase positive ER was associated with autophagic vacuoles that accumulated in Herring bodies exhibiting a degenerating appearance. Our speculation is that the ER contributes membrane for the formation of the autophagic vacuoles, while elongated tubular profiles of lysosomes, presumably derived from dense-body lysosomes in the perikarya (4, 7, 12), contribute the acid hydrolase enzymes to the same vacuoles. These observations suggest a direct correlation between G-6-Pase activity and energy metabolism in the hypothalamo-neurohypophysial system.

REFERENCES

1. G. Alonso and I. Assenmacher, *Cell Tiss. Res.* **199**, 415 (1979).
2. T. Barka and P. J. Anderson, *J. Histochem. Cytochem.* **10**, 714 (1962).
3. J. M. Boutry and A. B. Novikoff, *Proc. Natl. Acad. Sci. USA* **72**, 508 (1975).
4. R. D. Broadwell, *J. Histochem. Cytochem.* **28**, 87 (1980).
5. R. D. Broadwell and R. H. Bleier, *J. Comp. Neurol.* **167**, 315 (1976).
6. R. D. Broadwell and M. W. Brightman, *J. Comp. Neurol.* **166**, 257 (1976).
7. R. D. Broadwell and M. W. Brightman, *J. Comp. Neurol.* **185**, 31 (1979).
8. R. D. Broadwell and A. Cataldo (1983, in press), but see abstracts in *Soc. Neurosci. Abst.* **8**, 789 (1982) and *J. Cell Biol.* **95**, 405a (1982).
9. R. D. Broadwell and C. Oliver, "Morphological Basis for the Synthesis and Packaging of Neuronal Peptides," in J. L. Barker and T. G. Smith, Eds., *Role of Peptides in Neuronal Function*, Marcel Dekker, New York, 1980, p. 23.
10. R. D. Broadwell and C. Oliver, *J. Histochem. Cytochem* **31**, 325 (1983).
11. R. D. Broadwell and C. Oliver, *J. Cell Biol.* **90**, 474 (1981).

12. R. D. Broadwell, C. Oliver, and M. W. Brightman, *J. Comp. Neurol.* **190**, 519 (1980).
13. R. D. Broadwell, C. Oliver, and M. W. Brightman, *Proc. Natl. Acad. Sci. USA* **76**, 5999 (1979).
14. M. J. Brownstein, J. T. Russell, and H. Gainer, *Science* **207**, 373 (1979).
15. C. Bugnon, D. Fellmann, and B. Bloch, *Cell Tiss. Res.* **183**, 319 (1977).
16. L. G. Caro and G. E. Palade, *J. Cell Biol.* **20**, 473 (1964).
17. M. Castel and H.-Dieter Dellmann, *Cell Tiss. Res.* **210**, 205 (1980).
18. D. Coulter and R. Elde, personal communication.
19. N. Dauwalter, W. G. Whaley, and J. F. Kephart, *J. Cell Sci.* **4**, 455 (1969).
20. S. B. Doty, C. E. Smith, A. R. Hand, and C. Oliver, *J. Histochem. Cytochem.* **25**, 1381 (1977).
21. M. P. Dubois and E. Kolodziejczyk, *C. R. Acad. Sci., Paris, Ser. D*, **281**, 1737 (1975).
22. R. E. J. Dyball, *J. Physiol.* (London) **214**, 245 (1971).
23. E. Essner, "Phosphatases," in M. A. Hayat, Ed., *Electron Microscopy of Enzymes*, Van Nostrand Reinhold, New York, 1973, p. 44.
24. E. Essner and H. Haimes, *J. Cell Biol.* **75**, 381 (1977).
25. E. Essner and C. Oliver, *Lab. Invest.* **30**, 596 (1974).
26. H. Gainer, J. T. Russell, and D. J. Fink, "The Hypothalamo-Neurohypophysial System: A Cell Biological Model for Peptidergic Neurons," in A. M. Gotto, Jr., E. J. Peck, and A. E. Boyd III, Eds., *Brain Peptides: A New Endocrinology*, Elsevier/North-Holland Biomedical Press, 1979.
27. H. Gainer, Y. Sarne, and M. Brownstein, *J. Cell Biol.* **73**, 366 (1977).
28. J. George, *Science* **193**, 146 (1976).
29. A. Goldstein and J. E. Ghazarossian, *Proc. Natl. Acad. Sci. USA* **77**, 6207 (1980).
30. R. C. Graham and M. J. Karnovsky, *J. Histochem. Cytochem.* **14**, 291 (1966).
31. A. R. Hand, *Am. J. Anat.* **130**, 141 (1971).
32. A. R. Hand and C. Oliver, *Histochem. J.* **9**, 375 (1977).
33. A. R. Hand and C. Oliver, *J. Cell Biol.* **74**, 399 (1977).
34. E. Holtzman, *Cell Biology Monographs*, Vol. 3, Springer-Verlag, New York, 1976.
35. E. Holtzman and R. Dominitz, *J. Histochem. Cytochem.* **16**, 320 (1968).
36. E. Holtzman, A. B. Novikoff, and H. Villaverde, *J. Cell Biol.* **33**, 419 (1967).
37. W-Y. Huang, K. Chang, A. J. Kastin, D. H. Coy, and A. V. Schally, *Proc. Natl. Acad. Sci. USA* **76**, 6177 (1979).
38. K. Inoue and K. Kurosumi, *Cell Struct. Funct.* **2**, 171 (1977).
39. J. D. Jamieson and G. E. Palade, *J. Cell Biol.* **34**, 577 (1967).
40. J. D. Jamieson and G. E. Palade, *J. Cell Biol.* **34**, 597 (1967).
41. J. D. Jamieson and G. E. Palade, *J. Cell Biol.* **39**, 580 (1968).
42. J. D. Jamieson and G. E. Palade, *J. Cell Biol.* **39**, 589 (1968).
43. C. W. Jones and B. T. Pickering, *J. Physiol.* (London) **203**, 449 (1969).
44. M. L. Karnovsky, *The Neural Basis of Behavior*, Spectrum, New York, 1982, p. 47.
45. C. Kent and M. A. Williams, *J. Cell Biol.* **60**, 554 (1974).
46. G. P. Kozlowski, S. Frenk, and M. S. Brownfield, *Cell Tiss. Res.* **79**, 467 (1977).
47. R. E. Mains, B. A. Eipper, and N. Ling, *Proc. Natl. Acad. Sci. USA* **74**, 3014 (1977).
48. D. J. Morre, T. W. Keenan, and C. M. Huang, *Adv. Cytopharmacol.* **2**, 107 (1974).

49. M. Neutra and C. P. Leblond, *J. Cell Biol.* **30**, 119 (1966).

50. R. S. Nishioka, D. Zambrano, and H. A. Bern, *Gen. Comp. Endocrinol.* **15**, 477 (1970).

51. A. B. Novikoff, in A. V. S. de Reuck and M. P. Cameron, Eds., *Ciba Foundation Symposium on Lysosomes*, Little, Brown, Boston, 1963, p. 36.

52. A. B. Novikoff, *Biol. Bull.* (Woods Hole) **127**, 358 (1964).

53. A. B. Novikoff, in H. Hyden, Ed., *The Neuron*, Elsevier, Amsterdam, 1967, p 255.

54. A. B. Novikoff, in H. G. Hers and F. VanHoff, Eds. *Lysosomes and Storage Diseases*, Academic Press, New York, 1973, p. 1.

55. A. B. Novikoff, *Proc. Natl. Acad. Sci. USA* **73**, 2781 (1976).

56. A. B. Novikoff, E. Essner, and N. Quintana, *Fed. Proc.* **23**, 1010 (1964).

57. A. B. Novikoff and S. Goldfischer, *Proc. Natl. Acad. Sci. USA* **47**, 802 (1961).

58. A. B. Novikoff, M. Mori, N. Quintana, and A. Yam, *J. Cell Biol.* **75**, 148 (1977).

59. A. B. Novikoff and P. M. Novikoff, *Histochem. J.* **9**, 1 (1977).

60. A. B. Novikoff, P. M. Novikoff, M. Ma, W. Shin, and N. Quintana, *Adv. Cytopharmacol.* **2**, 349 (1974).

61. A. B. Novikoff, A. Yam, and P. M. Novikoff, *Proc. Natl. Acad. Sci. USA* **72**, 4501 (1975).

62. P. M. Novikoff, A. B. Novikoff, N. Quintana, and J-J. Hauw, *J. Cell Biol.* **50**, 859 (1971).

63. C. Oliver, *J. Histochem. Cytochem.* **28**, 78 (1980).

64. C. Oliver, R. E. Auth, and A. R. Hand, *Amer. J. Anat.* **158**, 275 (1980).

65. L. Paavola, *J. Cell Biol.* **79**, 45 (1978).

66. L. Paavola, *J. Cell Biol.* **79**, 59 (1978).

67. G. E. Palade, *J. Am. Med. Assoc.* **198**, 815 (1966).

68. G. Palade, *Science* **189**, 347 (1975).

69. G. E. Palade, P. Siekevitz, and C. G. Caro, in A. V. S. de Reuck and M. P. Cameron, Eds., *Ciba Foundation Symposium on The Exocrine Pancreas*, Churchill, London, 1962, pp. 23–49.

70. G. Pelletier, *J. Ultrastruct. Res.* **43**, 445 (1973).

71. G. Pelletier and A. B. Novikoff, *J. Histochem. Cytochem.* **20**, 1 (1972).

72. J. Quatacker, *Histochem. J.* **13**, 109 (1981).

73. C. M. Redman and D. Banerjee, in D. M. Prescott and L. Goldstein, Eds., *Cell Biology: A Comprehensive Treatise*, Vol. 4, Academic Press, New York, 1980, p. 444.

74. C. M. Redman, P. Siekevitz, and G. E. Palade, *J. Biol. Chem.* **241**, 1150 (1966).

75. J. Rossier, E. Battenberg, A. Bayon, R. J. Miller, R. Guillemin, and F. E. Bloom, *Nature* **277**, 653 (1979).

76. H. Sachs, P. Fawcett, Y. Takabatake, and R. Portanova, *Recent Prog. Horm. Res.* **25**, 447 (1969).

77. H. Sachs and Y. Takabatake, *Endocrinology* **75**, 943 (1964).

78. M. Sar, W. F. Stumpf, R. J. Miller, K-J. Chang, and P. Cuatrecasas, *J. Comp. Neurol.* **182**, 17 (1978).

79. A. P. Scott, J. G. Ratcliffe, L. H. Rees, J. Landon, H. P. J. Bennett, P. J. Lowry, and C. McMartin, *Nature New Biol.* **244**, 65 (1973).

80. A. J. Silverman, *J. Histochem. Cytochem.* **24**, 816 (1976).

81. M. Stachura and L. A. Frohman, *Endocrinology* **94**, 701 (1974).

82. D. F. Steiner, W. Kemmler, H. S. Tager, and J. D. Peterson, *Fed. Proc.* **33**, 2105 (1974).
83. L. Sternberger, *Immunocytochemistry*, 2nd ed., Wiley, New York, 1979.
84. A. C. Trakatellis, K. Tada, K. Yamaji, and P. Gardiki-Kouidou, *Biochemistry* **14**, 1508 (1975).
85. F. Vandesande, K. Dierickx, and N. Goossens, *J. Histochem. Cytochem.* **28**, 469 (1980).
86. S. J. Watson, H. Akil, V. E. Ghazarossian, and A. Goldstein, *Proc. Natl. Acad. Sci. USA* **78**, 1260 (1981).
87. W. G. Whaley, N. Dauwalter, and J. E. Kephart, in J. Reinert and H. Ursprung, Eds., *Origin and Continuity of Cell Organelles*, Springer-Verlag, Heidelberg, Germany, 1972, p. 1.
88. P. Whur, A. Herscovics, and C. P. Lebond, *J. Cell Biol.* **43**, 289 (1969).
89. E. A. Zimmerman, in L. Martini and W. F. Ganong, Eds., *Frontiers in Neuroendocrinology*, Vol. 4, Raven Press, New York, 1976, p. 25.

Chapter 6

Simultaneous Analysis of Monoaminergic and Peptidergic Neurons

John R. Sladek, Jr. and Gloria E. Hoffman

Department of Anatomy and Center for Brain Research
University of Rochester School of Medicine
Rochester, New York

1 CHEMICALLY IDENTIFIED NEURONS

In 1962, Bengt Falck and Nils-Åke Hillarp introduced a technique for the intracellular demonstration of monoamine neurotransmitters (1). This method, which has become known as the Falck–Hillarp or formaldehyde-induced fluorescence technique, enables the direct visualization of the catecholamines norepinephrine and dopamine and the indoleamine serotonin within the central and peripheral nervous system. A number of examinations quickly followed and delineated major ascending and descending monoaminergic pathways, the perikarya of which originated in the brain stem and hypothalamus (2, 3). These systems were characterized by far-reaching projections to virtually all parts of the neuraxis with prominent innervation patterns in the mammalian hypothalamus and preoptic area (3–10). The significance of these findings to the concept of neuron interactions was that some of these areas of innervation, even in the early 1960s, were considered prime sites for the location of peptidergic neurons.

With the advent of immunohistochemistry in the early 1970s, our knowledge of chemical neuroanatomy of the central nervous system expanded very rapidly. To date, more than 20 different central peptidergic neuron systems have been discovered (11) and many of these appear within regions which also house monoamine terminals and/or cell bodies. Further immunocytochemical techniques made it possible to examine monoaminergic systems by localizing their synthetic enzymes as well as the monoamine itself (12, 13).

The independent utilization of histofluorescence and immunocytochemical techniques for the localization of centrally occurring transmitters led to considerable speculation that these neuron systems were functionally interactive. For example, noradrenergic terminals are distributed abundantly in the supraoptic and paraventricular nuclei of the hypothalamus (4, 8, 10). These nuclei are composed of neurons that contain the peptides vasopressin and oxytocin and their associated neurophysins (14, 15). These morphological data suggested that norepinephrine could have an action on the magnocellular system. Subsequently, norepinephrine was shown to inhibit vasopressin release under a variety of conditions, lending credence to the notion that the anatomical relationship suggested is functional (16–20).

In order to establish the morphological basis for interactions between monoamine and peptide systems, histofluorescence and immunocytochemical techniques have been modified to allow the simultaneous visualization of these different substances (21–26). One striking advantage of a colocalization technique over independent analyses performed either on separate brains or on opposite sides of a single brain is the elimination of potential asynchronies or brain asymmetries as technical problems. For example, careful analysis of Björklund's initial study on the ontogenetic aspect of dopamine development in the hypothalamus of the mouse (27) revealed that even among littermates, the onset and maturation of fiber fluorescence in the median eminence was variable. If one attempted to localize and correlate the appearance of LHRH or another releasing factor within this same region and utilized littermates, it would in all probability be difficult to establish a precise pattern of developmental interactions between dopamine and LHRH. This is especially important when one considers, for example, the hypothesis proposed by Lauder and coworkers concerning the possible effect of the early differentiation of monoamine neurons on the promotion of differentiation of their subsequent targets (28–30). Through a series of extremely novel experiments, they were able to demonstrate that interruption of the early prenatal differentiation of catecholamine, as well as indoleamine neurons, resulted in a substantial delay in the differentiation of the majority of target neurons examined with birthdating techniques. A colocalization method would permit the further testing of this possible interaction by the simultaneous examination of afferent fibers and their target neurons as indexed by their chemical content.

For this and other reasons, an attempt was made to modify existing techniques; this resulted, in the summer of 1977 (22), in the presentation of the colocalization of monoamines and neuropeptides in a single microscope field from a single tissue preparation comparing adjacent tissue sections. These data, first presented at the Third Brain–Endocrine Interaction Symposium in Wurzburg, West Germany, graphically illustrated the juxtaposition of catecholamine varicosities to oxytocin and vasopressin neurons, as indexed by staining for neurophysin within the supraoptic and paraventricular nuclei of the hypothalamus. Subsequently, more extensive examination revealed the precise innervation pattern of vasopressin and oxytocin neurons, particularly in rodent and primate brain (31–35), and further analyzed the colocalization of dopamine and LHRH (21, 36–37) as well as dopamine and another parvicellular peptide, somatostatin (38). These studies have been extended into developmental (39), gerontological (40–42), and neural transplantation experiments (43, 44).

The methodological approach applied to the examination of magnocellular peptides is favored by the relatively large size of oxytocin and vasopressin nerve cell bodies. This, coupled with the excellent staining affinity of neurophysin in freeze-dried tissue (45) has made examination of the magnocellular system much simpler than that of many parvicellular systems. In part because of the difficulty of applying this approach to perikarya of parvicellular systems, a second colocalization technique was developed, using the peroxidase-antiperoxidase (PAP) technique with dual, primary antisera and chromogens, and it now affords a much clearer pattern of the colocalization of the catecholamines and any of a number of parvicellular peptides (41).

The LHRH system, for example, is a widely dispersed neuronal network (46, 47) that spreads throughout the rostral diencephalon. Its neurons, unlike the magnocellular system, are extremely small and do not localize well in freeze-dried tissue, possibly due to the extraction of the peptide during exposure of the tissue to hot paraffin and organic solvents (45). Yet the need for studying interactions of this system is great. There had been a growing body of evidence that indicated norepinephrine was the excitatory trigger for release of LHRH (49–60). In addition, dopamine appeared to influence LH and FSH release but for this monoamine, the action could be either facilitatory or inhibitory (52, 57, 61–79). The possible dual action of dopamine suggested that dopaminergic systems might innervate the LHRH system at two different anatomical sites, at the level of terminals in the median eminence and/or directly upon perikarya and dendrites in the medial preoptic area-septal LHRH field. What was known about the distribution of LHRH and the catecholamines provided little predictive value for possible interactions either because the catecholamine patterns in the regions of the LHRH cells were sparse or in transition (48). Even within the median eminence where dense patterns of both LHRH and dopamine occur, the patterns appeared sufficiently different to raise some uncertainty concerning possible axonal interaction. Although description of the relationship of dopamine and LHRH in the median eminence was possible with the freeze-dried paraffin-embedded material as described for the magnocellular interactions (21–24), attempts to colocalize catecholamine varicosities and LHRH cell bodies in the medial preoptic area of the guinea pig, for example, resulted in the tedious reconstruction of these systems (36).

What follows is a detailed methodology for each of these two approaches, with a brief summary of some of the advantages and disadvantages of each technique.

2 THE COMBINED FORMALDEHYDE HISTOFLUORESCENCE AND PAP IMMUNOCYTOCHEMICAL TECHNIQUE

This approach is basically a modification of the Falck–Hillarp histofluorescence technique for monoamines (1). It essentially involves the usual paradigm of tissue collection, immediate rapid freezing, and subsequent freeze-drying. Formaldehyde gas vapors create a fluorogenic substance from the endogenous monoamines, and following paraffin-embedding these blocks are rotary microtomed and examined in an adequately equipped fluorescence microscope. Our addition to this methodology is the ability to stain these same sections for any of a number of different peptides or proteins with PAP immunohistochemistry. Two modifications of the technique have been developed. One permits the simultaneous examination of peptides and monoamines from two adjacent tissue sections in an instrument designated as a comparator bridge microscope (22, 80), and the second allows a sequential visualization of first, histofluorescence, and then immunohistochemistry on a single tissue section with a photographic reconstruction of the colocalization patterns (33).

The methodological approach follows. Animals are decapitated to provide a bloodless field for dissection. Brains are removed rapidly, blocked to 2–5 mm slabs, and placed on dental wax for immediate freezing in Freon 22 prechilled to −100°C in liquid nitrogen. Each tissue block is sealed within a paper embedding bag and then can either be stored in liquid nitrogen indefinitely or processed for freeze-drying for a period of time from one to several weeks depending upon the size of the sample. Following freeze-drying, blocks are slowly equilibrated to room temperature over a 1 hr period and then are superheated with the use of a 50°C waterbath surrounding the freeze-drying tissue chamber. The freeze-drying system and its utilization are described elsewhere (81). Brain samples are then treated for 2 hr at 80°C with hot paraformaldehyde vapor that has been generated from about 5g of paraformaldehyde which was equilibrated for at least two weeks at a relative humidity of 65% (82). Tissue blocks are then removed from the fluorophor-generating chamber and are paraffin-embedded according to a modification of the procedure introduced by Björklund and Falck (83). This involves a separate evacuation of the hot paraffin and the tissue samples followed by a short exposure in paraffin. Failure to observe this step will result in the extraction of fluorophors, especially from fine-sized fibers (which accounts for their absence in some histochemical preparations using the formaldehyde technique). Sections then are ready for serial microtomy, usually performed on a rotary microtome. We have been successful at cutting sections as thin as 3–4 μm although the usual thickness is in the 6–10 μm range. Thicker sections produce a busier microscopic field with higher background fluorescence than do thinner sections. Thin sections optimally prepared with the modified embedding technique reveal fine-sized varicosities in apposition to suspected peptidergic neurons (Figure 1).

Sections are stored as ribbons in light-tight, dry containers and will usually retain histofluorescence for several months in reasonably dry climates. Peptide antigenicity remains much longer; we have successfully stained sections that have been in storage for several years. For fluorescence microscopy, sections are pressure mounted onto alcohol-cleaned slides. Immediately before examination in the fluorescence microscope, they are gently heated on a warming tray until a point where paraffin begins to liquefy. At this time they are covered with nonfluorescent immersion oil and are viewed with oil immersion lenses in a properly equipped fluorescence microscope. Optimal fluorescence intensity and contrast is achieved with epi-illumination utilizing the Leitz D-cube system. This is a narrow-band excitation (405 nm) which, coupled with low-band barrier filters (460), allows the true blue color of the catecholamine fluorophor to be easily distinguished from the yellow hue of serotonin (84, 85).

For the purpose of colocalization of peptides, one of two approaches may be followed. The first is the analysis of the same section utilized for histofluorescence, which has been described in detail elsewhere (33).

With this method, sections are photographed and then cleared of immersion oil with the use of xylene. They then are rehydrated through alcohols into phosphate-buffered saline. Each section is stained immunohistochemically according to the PAP technique (86) described in detail below. Following immunohistochemical staining the fluorescence micrographs are then compared with the microscopically observed peptidergic perikarya or fibers, the same field is located, and subsequent photographs are taken for subsequent analysis. This approach allows the identification of a peptide within a neuron and the analysis of the degree to which catecholamine or indoleamine fibers may juxtapose to that chemically identified target neuron. Although this technique offers the security of coidentification in a single tissue section, it is an exceptionally time-consuming and expensive process. For example, colocalizations within extremely large neural loci or within peptidergic areas that are widely scattered throughout the brain such as those for somatostatin and LHRH would, of necessity, involve extensive photographic reconstructions. Also, the best resolution and detail is achieved with the higher magnification objective lenses that reduce the size of the overall field of view. Thus, an analysis of LHRH or somatostatin neurons would require large numbers of photographic montages to achieve the identification of the colocalizations. For this and other reasons, a second approach was developed that involves the analysis of adjacent tissue sections through the use of a comparator bridge microscope (22). Once a target area is located through routine staining with Cresyl violet and Luxol-fast blue, sections are taken for fluorescence and immunochemistry. A single section, mounted as described above, is placed in a fluorescence microscope. The fluorescence section, adjacent to that which has been immunohistochemically stained for a particular peptide (as described below), is placed in another, optically identical microscope. The two microscopes are linked by a comparator bridge which yields a single field of view of either the two adjacent images or, when properly aligned, a superimposed view of the two images (Figure 1). The usual procedure follows. The immunocytochemical field is located with brightfield illumination. Then, in order to reduce background to allow the visualization of a fluorescence overlay, a darkfield illumination is employed. This imparts an orange color to the peptidergic cell bodies, processes, and terminals. Backgrounds are dark. The same field is found in the histofluorescence preparation utilizing white-light, darkfield illumination. This allows identification of the field and minimizes any blue-light fading during this illumination procedure. Proper alignment involves not only the rotation of one slide through a rotary stage into the plane of view of the other field, but also a compensation for differences in the size of each section. Fluorescence

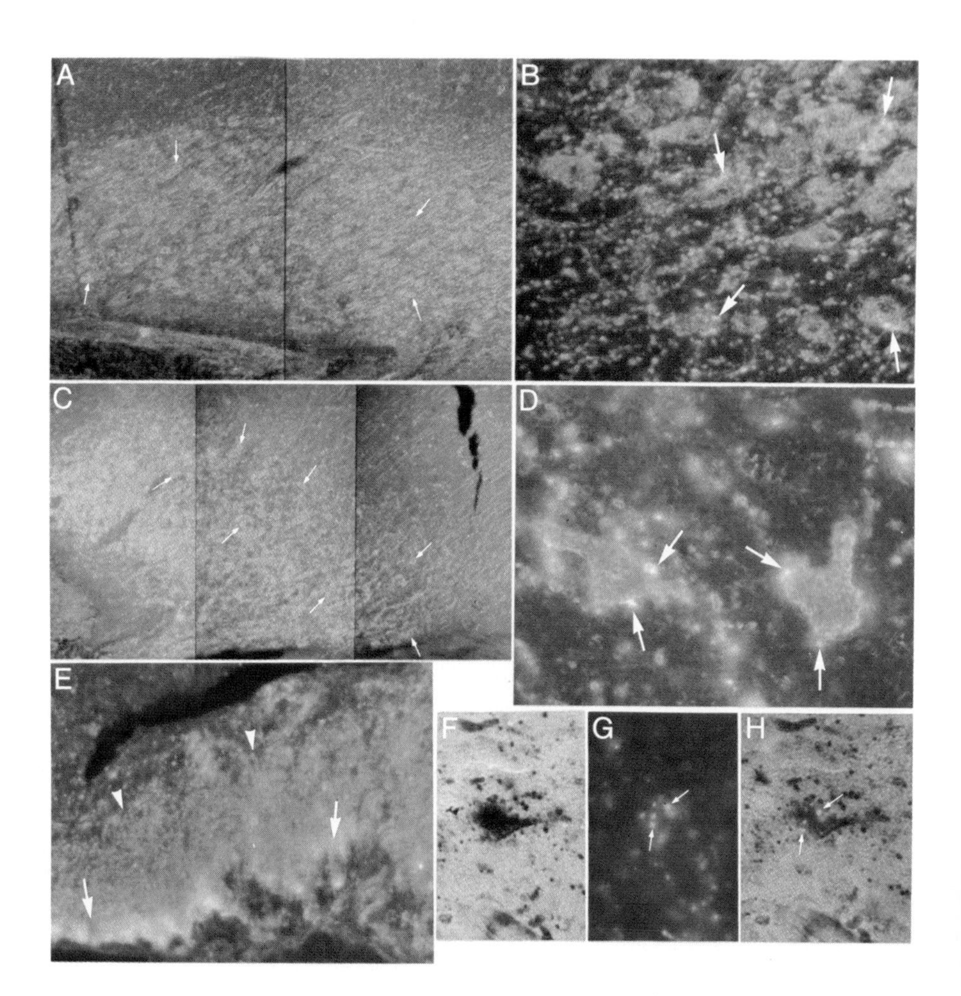
A
B
C
D
E
F
G
H

Figure 1. (A) Vasopressin-containing neurons (→) are distributed within a dense field of noradrenergic varicosities in the supraoptic nucleus of the rhesus monkey. This combined view illustrates the widespread distribution of vasopressin in neurons throughout most parts of the noradrenergic field. Magnification × 130. (B) At a higher magnification, vasopressin neurons are filled with dense reaction product, seen here as an orange material as illustrated with darkfield illumination. Catecholamine varicosities are abundant and often appear in juxtaposition to the vasopressin-containing perikarya (→). Magnification × 380. From Sladek and Zimmerman (35). (C) In this low-power photomontage, the supraoptic nucleus of the rhesus monkey is seen to contain a group of oxytocin neurons located in the dorsomedial portion of the nucleus. This group (→) appears somewhat apart from the dense noradrenergic zone in the lateral portion of the nucleus. Magnification × 80. (D) Oxytocin-containing neurons also appear to be innervated by noradrenergic varicosities (→). Magnification × 550. From Sladek and Zimmerman (35). (E) This combined view in the median eminence of the rat illustrates somatostatin (▶), seen as a dense band of orange reaction product in the fibrous zone of the median eminence, and dopamine (→), which appears as a band of blue histofluorescence in the external zone of the median eminence. This parvicellular peptide as well as another, LHRH, does not show significant overlap with the dopaminergic zone. Magnification × 375. (F) An LHRH-containing cell body is seen in a 10-μm paraffin-embedded section of the medial preoptic-septal area in guinea pig. The nucleus of the neuron is not evident because of the small size of the perikaryon, which is probably contained in its entirety within this tissue section. Magnification × 670. Photomicrograph from an unpublished study by A.-J. Silverman and J. Sladek (36). (G) On an adjacent tissue section to that seen in (F), catecholamine fluorescence (→) appears over the polar region of the LHRH neuron. Intense catecholamine varicosities are evident. Magnification × 670. (H) In a combined view of (E) and (F), the intense varicosities (→) are seen to reside over the polar cap of the LHRH neuron. This may suggest a morphological site of interaction between catecholamines and LHRH at the level of the perikaryon in addition to the suspected interaction in the median eminence. Magnification × 670.

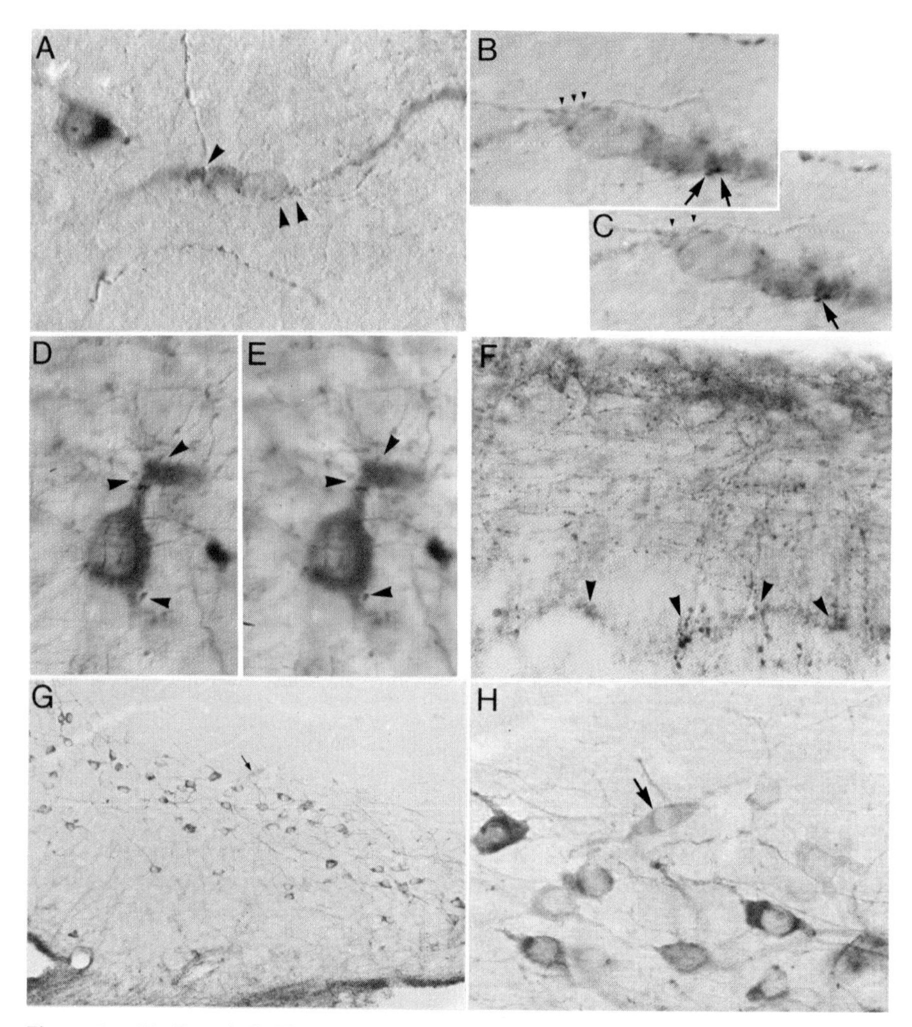

Figure 2. (A, B and C) This tissue was processed sequentially for the localization of LHRH (diaminobenzidine) and TH (4 Cl naphthol). Catecholamine fibers (blue) were found in apposition to an LHRH perikaryon (brown). The sites of possible contact are indicated by arrowheads. (B) and (C) show this interaction at a higher magnification in two different focal planes. One of the catecholamine axons appears to contact the LHRH cell with boutons en passant (▶). This pattern of interaction was characteristic of the majority of TH axons seen in juxtaposition to LHRH neurons. Magnification (A) × 380, (B) and (C) × 960. (D and E) With reversal of the sequence of antigen colocalization, i.e., when TH localization (diaminobenzidine) preceded LHRH localization (4 Cl naphthol), the staining intensity of the LHRH neurons is reduced, while that seen for the fibers is still intense. In the arcuate nucleus, which contains dopamine, LHRH axons (blue) were found in apposition to catecholamine neurons (brown) (▶). Comparing panels (D) and (E), one can see how Nomarski optics (D) facilitates the visualization of apparent contacts between substances. Magnification × 960. (F) The distribution of LHRH (brown) and TH (blue) is characterized by a separate organization within the contact zone of the median eminence in rat. In a few sites, overlap in the staining patterns was noted (▶). Magnification × 960. (G and H) Generally the arcuate nucleus in rat was devoid of LHRH neurons; however, an occasional LHRH neuron (brown) was seen (→). TH was stained blue. Panel (H) shows the isolated LHRH neuron at a higher magnification. Note that mixing of the two colors does not occur. Magnification (G) × 100, (H) × 240.

sections tend to be 1.25 times greater in size than the rehydrated, immunocytochemically stained sections. Compensation is achieved with a Leitz Variotube, which has a continual zoom lens from 1.0 to 3.2 × magnification. Once the field is properly aligned and compensated for magnification, fluorescence illumination can be applied to the unstained section to produce a side-by-side image of histofluorescence and peptide staining. These fields of view then can be superimposed by the simple realignment of the two sections. Increasing magnifications reveal greater detail and are easily achieved.

Immunocytochemical staining basically follows the PAP technique. Sections are rehydrated in phosphate-buffered saline and the unlabeled antibody immunoperoxidase technique of Stenberger (86) is applied. Several different antisera have shown positive staining in our studies; these include LHRH, somatostatin, vasopressin, oxytocin, and several neurophysins. All antisera are initially used at a dilution of 1:1000 for 24 hr. The neurophysin antisera are usually used at a dilution of 1:3000 for 24 hr. In subsequent steps of the immunoperoxidase bridge technique, sections are incubated with sheep antiserum to rabbit gamma globulin (anti-rabbit gamma globulin) and rabbit PAP purchased commercially. Sections are immersed in a solution of $3,3^1$-diaminobenzidine tetrahydrochloride and 0.003% H_2O_2. Sections are then dehydrated, cleared, and cover-slipped according to standard protocol.

3 THE DUAL IMMUNOPEROXIDASE TECHNIQUE

The previous technique is difficult and sometimes impractical to use for studying catecholamine–peptide interactions when (i) the innervated neurons are too small for serial section reconstruction, (ii) they are too widely dispersed to permit single section analysis, or (iii) they lose a significant amount of perikaryal peptide through extraction. Instead, an adaptation of the basic immunocytochemical method for dual localization can be used (46) to provide simultaneous localization in a single preparation (Figure 2).

Animals are anesthetized with pentobarbitol at 60–80 mg/kg i.p. Heparin is administered intracranially following the attainment of surgical anesthesia. The descending abdominal aorta is clamped and a Teflon cannula is inserted through the left cardiac ventricle through the aorta to ensure optimal perfusion. Saline is utilized as a flush and is immediately followed by Zamboni fixative for approximately 20 min. Each brain is then removed and post-fixed in Zamboni for 2 hr and subsequently blocked into 8 mm slabs in the anatomical plane of choice. Each block is glued to an aluminum chuck with a superglue such as Eastman 910 or Elmer's Wonderbond and is ready for sectioning with a vibrating microtome. Sections are cut at a thickness from 20–30 μm and are floated onto a bath of potassium phosphate-buffered saline (0.05 *M*, pH 7.4). These sections are rinsed several times in this saline and then are ready for immunohistochemical staining. Sections are processed according to a modification of the procedure of Grzanna et al. (87), utilizing any of a number of antisera; the catecholamine-synthesizing enzyme tyrosine hydroxylase (TH) is used for the localization of catecholamines. In order to simultaneously examine the distribution of catecholamine and peptide neurons, the tissue sections are processed sequentially for the localization of each particular antigen. Sections are treated for one antigen at an appropriate dilution (usually 1:1000 for anti-TH or 1:40,000 for our anti-LHRH) in phosphate-buffered saline containing 0.4% to 1.0% Triton X-100. Sections are incubated at 4°C for 16–24 hr and are subsequently rinsed in phosphate-buffered saline containing 0.02% Triton X-100. Ten rinses at 10 min each are performed. Sections then are incubated for 1 hr in sheep anti-rabbit IGG at a dilution of 1:40 in phosphate-buffered saline with 0.02% Triton X-100. After four 15 min rinses in this buffer, sections are incubated for 1 hr in PAP complex at 1:100 in phosphate-buffered saline at 0.02% Triton X-100. Four rinses later the sections are rinsed twice more in Tris-buffered saline (0.05 *M*, pH 7.2) and are transferred to 1 mL Tris-buffered saline to which 1 mL of 3,3^1-diaminobenzadine HCL solution (0.5g/mL in Tris buffer) containing 0.045% H_2O_2 has been added and incubated for 3–6 min at room temperature. For dual antibody staining, sections are rinsed in phosphate-buffered saline and then transferred to the primary antibody for the second antigen, which has been diluted in phosphate-buffered saline containing 1% Triton X-100. The sections are then incubated for 16–24 hr at 4°C. The same procedure described above for single antigen immunohistochemistry is then followed to the staining step. For visualization of the second antibody complex, sections are placed in 20–25 mL of a solution containing 0.435mg/mL 4-Cl naphthol in Tris-buffered saline. This solution is prepared by dissolving 100 mg 4-Cl naphthol in 5 mL absolute alcohol, adding 225 mL Tris-buffered saline and filtering just before use. After the addition of 0.1 to 0.15 mL of 3% H_2O_2, the sections are incubated for 6 min and are transferred into saline with several rinses and ultimately mounted on subbed glass slides. Sections are then air dried overnight and mounted with Apathie's water-soluble mounting media and are cover-slipped. To ensure that the color of the 4-Cl naphthol reaction is not caused by unreacted peroxidase from the first antibody action, or the attachment of anti-rabbit immunoglobulin with the primary antibody, or antiperoxidase from the first series of reactions, a control is employed as follows. A series of sections are analyzed in which the tissue was processed for the first antigen with 3,3′-diaminobenzidine and subsequently the second series was performed with the elimination of the second primary antibody or application of the first primary antibody. In this way it is possible to verify that the 4-Cl naphthol reaction, which appears blue in the double-staining procedure, is due only to the activity of the second antigen. Specificity of the antisera is tested by the addition of purified GH (2–5 μg/mL diluted antisera) to the antisera before incubating the tissue. Preimmune serum is used to test for selectivity of the TH antibody. Complete blockage of staining is the criterion for specificity.

4 EVALUATION OF THE METHODS

The two methods described offer the investigator a means of determining the light microscopic, morphological basis for interactions between monoamines and neuropeptides. In certain respects, the information that can be gained differs for each approach. The histofluorescence-immunocytochemical method assesses the presence of catecholamines per se within their neuroanatomical loci. The dual immunocytochemical technique localizes the catecholamine synthetic enzyme as a marker for the catecholamine system. Thus, studies aimed at assessing differences in the dynamics of the catecholamine–peptide interaction probably are best approached with the direct visualization of the transmitter instead of the synthetic enzyme, as a functionally induced change in turnover is more likely to be manifested in the transmitter. Also, histofluorescence is more easily coupled to quantitative assessments such as microspectrofluorometric or stereological analyses of innervation patterns (88–91) than is immunohistochemistry. For studies in which a qualitative analysis of the innervation is desired, the use of synthetic enzyme marker is preferred. Likewise, when the peptide under investigation is labile in conditions of paraffin-embedding, the dual immunocytochemical approach succeeds by eliminating the need for exposure of the tissue to heat or organic solvents. A major drawback in selecting the dual immunocytochemical approach is the lack of general availability of antisera to the enzyme markers. As a result, most data concerning the interactions of catecholamines and peptides has relied on the histofluorescence method described. The technique has been used extensively to map the colocalization of these substances during ontogeny and aging, with findings of concurrent changes in each system examined. For example, dopamine histofluorescence in the arcuate nucleus increases in aged Fischer 344 rats, whereas dopamine fluorescence in the median eminence as well as LHRH terminal staining in the median eminence decreases (41, 92). While these changes may be due to a cause and effect relationship between peripheral circulating hormone levels that feed back onto the dopaminergic neurons of the arcuate nucleus, it is equally probable that the changes represent inherent, age-related declines in the ability of dopamine neurons to transport newly synthesized perikaryal transmitter.

The application of the first method to the study of interactions during ontogeny revealed that some peptidergic neurons of the hypothalamus appear to synthesize and store their messengers at an early prenatal age (93), prior to the ingrowth of aminergic axons (39). This has led to an hypothesis that the peptidergic neurons, or possibly the peptide or something associated with it, may be responsible, in part, for the attraction of the adrenergic fiber to the target neuron. While this point awaits proof, it is interesting to note that the Brattleboro rat, a genetic mutant that lacks vasopressin but contains a full complement of neurons in the combined vasopressin-free, oxytocin-rich target region of the hypothalamus, exhibits a sharply reduced noradrenergic innervation pattern in that region (94, 95). This may suggest further that the peptidergic neuron is responsible for the maintenance of an appropriate innervation pattern. This has also been noted in the age-depleted normal rat, which contains reduced vasopressin staining and a reduction in the associated noradrenergic innervation pattern (42).

The formaldehyde histofluorescence-immunocytochemical technique also has been applied to the study of neuronal development following grafting through the work of Gash and colleagues. Here, vasopressin neurons transplanted into a host brain show a remarkable growth of neuronal processes, not only within the transplanted tissue but also into the host tissue, especially in the vicinity of blood vessels of the hypothalamo-hypophysial portal system (43, 96). Utilizing the colocalization technique, it also has been noted that catecholaminergic fibers appear to grow into the transplant from the host median eminence, which is rich in dopamine and to some extent norepinephrine, and from the host periventricular stratum, which contains a dense, noradrenergic innervation primarily arising from the locus coeruleus. With adequate post-operative time, these fibers have been seen in juxtaposition to neurophysin-containing perikarya in the donor tissue (44). Without a colocalization technique, such kinds of analyses in this experimental situation could not be made.

Although the dual immunocytochemical method has not yet been widely used, it, too, offers the promise of broad application to the area of monoaminergic interactions with peptidergic systems. The recent successes in generating antibodies against the monoamines themselves, as was done for serotonin (13), additionally have provided a possible means of examining specific monoamines in conjunction with peptide systems without having to contend with fluorophore fading. Also, a great deal of interest has focused recently on the possibility that two neurotransmitters are present within the same neuron. Ljungdahl and coworkers recently utilized a double immunocytochemical approach on adjacent frozen sections that revealed neurons containing serotonin and substance-P (97). Similar approaches have revealed the coexistence of various other combinations of peptides or monoamines (25, 98). The techniques described in this study are likewise adaptable for that purpose.

In summary, it appears that during the past 20 years the central nervous system has been further divided based upon the chemical constituents used for communication by neuron systems. This often has led to speculation of neuron interactions based on the observation that different substances occur within the same nuclear region. Complicating the issue is that chemically identified neurons do not always follow the known neuroanatomy of the brain—they may be located within and around any particular nucleus. Such is the case of norepinephrine in the A1 region of the lateral reticular nucleus and adjacent medullary reticular formation. Clearly, the lateral reticular nucleus is a heterogeneous group with only a small proportion of its neurons containing norepinephrine. Therefore, if another substance were localized in terminals in that region or in other nerve cell bodies in that region, it would be difficult, if not impossible, to predict a colocalization pattern unless each substance could be visualized simultaneously. Likewise, the paraventricular and to a lesser extent the supraoptic nuclei are extremely heterogeneous loci with regard to their neurochemistry (99). A number of different peptides and monoamines have been localized to perikarya and terminals in and around each nucleus. Other hypothalamic sites such as the median eminence also contain a host of monoamines and peptides. Of particular relevance, the LHRH neurons are widely dispersed, unlike more compact and chemically defined loci such as the locus coeruleus and supraoptic nucleus. This feature prohibits the accurate prediction of chemical interactions, which can be made more reliably with a combined localization technique.

ACKNOWLEDGMENT

This work was supported by USPHS grants AG 00847, NS 15816, NS 13725, Research Career Development Award NS 00321, and NSF grant

BNS 78 11153. We are indebted to Susan Wray for valuable assistance in the preparation of Figure 2 and to Judy VanLare, Barbara Blanchard, Julie Fields, and Laurie Koek for skilled technical assistance. Neurophysin antisera were generously supplied by Drs. Earl Zimmerman and Alan Robinson.

REFERENCES

1. B. Falck, N.-A. Hillarp, G. Thieme, and A. Torp, *J. Histochem. Cytochem.* **10,** 348–354 (1962).
2. A. Dahlström and K. Fuxe, *Acta Physiol. Scand.* (Suppl. 232) **62**, 1–55 (1965).
3. L. Fuxe, T. Hökfelt, and U. Ungerstedt, *Adv. Pharmakol.* **6,** 235–251 (1968).
4. K. Fuxe, *Acta Physiol. Scand.* (Suppl. 247), **64**, 39–85 (1965).
5. N. E. Andén, A. Dahlström, K. Fuxe, K. Larsson, L. Olson, and U. Ungerstedt, *Acta Physiol. Scand.* **67,** 313–326 (1966).
6. A. Björklund, *Z. Zellforsch.* **89,** 573–589 (1968).
7. U. Ungerstedt, *Acta Physiol. Scand.* (Suppl. 367) 1–48 (1971).
8. Y. Cheung and J. R. Sladek, Jr., *J. Comp. Neurol.* **164,** 339–360 (1975).
9. D. L. Felten, *J. Neuraltransm.* **39**, 269–280 (1976).
10. G. E. Hoffman, D. L. Felten, and J. R. Sladek, Jr., *Am. J. Anat.* **147,** 501–514 (1976).
11. T. Hökfelt, O. Johansson, Å. Ljungdahl, J. Lundberg, and M. Schultzbert, *Nature* **284**, 515–521 (1980).
12. T. Hökfelt, R., Elde, K. Fuxe, O. Johansson, A. Ljungdahl, R. Goldstein, S. Luft, G. Efendic, L. Nilsson, L. Terenius, D. Ganten, S. L. Jeffcoate, J. Rehfeld, S. Said, M. Perez de la Mora, L. Possani, R. Tapia, L. Teran, and R. Palacios, "Aminergic and Peptidergic Pathways in the Nervous System with Special Reference to the Hypothalamus," in S. Reichlin, R. J. Baldessarini, and J. B. Martin, Eds., *The Hypothalamus*, Research Publications, Vol. 56, Association for Research in Nervous and Mental Disease, Raven Press, 1978, pp. 69–135.
13. H. W. M. Steinbusch, *Neuroscience* **6,** 557–618 (1981).
14. E. A. Zimmerman, A. G. Robinson, M. K. Husain, A. Acosta, A. Frantz, and W. H. Sawyer, *Endocrinology* **95,** 931–936 (1974).
15. D. Swaab, C. Pool, and F. Nijveldt, *J. Neural Transmission* **36,** 195–215 (1975).
16. J. L. Barker, J. W. Crayton, and R. A. Nicoll, *Science* **171,** 208–210 (1971).
17. F. E. Hoffman, M. I. Phillips, and P. Schmid, *Neuropharmacology* **16,** 563 (1977).
18. A. Urano and H. Kobayashi, *Exp. Neurol.* **60,** 140 (1978).
19. W. E. Armstrong, C. D. Sladek, and J. R. Sladek, Jr., *Endocrinology III*, **111**, 273–279 (1982).
20. C. D. Sladek, W. E. Armstrong, and J. R. Sladek, Jr., "Norepinephrine Control of Vasopressin Release," in Y. Ibata and E. A. Zimmerman, Eds., *Structure and Function of Peptidergic and Aminergic Neurons*, Wiley, New York, 1983, in press.
21. T. H. McNeill and J. R. Sladek, Jr., *Science* **200,** 72–74 (1978).
22. J. R. Sladek, Jr., C. D. Sladek, T. H. McNeill, and J. G. Wood, "New Sites of Monoamine Localization in the Endocrine Hypothalamus As Revealed by New Methodological

Approaches," in D. E. Scott, G. P. Kozlowski, and A. Weindl, Eds., *Brain-Endocrine Interaction III: Neural Hormones and Reproduction*, Karger, Basel, 1978, pp. 154–171.

23. K. Ajika, *J. Anat.* **128,** Part 2, 331–348 (1979).
24. Y. Ibata, K. Watanabe, H. Kinoshita, S. Kubo, Y. Sano, S. Sin, E. Hashimura, and K. Imagawa, *Neurosci. Lett.* **11,** 181–186 (1979).
25. T. Hökfelt, L. Skirboll, J. Rehfeld, M. Goldstein, K. Markey, and O. Dann, *Neuroscience* **5** (12) 2093 (1980).
26. R. P. Johnson, M. Sar, W. E. Stumpf, *Brain Res.* **194** (2), 566 (1980).
27. A. Björklund, A. Enemar, and B. Falck, *Z. Zellforsch.* **89,** 590–607 (1968).
28. J. M. Lauder and F. E. Bloom, *J. Comp. Neurol.* **155,** 469–482 (1974).
29. J. M. Lauder and F. E. Bloom, *J. Comp. Neurol.* **163,** 251–264 (1975).
30. J. Lauder and H. Krebs, *Brain Res.* **107,** 638–644 (1976).
31. J. R. Sladek, Jr., E. A. Zimmerman, J. L. Antunes, and T. H. McNeill, "Integrated Morphology of Catecholamine and Neurophysin Neurons," in E. Usdin, I. J. Kopin, and J. Barchas, Eds., *Catecholamines: Basic and Clinical Frontiers*, Pergamon Press, New York, 1978, pp. 1325–1327.
32. T. H. McNeill and J. R. Sladek, Jr., *J. Comp. Neurol.* **193,** 1023–1033 (1980).
33. J. R. Sladek, Jr., and T. H. McNeill, *Cell Tiss. Res.* **210,** 181–189 (1980).
34. J. R. Sladek, Jr., T. H. McNeill, H. Khachaturian, and E. A. Zimmerman, "Chemical Neuroanatomy of Monoamine-Neuropeptide Interactions in the Hypothalamic Magnocellular System," in S. Yoshida, L. Share, and K. Yagi, Eds., *Antidiuretic Hormone*, Japan Scientific Societies Press, Tokyo, 1980, pp. 3–17.
35. J. R. Sladek, Jr., and E. A. Zimmerman, *Brain Res. Bull.* **9**, 431–440 (1982).
36. A. J. Silverman and J. R. Sladek, Jr., *Soc. Neurosci. Abstr.* **4,** 411 (1978).
37. T. H. McNeill, D. E. Scott, and J. R. Sladek, Jr., *Peptides* **1,** 59–68 (1980).
38. L. Koek and J. R. Sladek, Jr., *Anat. Rec.* **202,** 101A (1982).
39. H. Khachaturian and J. R. Sladek, Jr., *Peptides* **1,** 77–95 (1980).
40. J. R. Sladek, Jr., T. H. McNeill, P. Walker, and C. D. Sladek, "Age-related Alterations in Monoamine and Neurophysin Systems in Primate Brain," in D. M. Bowden, Ed., *Aging in Non-Human Primates*, Van Nostrand Reinhold, New York, 1979, pp. 80–99.
41. G. E. Hoffman and J. R. Sladek, Jr., *Neurobiol. Aging* **1,** 27–38 (1980).
42. J. R. Sladek, Jr., H. Khachaturian, G. E. Hoffman, and J. Schöler, *Peptides* **1,** Suppl. 1, 141–157 (1980).
43. D. M. Gash, J. R. Sladek, Jr., and C. D. Sladek, *Science* **210,** 1367–1369 (1980).
44. D. Gash, C. D. Sladek, and J. R. Sladek, Jr., *Peptides* **1,** Suppl. 1, 125–134 (1980).
45. C. J. Clayton, T. H. McNeill, and J. R. Sladek, Jr., *Cell Tiss. Res.* **220,** 223–230 (1981).
46. G. E. Hoffman, in Y. Ibata, Ed., *LHRH Neurons and Their Projections*, in Y. Ibata and E. A. Zimmerman, Eds., *Structure and Function of Peptidergic and Aminergic Neurons*, Wiley, New York, 1983, in press.
47. G. E. Hoffman and F. P. Gibbs, *Neuroscience* **7**, 1979–1993 (1982).
48. G. E. Hoffman, S. Wray, and M. Goldstein, *Brain Res. Bull.* **9**, 417–430 (1982).
49. S. P. Kalra, *J. Reprod. Fertil.* **49,** 371–373 (1977).
50. C. Kordon and J. Glowinski, *Neuropharmacology* **11,** 153–162 (1972).
51. P. C. K. Leung, G. W. Arendash, D. I. Whitmoyer, R. A. Gorski, and C. H. Sawyer, *Neuroendocrinology* **34,** 207–214 (1982).

52. A. Lofstrom, P. Eneroth, J. A. Gustafsson, and P. Skett, *Endocrinology* **101,** 1559–1569 (1977).
53. J. Markee, J. Everett, and C. Sawyer, *Recent Prog. Horm. Res.* **7,** 139–157 (1952).
54. S. M. McCann and R. L. Moss, *Life Sci.* **16,** 833–852 (1975).
55. S. R. Ojeda, A. Negro-Vilar, and S. M. McCann, *Endocrinology* **104,** 617–624 (1979).
56. M. Palkovits, C. Léranth, J. Y. Jow, and T. H. Williams, *Proc. Natl. Acad. Sci. USA,* **298**, 2705–2708 (1982).
57. D. K. Sarkar and G. Fink, *Endocrinology* **108,** 862–867 (1981).
58. C. H. Sawyer, H. M. Radford, R. J. Krieg, and H. T. Carrer, "Control of Pituitary-Ovarian Function by Brain Catecholamines and LH-Releasing Hormone," D. E. Scott, G. P. Kozlowski, and A. Weindl, Eds., *Neural Hormones and Reproduction,* Karger, Basel, 1978, pp. 263–273.
59. E. Vijayan and S. M. McCann, *Neuroendocrinology* **25,** 150–165 (1978).
60. C. Wilson, *Adv. Drug Res.* **8,** 110–294 (1974).
61. L. Agnati, K. Fuxe, A. Löfström, and T. Hökfelt, *Adv. Biochem. Psychopharmacol.* **16,** 159–168 (1977).
62. K. Ahrén, K. Fuxe, L. Hamberger, and T. Hökfelt, *Endocrinology* **88,** 1415–1424 (1971).
63. W. Beck and W. Wuttke, *J. Endocrinol.* **74,** 67–74 (1977).
64. W. Beck, J. L. Hancke, and W. Wuttke, *Endocrinology* **102,** 837–843 (1978).
65. J. A. Clemens, F. C. Linsley, and R. W. Fuller, *Acta Endocrinol.* **85,** 18–24 (1977).
66. J. A. Coppola, *Neuroendocrinology* **5,** 75–80 (1969).
67. A. O. Donoso and R. C. Santolaya, *Experientia* **25,** 855–857 (1969).
68. A. O. Donoso, W. Bishop, C. P. Fawcett, L. Krulich, and S. M. McCann, *Endocrinology* **89,** 774–784 (1971).
69. G. A. Gudelsky, J. Simpkins, G. P. Mueller, J. Meites, and K. E. Moore, *Neuroendocrinology* **22,** 206–215 (1976).
70. J. S. Kizer, M. Palkovits, J. Zivin, M. Brownstein, J. M. Saavedra, and I. J. Kopin, *Endocrinology* **95,** 799–812 (1974).
71. L. Krulich, *Ann. Rev. Physiol.* **41,** 603–615 (1979).
72. G. C. Lachelin, H. Leblanc, and S. S. Yen, *J. Clin. Endocrinol. Metab.* **44,** 728–732 (1977).
73. H. Leblanc, G. C. L. Lachelin, S. Abu-Fadil, and S. S. C. Yen, *J. Clin. Endocrinol. Metab.* **43,** 668–674 (1976).
74. R. E. Owens, J. L. Fleeger, and P. G. Harms, *Endocr. Res. Comm.* **7**(2), 99–105 (1980).
75. H. P. G. Schneider and S. M. McCann, *Endocrinology* **87,** 249–253 (1970).
76. W. Schulz and H. G. Hartwig, *Exp. Brain Res.* **38,** 293–298 (1980).
77. E. Vijayan and S. M. McCann, *Neuroendocrinology* **25,** 221–235 (1978).
78. R. F. Walker, R. L. Cooper, and P. S. Timiras, *Endocrinology* **107,** 249–255 (1980).
79. M. M. Wilkes, R. M. Kobayashi, S. S. C. Yen, and R. Y. Moore, *Neurosci. Lett.* **13,** 41–46 (1979).
80. T. H. McNeill and J. R. Sladek, Jr., *Brain Res. Bull.* **5,** 599–608 (1980).
81. D. L. Felten, A. M. Laties, and J. R. Sladek, Jr., *J. Histochem. Cytochem.,* **30**, 744–749 (1982).
82. B. Hamberger, T. Malmfors, and C. Sachs, *J. Histochem. Cytochem.* **13,** 147 (1965).

83. A. Björklund and B. Falck, *J. Histochem. Cytochem.* **16,** 717–720 (1968).

84. J. S. Ploem, "The Microscopic Differentiation of the Colour of Formaldehyde-Induced Fluorescence," in O. Eränkö, Ed., *Progress in Brain Research, Vol. 34: Histochemistry of Nervous Transmissions*, Elsevier, Amsterdam, 1971, pp. 27-37.

85. J. Ochi and Y. Hosoya, *Histochemistry* **40,** 263–266 (1974).

86. L. A. Sternberger, *Immunocytochemistry*, Wiley, New York, 1979.

87. R. Grzanna, M. E. Molliver, and J. T. Coyle, *Proc. Natl. Acad. Sci. USA* **75,** 2502–2506 (1978).

88. A. Björklund, B. Falck, and Ch. Owman, "Fluorescence Microscopic and Microspectrofluorometric Techniques for the Cellular Localization and Characterization of Biogenic Amines," *Methods of Investigative and Diagnostic Endocrinology, Vol. 00: The Thyroid and Biogenic Amines*, J. Rall and I. Kopin, Eds., North-Holland, Amsterdam, 1972, pp. 318–368.

89. J. R. Sladek, Jr. and C. D. Sladek, *Adv. Exp. Med. Biol.* **113,** 231–240 (1979).

90. L. F. Agnati, K. Andersson, F. Wiesel, and K. Fuxe, *J. Neurosci. Methods* **1,** 365–373 (1979).

91. J. Schipper, F. J. H. Tilders, and J. S. Ploem, *Brain Res.* **190,** 459–472 (1980).

92. L. D. Selemon and J. R. Sladek, Jr., *Brain Res. Bull.* **7,** 585–594 (1981).

93. C. Sinding, A. G. Robinson, S. M. Seif, and P. G. Schmid, *Brain Res.* **195,** 177-186 (1980).

94. J. Schöler and J. R. Sladek, Jr., *Science* **214,** 347–349 (1981).

95. J. Schöler and J. R. Sladek, Jr., *Ann. N. Y. Acad. Sci.*, **394**, 718–728 (1982).

96. J. R. Sladek, Jr., J. Schöler, M. F. Notter, and D. M. Gash, *Ann. N. Y. Acad. Sci.*, **394**, 102–117 (1982).

97. A. Ljungdahl, T. Hökfelt, G. Nilsson, and M. Goldstein, *Neuroscience* **3,** 945-976 (1978).

98. S. J. Watson, H. Akil, W. Fischli, A. Goldstein, E. A. Zimmerman, G. Nilaver, and T. B. van Wimersma, *Science* **216,** 85–87 (1982).

99. L. W. Swanson and P. E. Sawchenko, *Neuroendocrinol.* **31,** 410–417 (1980).

Chapter 7

The Use of Retrogradely Transported Fluorescent Markers in Neuroanatomy

L. W. Swanson

The Salk Institute for Biological Studies
San Diego, California
and
The Clayton Foundation for Research, California Division
La Jolla, California

1 INTRODUCTION

In the past decade, new methods based on the axonal transport of markers in the anterograde as well as in the retrograde direction, and on the immunohistochemical (IHC) localization of specific antigens, have revolutionized the field of neuroanatomy, and in particular the study of neuronal circuitry. The purpose of this chapter is to review the use of one of the newest family of markers, the retrogradely transported fluorescent dyes introduced by Kuypers and his colleagues. Thus far, the dyes have been used successfully to determine the cells of origin of an injected terminal field, much like horseradish peroxidase (HRP); to characterize biochemically the retrogradely labeled cells when used in combination with IHC methods; and to determine whether individual cells send axon collaterals to more than one discrete terminal field. Before describing these methods in detail, some of the major problems encountered in the study of central pathways are outlined. This may help put in some perspective the problems that can be approached with fluorescent markers as well as some of the limitations that must be dealt with when they are used. Because the fluorescent tracer methods are so new, a great deal undoubtedly remains to be learned about their use, as well as about the cell biology of their uptake, transport, and storage.

2 PATHWAY MARKING

A hypothetical set of pathways in the central nervous system (CNS) is shown in Figure 1. The basic problem illustrated here is to demonstrate experimentally that cell group *A* projects to cell group *B*. Until quite recently, two approaches were commonly used. (i) A lesion was placed in *A* and degenerating axons were traced to *B* in silver-impregnated material, prepared, for example, by one of the variants of the Nauta method. Although quite useful, this approach was limited because (*a*) fibers-of-passage (from cell group *E*) are also interrupted and the distal ends of the axons ending in *B* degenerate, (*b*) the lesion destroys the projection neurons of interest, and (*c*) small-diameter axons are often not impregnated. (ii) A lesion was placed in region *B*, often in young animals, and chromatolytic changes were looked for in cell group *A*. This approach is also plagued by the fiber-of-passage problem (cell *3* in group *A*), but more important, it often fails if the projection neurons maintain an intact "sustaining collateral" (cell *1* in group *A*).

The fiber-of-passage problem was effectively solved with the introduction of the autoradiographic method for tracing pathways (1, 2), and it is still the only pathway-tracing method that unequivocally avoids this problem (see ref. 3 for a recent discussion of this method). The autoradiographic method is based on the anterograde axonal transport of

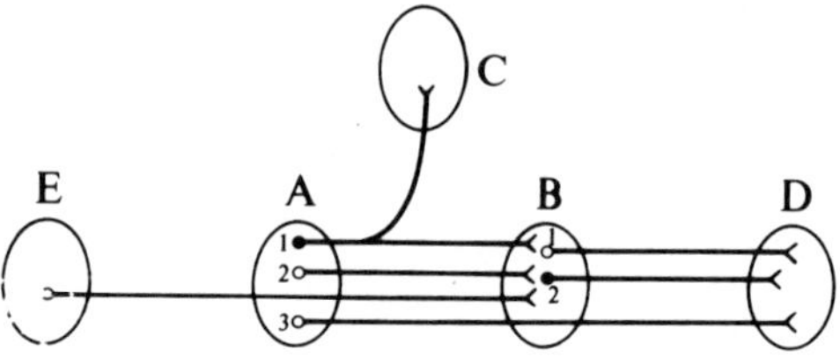

Figure 1. A hypothetical circuit diagram to illustrate some of the problems associated with marking central pathways. See text for details.

macromolecules that have incorporated radiolabeled amino acids (injected into the vicinity of the projection neurons). A different experimental approach, introduced at about the same time (4, 5), relies on the retrograde axonal transport of markers, such as the enzyme horseradish peroxidase (HRP) and conjugates of the fluorescent dye Evans blue, from an injected terminal field to the cell bodies of the projection neurons. This approach has proved much more sensitive than that based on the detection of retrograde degenerative changes, but as discussed more fully below, damaged or intact fibers-of-passage may take up label and transport it retrogradely as well as anterogradely. Nevertheless, the concurrent use of anterograde and retrograde transport methods provides a powerful approach to studying neural circuitry.

Another recently appreciated problem is illustrated in Figure 1, namely, that cell groups in the brain (and in the periphery) often contain more than one type of projection neuron, defined on the basis of different neurotransmitters (cells *1* and *2* in group *A*). This has become particularly clear with the introduction of IHC methods. Unfortunately, the use of IHC methods alone to unravel neural circuitry has been difficult and fraught with all the vagaries of interpretation faced by 19th-century neuroanatomomists working with normal material. The combined use of lesions and IHC has been useful in some systems, but, as discussed below, methods based on IHC and axonal transport methods, although tedious, hold a great deal more promise.

As already alluded to, many neurons in the CNS have one or more collaterals, as the Golgi method has made abundantly clear. It has proved difficult to trace the longer collaterals to their sites of termination in Golgi preparations, but recently experimental anatomical methods have been introduced that make the task easier. One approach is based on the injection of different, readily distinguishable, retrogradely transported markers into separate terminal fields (*B* and *C* in Figure 1) and the subsequent examination of labeled cells in the region of interest (*A*). For example, in Figure 1 some doubly labeled cells (cell type *1* in *A*) and some singly labeled cells (cell type *2* in *A*) would be found. This approach has the distinct advantage over the Golgi method that whole populations of projection neurons, rather than one cell at a time, are observed. It is important to point out, however, that at least for semiquantitative studies, markers with approximately the same sensitivity should be used, otherwise many cells will be singly labeled with the more sensitive marker even if they send a collateral to the site injected with the less sensitive marker. A number of different retrograde markers have been used with some success in conjunction with HRP; these include ^{3}H-HRP (6), ^{3}H-Apo-HRP (7), ^{3}H-bovine serum albumen (BSA) (8), iron dextran complex (9), and Evans blue (10). But only the fluorescent dyes are considered here because they are highly sensitive, because in general the two markers are easier to differentiate than in the other methods, and because histological processing is quite simple and rapid. A second approach is based on the intracellular injection of markers, such as HRP (see, for example, ref. 11), which are transported throughout the axonal and dendritic processes of the cell. This technique is similar in many ways to the Golgi method, but it has the major advantage that the inputs and outputs of the injected cell can be characterized electrophysiologically.

A complete description of the circuitry outlined in Figure 1 would include an account of synaptic relationships. Because most regions contain more than one cell type, it is necessary to determine what type of cell is innervated by each input. For example, in Figure 1, does cell *1* or cell *2* (or both) in group *A* innervate cell *1* in group *B*? For this, only electron microscopy can provide definitive answers, and the fluorescence methods described here are of little use because they are not electron-dense.

3 FLUORESCENT DYES AS RETROGRADE MARKERS

A fluorescent retrograde marker, Evans blue bound to albumin, was actually described (12) before HRP for the same purpose (4). However, it remained for Kuypers and his colleagues (10) to show that Evans blue alone was a more sensitive tracer than Evans blue–albumin, and to screen a large number of fluorescent compounds to find other suitable retrograde markers. Although the primary goal was to develop multiple-labeling methods for the determination of possible collateralization, several of these markers have proved quite useful as simple retrograde markers in their own right. Therefore, the advantages and limitations of the more useful markers will be considered, as a prelude to a discussion of their combined application in multiple-labeling studies.

The retrogradely transported fluorochromes identified to date by Kuypers and hs colleagues are listed in Table 1, along with their major features. A quantitative comparison of the relative sensitivity of all of these markers has not been carried out, but careful examination of the descending projections from the paraventricular nucleus of the hypothalamus (PVH) to the spinal cord indicates that the order in which the markers are listed in Table 1 roughly corresponds to their sensitivity, with true blue being the most sensitive (17, 18; unpublished observations). A quantitative comparison of the sensitivity of true blue, bisbenzimide, HRP-polyacrylamide gel implants, and HRP in this system showed that 88%, 58%, 39%, and 24%, respectively, of the total number of cells in a

Table 1. Fluorescent Compounds Suitable for Retrograde Axonal Transport Studies of Neural Pathways

Name (Concentration)	Filter System[a]	Excitation	Fluorescence	Comments	Reference
True blue (2%)	A	UV	Bright blue (cytoplasm, nucleolus)	Quite sensitive; combine with IHC; stays in cell for many weeks	13
Granular blue (2%)	A	UV	Blue (cytoplasm, nucleolus)	Somewhat less bright than true blue	13
Fast blue (2%)	A	UV	Silvery blue (cytoplasm, nucleolus)	Most useful for long pathways in large animals	14
Bisbenzimide (2%)	A	UV	Yellow (nucleus)	Quite sensitive; short survival time necessary	15, 16
Nuclear yellow (1–5%)	A	UV	Yellow (nucleus)	Brilliant yellow fluorescence; labels longer pathways than bisbenzimide	14
Propidium iodide (5%)	$N_{2.1}$	green	Orange (cytoplasm, nucleolus)	Labeling can be confused with autofluorescence; short survival time necessary	10
Evans blue (10%)	$N_{2.1}$	green	Red (cell)	Not as sensitive as above markers; large injection sites	10
DAPI-primuline (2.5–10%)	A	UV	Blue cell with golden cytoplasmic granules	Less sensitive than most others	10

[a] The letters refer to Leitz Ploempak filter cube systems. A: wideband UV excitation, excitation filter BP 340-380, mirror RKP 400, barrier filter LP 430. $N_{2.1}$: narrow band green excitation, excitation filter RP 515-560, mirror RKP 580, barrier filter LP 580.

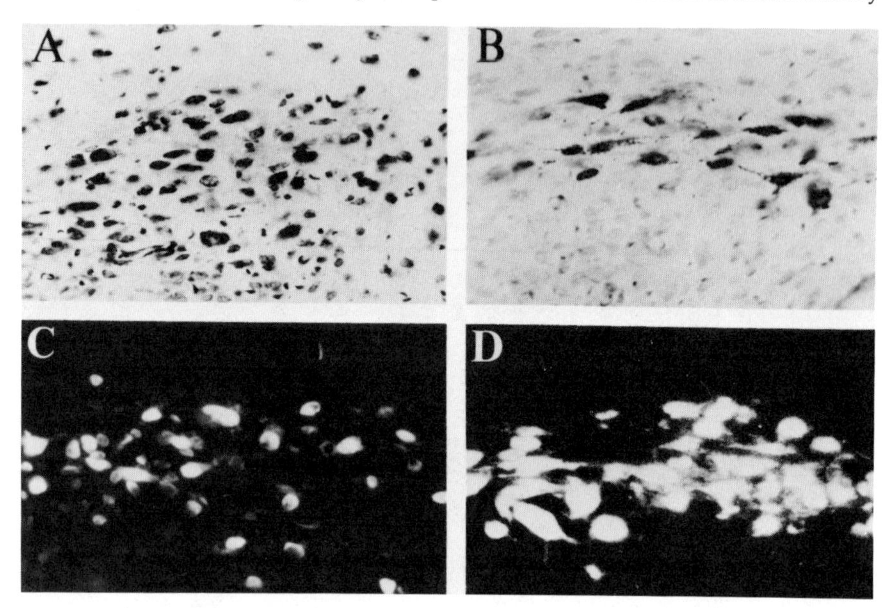

Figure 2. Photomicrographs through similar levels of the hypothalamus to show the appearance of retrogradely labeled cells after injections of HRP (B), bisbenzimide (C) and true blue (D) into upper thoracic levels of the spinal cord. Note labeled nuclei in (C) and labeled cytoplasm in (B) and (D). (A) shows the appearance of the paraventricular nucleus in a Nissl-stained section. Magnification ×160. From Sawchenko and Swanson (18), reproduced with permission.

specific part of the PVH were retrogradely labeled (18) (Figure 2). It appears, therefore, that true blue and bisbenzimide are more sensitive than HRP, a conclusion that is supported by the results of fluorescent marker studies on the connections of the hippocampal formation (19). This sensitivity, combined with the extremely simple and rapid methods required to process the tissue (described in the next section), are the principal advantages of certain fluorescent dyes when used as retrogradely transported markers. An additional feature, which may be quite useful, is that cells retrogradely labeled with true blue or with bisbenzimide (or nuclear yellow) can be counted accurately, since the former stains nucleoli and the latter clearly outlines the nucleus. Further studies are needed to determine if true blue is as sensitive as recently introduced HRP–lectin conjugates (20, 21).

There are, however, at least four limitations that must be taken into account:

1. None of the fluorescent markers adequately label fibers between the injection site and the parent cell bodies, as does HRP. In fact, several of the markers, in particular bisbenzimide, label glial cells along the course of the fibers, indicating leakage out of axons. This suggests that initially unlabeled axons may take up, and retrogradely transport, bisbenzimide from adjacent labeled axons, but the evidence to date suggests that this does not occur within the limits of detectability (18, 22).

2. It is clear that bisbenzimide and true blue, and most likely the other fluorescent markers, are readily taken up by damaged as well as by undamaged fibers-of-passage. This can be a major limitation for some studies of central connectivity, but it can be used to advantage for the injection of peripheral nerves (18).

3. Survival time is a critical variable for certain of the fluorescent tracers, especially bisbenzimide (16, 23). At longer survival times, bisbenzimide (as well as nuclear yellow, propidium iodide, and DAPI) appears to leak out of retrogradely labeled cell bodies and label adjacent glial and neuronal nuclei. When using these tracers, it is important to determine empirically a survival time that produces maximal retrograde labeling without glial labeling. On the other hand, significant glial labeling is usually not a problem with true blue (or with granular and fast blue); in fact, survival times of several weeks can be used with these tracers (Figure 3), and the retrograde labeling continues to become brighter with time. This suggests that, unlike HRP, the fluorescent markers are not effectively metabolized by the labeled cells, and this property has been used to advantage in developmental studies, as discussed below. Nothing

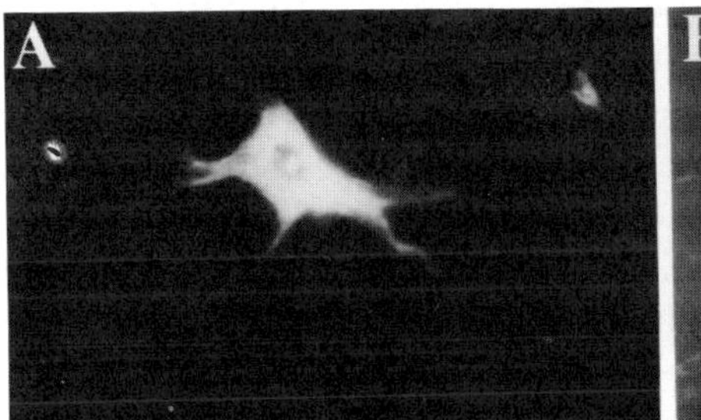

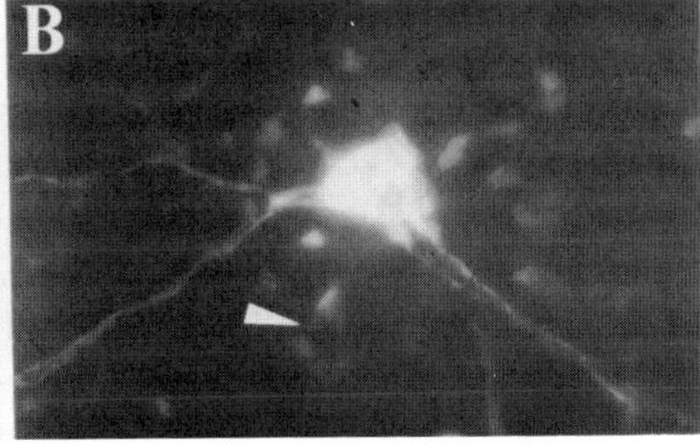

Figure 3. Photomicrographs of individual true blue-stained neurons in the reticular formation of the medulla following injections in upper thoracic levels of the spinal cord. (A) The large retrogradely labeled neuron in the center illustrates the morphological features that are typically stained by true blue. Note that the nucleolus is uniformly stained while the nucleus is not. A small retrogradely labeled cell is seen in the upper right corner. Survival time was 14 days. Magnification ×160. (B) An illustration of the true blue "halo" surrounding a retrogradely labeled neuron after an excessively long survival time. Dye has diffused from the labeled cell to label adjacent neuropil, glial cells, and neurons (arrowhead). This effect occurs much more rapidly with bisbenzimide. Survival time was 28 days. Magnification ×160. From Sawchenko and Swanson (18), reproduced with permission.

is known with certainty about the mechanism by which the fluorescent tracers are taken up and transported down the axon, although Björklund and Skagerberg (24) have shown that in freeze-dried material, where diffusion during histological processing has been greatly limited, both bisbenzimide and true blue are localized within small cytoplasmic granules in retrogradely labeled cells. This suggests that the nuclear labeling, and the glial labeling at longer survival times, found with bisbenzimide in conventional frozen sections is due to diffusion after the animal has been sacrificed. It also suggests that the fluorescent markers may be sequestered in perikaryal lysosomes *in vivo*. The pattern of labeling seen in frozen sections (true blue: cytoplasm and nucleolus; bisbenzimide: nucleus) may be caused by a combination of post-mortem diffusion from lysosomes and the relative affinities of the dyes for RNA (true blue) and DNA (bisbenzimide) under the histological conditions used (see refs. 25, 26).

4. The fluorescent markers are transported in the anterograde direction. This is not a problem with true blue and related tracers, but it can be troublesome when bisbenzimide and nuclear yellow are used. For example, when bisbenzimide is injected into the habenula, even after relatively short survival times, massive glial and neuronal labeling is found in and around the interpeduncular nucleus, which is due to the anterograde transport of label through the fasciculus retroflexus (27). This labeling makes it impossible to determine the origin of a pathway from the region of the ventral tegmental area to the lateral habenula, which is easily demonstrated with injections of true blue. Finally, there is no evidence that the fluorescent markers undergo transneuronal transport in either the anterograde or the retrograde direction when appropriate survival times are used (18).

In summary, true blue appears to be the most useful fluorescent retrograde marker for studies in smaller animals, while fast blue may be the most useful for larger animals since it is transported over greater distances (14).

4 MULTIPLE LABELING WITH FLUORESCENT MARKERS

In the first double-labeling experiments with fluorescent markers, Evans blue and DAPI-primuline were used to examine collateralization in the projections of the mammillary body of the rat (28). This system was chosen because previous studies with the Golgi method indicated that the axons of cells in the mammillary body that course through the mammillothalamic tract give off a collateral to the mammillotegmental tract, and this was fully confirmed with the double-labeling method. In addition, it was shown that most cells in the lateral mammillary nucleus project bilaterally to the anterior thalamic nuclei, while cells in the medial mammillary nucleus project only to the anterior thalamic nuclei on the same side of the brain. This report established the usefulness of the fluorescent retrograde double-labeling method.

Since that time, various combinations with each of the tracers listed in Table 1 have been tried in a variety of systems. At the present time, the most useful combination of markers for double-labeling experiments is true blue (or fast blue in larger animals) and bisbenzimide or nuclear yellow (29). There are four major reasons for this: (i) As mentioned above, these appear to be the most sensitive markers. (ii) Both markers are visible with light of the same (UV) excitation wavelength although they have different emission wavelengths, so that it is not necessary to switch filter systems to observe cells labeled with the two markers. (iii) The markers label different parts of the cell, i.e., the nucleus (bisbenzimide and nuclear yellow) and the cytoplasm and nucleolus (true blue family). Thus, when viewed with UV excitation, the nucleus and cytoplasm of doubly labeled cells are differentially stained, and they fluoresce in complementary (yellow and blue, respectively) colors (Figure 4). And fourth, these markers (especially fast blue and nuclear yellow) are transported over relatively long distances and are thus quite useful in larger animals such as the cat and monkey.

One problem with this combination of tracers is that injections of each usually must be made on separate days. It is often useful to allow a survival time of 3 days to 2 weeks after injections of the true blue family, so that bright labeling may be obtained. On the other hand, shorter survival times for bisbenzimide and nuclear yellow are necessary to avoid glial (and artifactual neuronal) labeling around retrogradely labeled cells (see above). The best combination of survival times can only be determined empirically, although we have found that, as a first approximation, survival times for experiments involving bisbenzimide can be estimated by multiplying the length of the pathway in mm by 2 hr. The most effective time can vary considerably, however, depending, for example, on cell density in the retrogradely labeled region. Thus, longer survival times can be used in the reticular formation, where the density of cells is rather low, than in the pyramidal layer of the hippocampus. The survival time for nuclear yellow experiments is not as critical and is usually on the order of days rather than hours. Other problems that may be encountered when using the markers in this way were mentioned in the preceding section, that is, the involvement of fibers-of-passage and the fact that fibers between the injection site and the parent cell bodies are usually not labeled.

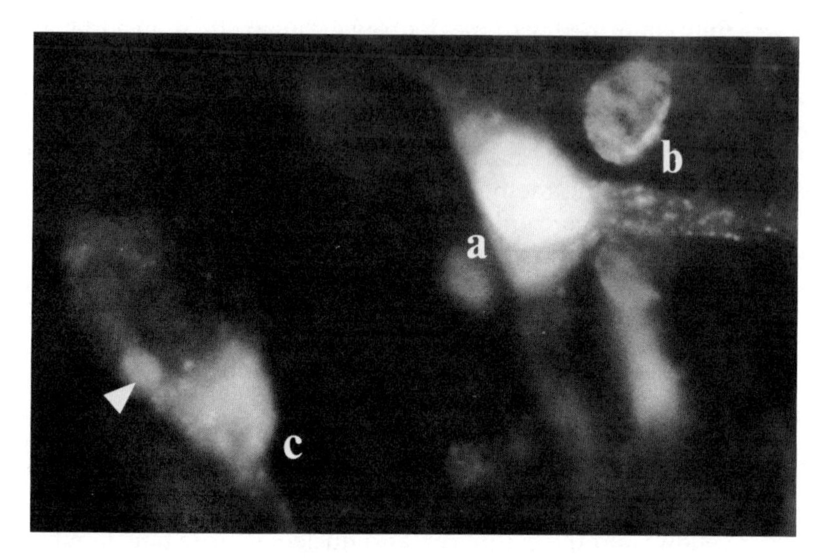

Figure 4. Three retrogradely labeled cells in the hilar region of the dentate gyrus in the rat. True blue was injected into the molecular layer of the dentate gyrus on one side of the brain and bisbenzimide was injected into the same region on the other side of the brain. Cell (*a*) contains label in the cytoplasm (true blue) and nucleus (bisbenzimide), cell (*b*) contains label only in the nucleus, and cell (*c*) contains label only in the cytoplasm and nucleolus (arrowhead). Magnification ×1300.

An ever present worry associated with any method is the interpretation of negative evidence, in this instance, a lack of doubly labeled cells. It is clear that in some systems, such as the mammillary body (28, 29) and Ammon's horn (19), a large proportion of the cells in specific regions can be doubly labeled. On the other hand, when the descending projections of the paraventricular nucleus to the dorsal vagal complex and to the spinal cord were examined with true blue and bisbenzimide, only some 15% of the retrogradely labeled cells contained both markers (17), and in a detailed examination of the ascending and descending connections of the ventral tegmental area with the same markers, involving some 15 different combinations of injection sites, almost no double labeling was found (27). In the latter experiment, and where quantitative estimates of the proportion of singly and doubly labeled projection neurons would be useful, it is important to consider possible reasons for not detecting doubly labeled cells, if cells with collateral projections are in fact present. First, one retrograde marker may be more sensitive than the other, and thus label more cells (the "extra" cells would, of course, be artifactually singly labeled). This possibility can be controlled for by reversing the sites into which the markers are injected (i.e., marker *A* in site 1, marker *B* in site 2; then, marker *A* in site 2, tracer *B* in site 1). This may be an important control, since it is unlikely that the commonly used markers are equally sensitive (18, 27). Second, one tracer may mask another. For example, if true blue labeling is particularly bright, it may mask somewhat dim bisbenzimide or nuclear yellow labeling, particularly in true-blue-labeled cells that lie entirely within a single section, since a layer of fluorescent cytoplasm surrounds the nucleus. Nuclear labeling (bisbenzimide or nuclear yellow) can be faint because the axon may have passed through the periphery of the injection site, because the terminal arborization is small and may have taken up only a small amount of label, because the pathway may be too long, or because an inadequate survival time may have been used. Third, even if both injection sites lie within the appropriate terminal fields, double labeling will not be found if the projections are topographically organized and inappropriate parts of the terminal fields have been labeled (19). Fourth, it is important to remember the trivial fact that true-blue-labeled cells (and other cytoplasmically labeled) cells should not be regarded as singly labeled unless an unlabeled nucleus is clearly visible. When tissue sections are examined, cells are often cut by the knife and end up in two adjacent sections, one of which may contain a great deal of labeled cytoplasm and proximal dendrites but not the nucleus.

5 PROTOCOL FOR RETROGRADE MULTIPLE-LABELING TECHNIQUE

The following method has been found useful for the demonstration of retrogradely transported fluorescent tracers in frozen sections. It is based on the work of Kuypers et al. (29) and Sawchenko and Swanson (18).

1. *Preparation of injection solutions.* The dyes are dissolved (e.g., bisbenzimide) or suspended (true blue family) in distilled water at the desired concentration (Table 1). Solutions of dyes that are not readily dissolved should be sonicated for about 5 min just before use. After use, solutions are stored at 4°C; solutions of true blue can be used for at least six months; solutions of bisbenzimide can be used safely for at least two weeks.

2. *Delivery of markers.* The markers can be delivered stereotaxically to selected areas in the CNS, or to peripheral nerves, with either a microsyringe or a glass micropipette. The volume to be delivered depends, of course, on the size of the cell group of interest; in the rat brain, for example, it is often necessary to inject as little as 20 to 50 nL. When suspensions of true blue and other relatively insoluble markers are delivered by pressure injections through a micropipette, a tip diameter of about 50–80 μm is necessary.

3. *Perfusion.* After an appropriate survival time (see above) the animal is anesthetized and perfused transcardially for 3–5 min with isotonic saline followed by 10% formalin in 0.1 *M* phosphate buffer at pH 7.5, for 20 min. The brain is then removed from the skull, blocked if desired, and placed in phosphate-buffered 10% formalin, containing 15% sucrose, for 18 to 48 hours. A variety of other fixatives may also be used, but glutaraldehyde in concentrations greater than 0.5% should be avoided because of increased nonspecific fluorescence, which reduces contrast.

4. *Tissue sectioning.* The piece of tissue to be cut is washed for 5–10 min in cold running tap water, and is then frozen in dry ice (or with a freezing stage on the microtome). Sections 20–30 μm thick are cut on a sliding microtome (or in a cryostat) and are collected in ice-cold distilled water. The sections should be mounted immediately on gelatin-chromalum-coated slides, although sections that contain only true blue can be stored in water at 4°C for several weeks.

5. *Microscopic observation.* The sections are examined with a fluorescence microscope that is equipped with an incident illumination system and oil immersion objectives, for optimal brightness. Appropriate filter systems are shown in Table 1, and the typical appearance of retrogradely labeled cells are shown in Figures 2–5. For purposes of orientation, major fiber tracts can be observed when the sections are viewed alternately with darkfield illumination and the sections can be counterstained with ethidium bromide (see Section 8).

6. *Photography.* Because the fluorescent markers fade when exposed to light of the appropriate (excitation) wavelength, it is necessary to use sensitive black-and-white or color film. Films with an ASA rating of 400 are normally used, with exposure times that range between 5 and 60 sec. Unfortunately, experience has shown that it is difficult, though not impossible, to obtain convincing black-and-white photomicrographs of cells doubly labeled with true blue and bisbenzimide or nuclear yellow.

6 RETROGRADE MARKERS COMBINED WITH IMMUNOHISTOCHEMISTRY

Several methods that permit the biochemical characterization of retrogradely labeled cells have been described. They include the combined use of HRP and IHC techniques (30), HRP and histofluorescence techniques for monoamines (31, 32), fluorescent markers and IHC techniques (18, 33, 34), and fluorescent tracers and histofluorescence techniques (24, 35). The double-labeling methods that rely on histofluorescence techniques for monoamines are quite useful for some studies, and the technical aspects of their application are clearly detailed in the references just cited. However, because IHC methods are much more generally applicable, their use with fluorescent markers will be considered in some detail. The first such method, based on the use of true blue, was developed by Hökfelt et al. (33). It is, however, somewhat inconvenient because the retrogradely labeled cells are photographed before the IHC procedure, after which the sections are rephotographed and individual cells in the two photographs are identified. This procedure is necessary because under the conditions used the true blue labeling fades during IHC processing.

It has been found, however, that with more complete fixation, true blue and IHC staining can be viewed simultaneously (18). Furthermore, the usefulness of this method for quantitative studies was examined in some detail. It was concluded that any such approach is based on four important considerations: (i) What is the sensitivity and reliability of the retrograde labeling method? As reviewed above, true blue is at least as sensitive and reliable as any known retrograde label. Nevertheless, it is important to bear in mind that still more sensitive tracers may well be found in the future. (ii) How well does the retrograde labeling survive IHC processing? Counts of true-blue-labeled cells before and after IHC processing indicated that only about 5% of the retrogradely labeled cells were no longer detectable following IHC processing. (iii) How clearly distinguishable is the retrograde and the IHC labeling? In this method the two are clearly differentiable: true blue fluoresces blue with wide-band UV excitation wavelengths (Leitz filter system A), and the IHC reaction is detected with fluorescein, which fluoresces yellow-green with blue excitation wavelengths (Leitz filter system I_2) (Figure 5). (iv) How sensitive and how specific is the IHC reaction? This, of course, is a particularly difficult problem (see ref. 36), and it is likely that neither part of the question can ever be fully answered. The best that can be done is to examine in detail, for each antigen, the best method of fixation (18, 37), and to carry out all of the proper controls for immunological specificity and nonspecific staining (36, 38).

This approach can provide a great deal of information about the morphology, location, and relative proportion of biochemically defined neurons in a cell group that projects to a specific terminal field. It is still necessary, however, to consider whether the injection of retrograde marker has labeled fibers-of-passage rather than a terminal field, a problem that

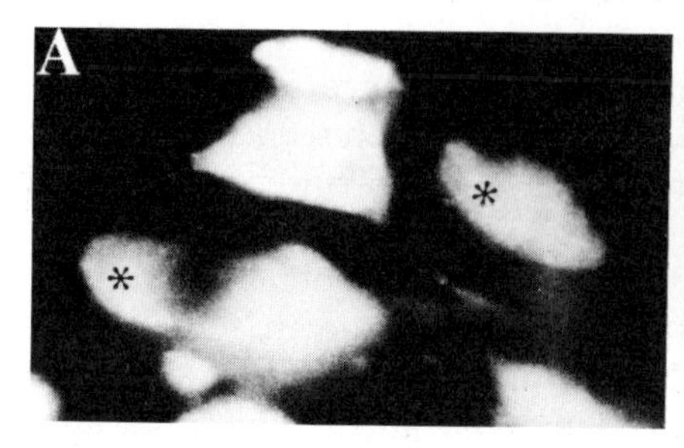

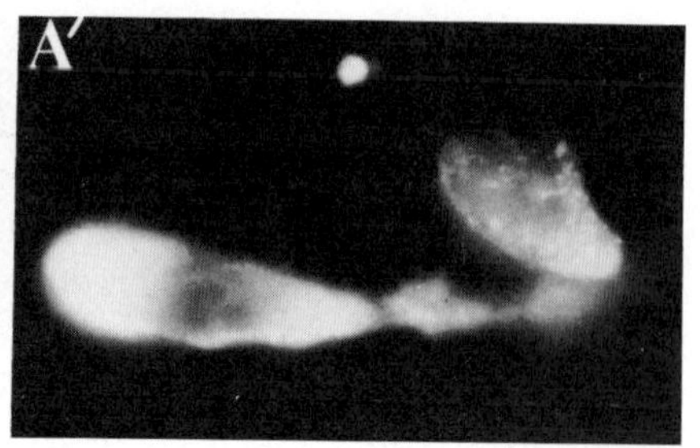

Figure 5. A pair of photomicrographs of the same field in the paraventricular nucleus, taken with two filter systems. (A) shows the appearance of cells labeled with true blue, after an injection in the spinal cord. (A′) shows the appearance of cells stained with an antiserum to oxytocin. Note two doubly labeled cells, indicated by asterisks. Magnification ×160.

may be avoided if the course of the antigen-containing pathway has also been determined immunohistochemically. The uptake of marker by fibers-of-passage may even be used to advantage in some circumstances. For example, injections at upper cervical levels may be used to label relatively completely certain fiber systems that descend from the brain to various levels of the spinal cord.

Two other advantages of the fluorescent marker–IHC method deserve comment. First, it has been used to identify what appear to be very limited projections that would be difficult, if not impossible, to demonstrate with other methods. For example, an average of about 20 dopaminergic neurons in the PVH on each side of the brain were found to project to the spinal cord in the rat (39). And second, the method can be used to demonstrate the distribution of antigen-specific fibers in the vicinity of retrogradely labeled neurons. Without ultrastructural evidence, this, of course, can only provide indirect evidence that the retrogradely labeled neurons are innervated by the IHC-labeled fibers or presumptive terminal field.

7 PROTOCOL FOR RETROGRADE MARKER IMMUNOHISTOCHEMISTRY

The following procedure has been found to yield adequate IHC staining for a wide variety of antigens, if proper fixation has been achieved, and is based on ref. 18.

1. *Fixation.* The animal (rat) is deeply anesthetized and is perfused transcardially for 3–5 min with isotonic saline followed by 10% formalin in 0.1 potassium phosphate-buffered saline (KPBS) at pH 7.5 for 20 min. The brain is removed from the skull, blocked, and placed immediately in cold (4°C) phosphate-buffered 10% formalin containing 15% sucrose, for 14–18 hr.

A variety of other fixation protocols are described in ref. 18. Recently, the following modification of the Bérod Hartman method (37) has been found to demonstrate a variety of enzymes, peptides, and neurotransmitters (serotonin) effectively. Deeply anesthetize the animal and place its head in crushed ice prior to perfusion. Begin with a brief (3–5 min) isotonic saline rinse, followed by a solution of ice-cold (2–4°C) 4% paraformaldehyde in carbonate buffer at pH 6.5 for 10–15 min.

Then switch to an ice-cold solution of 4% paraformaldehyde in borate buffer at pH 9.5 and perfuse for an additional 15–20 min. Remove the brain, block it, and place it immediately in a precooled solution of the second fixative, which contains 15% sucrose, and store overnight at 4°C.

2. *Tissue Sectioning.* Wash the brain briefly in cold running tap water, freeze a block of tissue with dry ice (or on the freezing stage of a microtome), and cut 20–30 μm thick sections on a sliding microtome. Collect the sections in ice-cold KPBS, and rinse for about 10 min.

3. *Incubation in Primary Antiserum.* Transfer the sections to the wells of porcelain spot dishes (1–3 sections/well) that contain primary (rabbit) antiserum (a dilution of 1:1000 is often useful, but initially a series of dilutions covering a wide range should be tested), 2% normal goat serum (which reduces staining due to nonspecific antibodies in the secondary antiserum), and 0.3–1.2% Triton X-100 (a detergent that reduces nonspecific staining and increases penetration of the antibodies by extracting lipids from membranes) in KPBS. The spot dishes are placed in closed containers (such as covered Petri dishes) and are stored at 4°C for 18 hr to 4 days.

4. *Incubation in Secondary Antiserum.* Rinse the sections in two changes (10 min each) of KPBS. Transfer the sections to the wells of spot dishes that contain the secondary antiserum (affinity purified goat anti-rabbit immunoglobulin G) that has been conjugated to fluorescein isothiocyanate (FITC-ARG; ref. 38) at an appropriate dilution (usually 1:200) in KPBS. Incubate at room temperature for 45 min.

5. *Mounting.* Wash the sections in two changes (15 min each) of KPBS at room temperature. Mount the sections from distilled water onto gelatin-chromalum-coated slides, let the sections dry in air, and coverslip with a nonfluorescent solution of glycerol at pH 8.6 (one part glycerol, one part 0.4 *M* carbonate buffer; for buffer add 2.0 g potassium bicarbonate to 50 mL DW, and adjust pH with KOH). Before mounting, the sections can be counterstained with ethidium bromide (see Section 8).

6. *Microscopy.* With the Leitz Ploem illumination system, filter system I_2 is used to observe FITC-labeling and filter system A is used to observe true blue labeling. Stained sections can be stored for several months at 4°C with only gradual deterioration of labeling.

8 COUNTERSTAINING

One major feature associated with the use of fluorescent markers is that brightly labeled cells and processes are viewed against a dark background. This is advantageous because it accounts for a major part of the sensitivity of fluorescence methods, but it is problematic when the precise localization of labeled cells must be determined. Commonly, this problem has been approached by the use of adjacent Nissl-stained and/or fiber-stained sections as a guide, or by the use of X-Y plotter systems. However, these methods are not completely accurate, and are tedious. Recently, several approaches that are more direct have been developed. The most useful involves the direct counterstaining of sections with a dye, ethidium bromide, that provides a fluorescent Nissl stain (26). The major advantages of ethidium bromide are that it can be used with a wide variety of fluorescence methods; that it is excited, and fluoresces, at different wavelengths (Leitz filter system $N_{2.1}$) than those used for the true blue family of markers, bisbenzimide and nuclear yellow, and IHC (when fluorescein is used), and thus does not mask the label; and that the counterstaining is rapid and simple. Thus, for example, the blue fluorescence of true blue is viewed with Leitz filter system A, the yellow-green fluorescence of fluorescein is viewed with Leitz filter system I_2, and the red fluorescence of ethidium bromide is viewed with Leitz filter system $N_{2.1}$ (see color plate in ref. 26). To counterstain fluorescent-marker labeled and/or IHC-labeled material, briefly dry the mounted sections in air to firmly attach them to the coated slide. Then, in Coplin jars, rehydrate the sections in distilled water for 2 min, stain the sections for 10–60 sec in a 0.0001–0.00001% aqueous solution of ethidium bromide, and wash the sections for 2 min in distilled water. The sections can be examined after they are dry, or they can be cleared in xylene and mounted with DPX (for retrograde markers only), or mounted with buffered glycerol (for IHC material). A second useful procedure is to view the sections alternately with fluorescence and darkfield illumination (27). With the latter, major fiber tracts are readily identifiable.

9 APPLICATIONS

Despite their recent introduction, fluorescent retrograde markers have been applied to a variety of neuroanatomical systems. It is not my intention here to review all of the studies, some of which have already been referred to in previous sections, that have used such markers. Instead, a flavor for the types of problem that can be approached with the fluorescent marker methods will be given by referring to a limited number of reports.

There seems little doubt that most of the connections in the mature CNS will eventually be reexamined with retrograde multiple-labeling techniques because the method is relatively simple and the results are not difficult to interpret. However, more sophisticated approaches that combine multiple-retrograde marker and electrophysiological methods will be of particular interest since they provide independent evidence for collateralization as well as information about conduction velocities. This approach has been used by Catsman-Berrevoets et al. (40) to show that separate neurons in layer V of the motor cortex in the rat project to the spinal cord and to the cerebral cortex on the opposite side of the brain.

New insights into the organization of the cerebral cortex, as well as into the reorganization of fiber systems in response to injury, have emerged from a retrograde multiple-labeling study of the hippocampal formation in the rat (19). In this study, three different tracers (bisbenzimide, true blue, and Evans blue) were injected, in the same animal, into one or another of the areas innervated by field CA3 of Ammon's horn, and the results indicated that the associational, commissural, and subcortical projections of this field arise from an essentially homogeneous population of pyramidal cells. This and other evidence (see ref. 19) indicates that the projection neurons in the hippocampal cortex are organized quite differently from those in the neocortex, where retrograde transport studies have clearly shown that pyramidal cells, which project to different sites, tend to be segregated in distinct sublaminae.

In the same study, it was shown that the commissural and associational projections to the molecular layer of the dentate gyrus arise as collaterals from individual cells in the hilar region of the dentate gyrus. This finding sheds new light on studies of plasticity in this system. When the commissural fibers are interrupted, the associational fibers appear to sprout, and eventually they occupy all of the vacated synaptic sites (see ref. 41). Although this response has been viewed as a form of "reactive synaptogenesis"—induced by degeneration in a neighboring fiber system (42)—the evidence from retrograde double-labeling experiments suggests a different interpretation. Since commissurotomy also partially axotomizes the cells that give rise to the associational system, it seems likely that at least part of the sprouting in the molecular layer is due to a "pruning" effect, as first described in the visual system by Schneider (43).

The fact that true blue labeling, unlike HRP labeling, persists for long periods of time (at least several months) in the intact animal has been used to advantage in studies of the developing nervous system. For example, Innocente and his colleagues (44) first demonstrated with the HRP method that early in development all parts of the primary visual cortex give rise to a massive commissural projection, only remnants of which are found along the representation of the vertical meridian, near the border between areas 17 and 18, in the adult. This observation raises the question whether cells that give rise to the exuberant early projection die during the course of development or whether commissural collaterals merely retract or die. This problem has been addressed directly in visual and somatosensory cortices with the use of fluorescent retrograde tracers (45–47). True blue was injected into the expanded commissural field early in development, and the animals were allowed to mature for several weeks before a second injection of bisbenzimide or nuclear yellow was made into the same region. The results of all of these studies are in general agreement: After tissue processing, widespread true-blue-labeled cells are found in the cerebral cortex on the uninjected side of the brain, and nuclear yellow labeling on the uninjected side is much more restricted (i.e., the adult pattern). It appears at the least, therefore, that cell death cannot account for all of the developmental remodeling that takes place in the commissural system. Instead, many commissural collaterals appear to retract or degenerate. This immediately raises the question, where do the cells that have lost a commissural collateral project in the adult? Further double-labeling studies indicate that one such projection from cells in the somatosensory cortex may be an associational projection to the rostrally adjacent motor cortex (46). Clearly, retrograde multiple-labeling methods, combined eventually with IHC techniques, offer an exciting new tool for studies in developmental neurobiology.

There is also no reason to believe that fluorescent retrograde tracers cannot be applied to a wide variety of problems in comparative neuroanatomy. For example, preliminary experiments have shown that true blue, bisbenzimide, and nuclear yellow are effectively transported in parts of the mature and developing avian (chick) visual system (D. D. M. O'Leary and W. M. Cowan, personal communication) and in the amphibian (frog) isthmotectal system (L. C. Schmued and H. Zakon, personal communication).

As a final example, recent work in the rat on hypothalamic circuitry that appears to integrate autonomic and neuroendocrine responses may be cited as an example of how fluorescent retrograde double-labeling methods, combined retrograde transport-IHC methods, and anterograde autoradiographic pathway tracing methods can be used in an integrated way (see 48, 49). First, the retrograde double-labeling method (with true blue and bisbenzimide) was used to show that essentially separate though partly overlapping cell groups in the PVH project to the posterior lobe of the pituitary, to the median eminence, where they presumably affect anterior lobe functions, and to autonomic cell groups in the brain stem and spinal cord, although some cells in the latter group project to both sympathetic and parasympathetic preganglionic cell groups. These results, combined with cytoarchitectonic criteria, served to demonstrate that the PVH in the rat consists of at least eight distinct subdivisions.

Next, combined retrograde transport-IHC methods were used to determine the biochemical specificity of (parvocellular) cells in the PVH that project to autonomic centers, as well as to determine their distribution within the PVH, and their relative contributions to the projection. It was found, using appropriate antisera, that oxytocin-, vasopressin-, somatostatin-, dopamine- and Leu-enkephalin and Met-enkephalin-stained cells, all with distinct distributions in the PVH, project to the dorsal vagal complex and to the spinal cord; the order in which the cell types are listed corresponds to their relative contributions to the descending projection, with oxytocinergic cells being the most numerous. Interestingly, these cell types together still account for only about one-quarter of the total number of cells that can be retrogradely labeled in the PVH after true blue injections in the dorsal vagal complex and the spinal cord. This suggests that still more cell types in the PVH give rise to descending autonomic projections.

Finally, since essentially separate groups of cells in the PVH appear to affect autonomic, anterior pituitary, and posterior pituitary functions, it seems reasonable to suggest that specific neural inputs to more than one group of cells in the PVH mediate coordinated autonomic and neuroendocrine responses. Since the major known input to the PVH is noradrenergic, the origin, course, and terminal distribution of this input was examined in detail. By using injections of true blue centered in the PVH, combined with IHC-staining for dopamine-B-hydroxylase (DBH; a specific marker for noradrenergic and adrenergic cells), it was found that three cell groups in the brain stem, in the A1, A2 and A6 (locus ceruleus) regions, project to the PVH, and that a vast majority of the cells in each region are DBH-positive. Because of this, the projections of each region were examined autoradiographically, and the results indicated that each of the three noradrenergic cell groups in the brain stem gives rise to a unique pattern of innervation in the PVH. For example, the locus ceruleus innervates primarily the median eminence effector zone, while only the A1 region projects to the magnocellular neurosecretory part of the nucleus, and in addition, only to the region in which vasopressinergic cell bodies are concentrated. The A2 group, centered in the nucleus of the solitary tract, projects to the autonomic and median eminence effector zones of the PVH. This and other evidence indicates that central noradrenergic pathways play a major role in the relay of visceral sensory information, which synapses first in the nucleus of the solitary tract, to the hypothalamus. This is of particular interest in view of the fact that the PVH in turn projects back to autonomic cell groups, including the nucleus of the solitary tract and the dorsal motor nucleus of the vagus nerve. These pathways may therefore play an important role in the integration of visceral responses.

In summary, although only recently introduced, fluorescent retrograde markers have already been applied to a wide variety of interesting problems in neurobiology. And in view of their sensitivity and versatility, they will undoubtedly be widely used, and more will be learned about how they are taken up, transported, and stored by neurons. It is also possible that new markers, which label axons between the injection site and the parent cell bodies, and which are not taken up by fibers-of-passage, will be found. Perhaps this can be achieved by conjugating fluorescent dyes to larger molecules, which ideally could be visualized in the electron microscope, so that ultrastructural studies could be carried out as well. In any case, it is clear that by combing retrograde double-labeling, IHC, and autoradiographic methods, a great deal can now be learned about previously intractable neuronal circuitry.

ACKNOWLEDGMENTS

The original work referred to here was supported in part by grant NS-16686 from the National Institutes of Health. Dr. Swanson is a Clayton Foundation Investigator.

REFERENCES

1. R. Lasek, B. S. Joseph and D. G. Whitlock, *Brain Res.* **8**, 931 (1968).
2. W. M. Cowan, D. I. Gottleib, A. E. Hendrickson, J. L. Price, and T. A. Woolsey, *Brain Res.* **37**, 21 (1972).
3. L. W. Swanson, *J. Histochem. Cytochem.* **29**, 117 (1980).
4. K. Kristensson, Y. Olsson, and J. Sjöstrand, *Brain Res.* **32**, 399 (1971).
5. J. H. LaVail and M. M. LaVail, *Science* **176**, 1416 (1972).
6. E. E. Geisert, *Brain Res.* **165**, 321 (1979).
7. N. L. Hayes and A. Rustioni, *Brain Res.* **165**, 321 (1979).

8. O. Steward, S. A. Scoville, and S. L. Vinsant, *Neurosci. Lett.* **5**, 1 (1977).
9. P. Cesaro, J. Nguyen-Legros, B. Berger, C. Alvarez, and D. Albe-Fessard, *Neurosci. Lett.* **6**, 127 (1977).
10. H. G. J. M. Kuypers, C. E. Catsman-Berrevoets, and R. E. Padt, *Neurosci. Lett.* **6**, 127 (1977).
11. S. T. Kitai and J. D. Kocsis, in R. W. Ryall and J. S. Kelly, Eds., *Iontophoresis and Transmitter Mechanisms in the Mammalian Central Nervous System*, Elsevier, Amsterdam, 1978, p. 17.
12. K. Kristensson, *Acta Neuropath. (Berlin)* **17**, 127 (1970).
13. M. Bentivoglio, H. G. J. M. Kuypers, C. E. Catsman-Berrevoets, and O. Dann, *Neurosci. Lett.* **12**, 235 (1979).
14. M. Bentivoglio, H. G. J. M. Kuypers, C. E. Catsman-Berrevoets, H. Loewe, and O. Dann, *Neurosci. Lett.* **18**, 25 (1980).
15. H. G. J. M. Kuypers, M. Bentivoglio, D. Van der Kooy, and C. E. Catsman-Berrevoets, *Neurosci. Lett.* **12**, 1 (1979).
16. M. Bentivoglio, H. G. J. M. Kuypers, and C. E. Catsman-Berrevoets, *Neurosci. Lett.* **18**, 19 (1980).
17. L. W. Swanson and H. G. J. M. Kuypers, *J. Comp. Neurol.* **194**, 555 (1980).
18. P. E. Sawchenko and L. W. Swanson, *Brain Res.* **210**, 31 (1981).
19. L. W. Swanson, P. E. Sawchenko, and W. M. Cowan, *J. Neurosci.* **1**, 548 (1981).
20. G. N. K. Gontas, C. Harper, T. Mizutani, and J. O. Gonatas, *J. Histochem. Cytochem.* **27**, 728 (1979).
21. W. A. Staines, H. Kimura, H. C. Fibiger, and E. G. McGeer, *Brain Res.* **197**, 485 (1980).
22. R.-B. Illing, *Neurosci. Lett.* **19**, 125 (1980).
23. L. W. Swanson and H. G. J. M. Kuypers, *Neurosci. Lett.* **17**, 307 (1980).
24. A. Björklund and G. Skagerberg, *J. Neurosci. Methods* **1**, 261 (1979).
25. M. Bentivoglio, H. G. J. M. Kuypers, C. E. Catsman-Berrevoets, and O. Dann, *Neurosci. Lett.* **12**, 235 (1979).
26. L. C. Schmued, L. W. Swanson, and P. E. Sawchenko, *J. Histochem. Cytochem.* **30**, 123 (1982).
27. L. W. Swanson, *Brain Res. Bull.* in press (1982).
28. D. Van der Kooy, H. G. J. M. Kuypers, and C. E. Catsman-Berrevoets, *Brain Res.* **158**, 189 (1978).
29. H. G. J. M. Kuypers, M. Bentivoglio, C. E. Catsman-Berrevoets, and A. T. Bharos, *Exp. Brain Res.* **40**, 383 (1980).
30. A. Ljungdahl, T. Hökfelt, M. Goldstein, and D. Park, *Brain Res.* **84**, 313 (1975).
31. W. W. Blessing, J. B. Furness, M. Costa, and J. P. Chalmers, *Neurosci. Lett.* **9**, 311 (1978).
32. A. J. Smolen, E. J. Glazer, and L. L. Ross, *Brain Res.* **160**, 353 (1979).
33. T. Hökfelt, L. Terenius, H. G. J. M. Kuypers, and O. Dann, *Neurosci. Lett.* **14**, 55 (1979).
34. M. R. Brann and P. C. Emson, *Neurosci. Lett.* **16**, 61 (1980).
35. D. Van der Kooy, D. V. Coscina, and T. Hattori, *Neuroscience* **6**, 345 (1981).
36. L. A. Sternberger, *Immunocytochemistry*, 2nd ed., Wiley, New York, 1979.
37. A. Bérod, B. K. Hartman, and J. F. Pujol, *J. Histochem. Cytochem.* **29**, 844 (1981).

38. B. K. Hartman, *J. Histochem. Cytochem.* **21**, 312 (1973).

39. L. W. Swanson, P. E. Sawchenko, A. Bérod, B. K. Hartman, K. B. Helle, and D. E. Van Orden, *J. Comp. Neurol.* **196**, 271 (1981).

40. C. E. Catsman-Berrevoets, R. N. Lemon, C. A. Verburgh, M. Bentivoglio, and H. G. J. M. Kuypers, *Exp. Brain Res.* **39**, 433 (1980).

41. D. D. M. O'Leary, R. A. Fricke, B. B. Stanfield, and W. M. Cowan, *Anat. Embryol.* **156**, 283 (1979).

42. C. W. Cotman and J. V. Nadler, in C. W. Cotman, Ed., *Neuronal Plasticity*, Raven Press, New York, 1978, p. 227.

43. G. E. Schneider, *Brain Behav. Evol.* **8**, 73 (1973).

44. G. M. Innocente, L. Fiore, and R. Caminiti, *Neurosci. Lett.* **4**, 237 (1977).

45. G. M. Innocenti, *Science* **212**, 824 (1981).

46. G. O. Ivy and H. Killackey, *J. Comp. Neurol.*, **195**, 367 (1981).

47. D. D. M. O'Leary, B. B. Stanfield, and W. M. Cowan, *Dev. Brain Res.* **1**, 607 (1981).

48. L. W. Swanson and P. E. Sawchenko, *Neuroendocrinol.* **31**, 410 (1980).

49. P. E. Sawchenko and L. W. Swanson, *Science*, **214**, 685 (1981).

Chapter 8

Apposition Techniques of Autoradiography for Microscopic Receptor Localization

James K. Wamsley

Department of Psychiatry
University of Utah Medical Center, Salt Lake City

José M. Palacios

Sandoz Ltd.
Preclinical Research
Basel, Switzerland

1 INTRODUCTION

Ideally, to know where a specific receptor is, is to know where a compound which can bind to that receptor may exert its physiological effects. Many attempts to localize receptor sites have produced questionable results or have simply met with failure. Within the last decade, however, advances in neurobiological techniques have made possible *in vitro* binding of radioactive ligands to specific binding sites (1). Improved procedures for washing the unbound radioactivity from the membranes, along with the discovery of new and more specific ligands, allow the determination of specific binding sites in minute quantities of tissue. These receptor binding studies have generated a wealth of information about drug and neurotransmitter receptors, from normal receptor distribution in the brain to effects on receptor density of chronic and acute drug treatments to the effects of various pathological conditions. These studies have also greatly enhanced our appreciation of the pharmacology and kinetics of receptor binding. Receptor binding techniques thus seem to have an infinite application to neurobiological research.

Still, from a neuroanatomical point of view, the limit of resolution of receptor binding techniques leaves something to be desired. Since receptor density can be determined in very small quantities of tissue, this limit appears to be dependent on the ability of the investigator to cleanly separate smaller and smaller portions of known areas of the brain. A more direct approach, which provides microscopic limits of resolution, has been used to localize muscarinic cholinergic receptors (2, 3), opiate receptors (4–12), and dopamine receptors (13–16). This method involves injecting a radioactive ligand into the animal's system and allowing the binding to occur *in vivo*. Although very important information is being derived from *in vivo* studies, the successful application of the technique is dependent on the existence of potent and specific ligands that are resistant to metabolism in the living organism. The investigator must also be concerned with changes in regional blood flow after administration of the compound, with the ability of the ligand to enter the brain parenchyma from its point of injection, with waiting the precise amount of time before sacrificing the injected animal so that the majority of the receptors are still occupied while most of the unbound radioactivity has moved through the animal's system, and so forth. These constraints greatly limit the applicability of *in vivo* labeling techniques for receptor study.

In 1979, Young and Kuhar (17) introduced a microscopic technique for opioid receptor localization that utilized *in vitro* labeling of receptors on slide-mounted tissue sections followed by autoradiographic localization of the bound radioactivity with an apposition technique originally described by Roth et al. (18). More recently, methods have been developed that utilize the same *in vitro* binding to tissue sections but localize the bound radioactivity by apposition of a tritium-sensitive film (19–21).

The ability to label slide-mounted tissue sections for microscopic autoradiography has resulted in the microscopic localization of the following receptors and probably other types as well: opioid (11, 17, 22–24), glycine (25), beta-adrenergic (26), alpha-adrenergic (27, 28), neurotensin (29, 30), insulin (31), cholecystokinin (32, 33), kainic acid (34, 35), GABA (36, 37), benzodiazepine (38–40), muscarinic cholinergic (41–43), dopamine (44), serotonin (44–46), histamine-H_1 (47, 48).

The key to the success, wide applicability, and versatility of apposition techniques for receptor autoradiography is that they employ *in vitro* binding in much the same fashion as it is accomplished in membrane homogenates. It is possible to study the pharmacology and kinetics of the binding directly on the slide-mounted tissue sections. This data can then be compared with that found in membrane homogenate studies. Receptor by definition refers to a binding site that demonstrates a certain pharmacology. In this context, receptor refers to a site that will bind certain classes of compounds known to be specific for a certain receptor but that will not bind unrelated compounds. A more rigorous definition of receptor would involve the demonstration that the specific binding site is coupled to a physiological response. For example, an electrophysiologist electrically stimulates an area of brain and records a certain evoked potential. He finds he can prevent the elicitation of the evoked potential by administering atropine or scopolamine but not tubocurare, norepinephrine, muscimol, strychnine, or other unrelated compounds. Next, he finds he can elicit the evoked potential in response to iontophoretically applied acetylcholine in the area that he previously stimulated electrically. Again, the investigator finds he can block this effect with atropine or scopolamine, but not tubocurare, and so on. Thus, he classifies the evoked potential as occurring in response to activation of a muscarinic cholinergic receptor.

Binding studies, on the other hand, measure the amount of radioactivity in tissue once a substantial number of the binding sites are labeled with a suitable radioactive ligand, for instance with {^{3}H}-QNB. Then, by including atropine or scopolamine in the incubation medium and measuring the amount of radioactivity incorporated into the tissue, it is possible to determine the amount of radioactivity originally attributable to the binding of {^{3}H}-QNB to atropine- or scopolamine-displaceable binding sites. If this level of binding is not affected by unrelated compounds, it is attributable to a pharmacologically defined muscarinic cholinergic receptor.

Binding studies performed in this fashion allow the investigator to carefully control the binding conditions and thus study the pharmacology and kinetics of the binding sites. By including small quantities of drugs that effectively displace or compete with the radioactive ligand for attachment to the binding site, it is possible to classify the binding of a radioactive ligand to a pharmacological receptor. With *in vitro* labeling techniques of autoradiography it is possible to replicate (within reason) the pharmacology and kinetics found in homogenate studies. Thus, it can be established that the radioactive ligand is bound to specific receptor sites on the slide-mounted tissue sections. Apposition techniques of receptor autoradiography localize the sites of radioactivity remaining in the tissue. Since the radioactivity is incorporated in the molecular structure of a ligand and since the ligand is bound to specific receptor sites, the presence of autoradiographic grains in a region implies the microscopic location of pharmacologically relevant *receptors*.

2 BIOLOGICAL TECHNIQUES OF AUTORADIOGRAPHY

Autoradiography is the process by which various concentrations of incorporated or bound radioactivity can be microscopically localized in tissue sections. This is accomplished by covering the labeled tissue with a nuclear emulsion. This emulsion is similar to that used in photography and consists of silver halide crystals suspended in gelatin. When a high-energy electron is emitted from the radioactive molecule present in the tissue section, it passes through the emulsion and strikes the silver halide crystals along its path. Subsequent development of the emulsion reduces the crystals to metallic silver. Unreacted halide crystals (those not contacted by a charged particle) are rinsed away in the development process. Thus, a black dot of silver is deposited in the region of the source of radioactivity. Microscopic visualization of these "dots" constitutes autoradiography. (For more information, see ref. 49).

Classical techniques of autoradiography involve dipping the tissue section containing the radioactive label in warm liquid emulsion. This procedure can be used for receptor autoradiography when employing irreversibly binding ligands (50–54). Since only a few of these compounds exist, it is imperative that other methods of emulsion application be devised. Roth and Stumpf (55) developed a technique for applying a labeled tissue section on a dry emulsion coated slide. This methodology has been employed for localizing receptors after *in vivo* labeling. *In vitro* labeling of the section is prohibited since the incubation procedure would

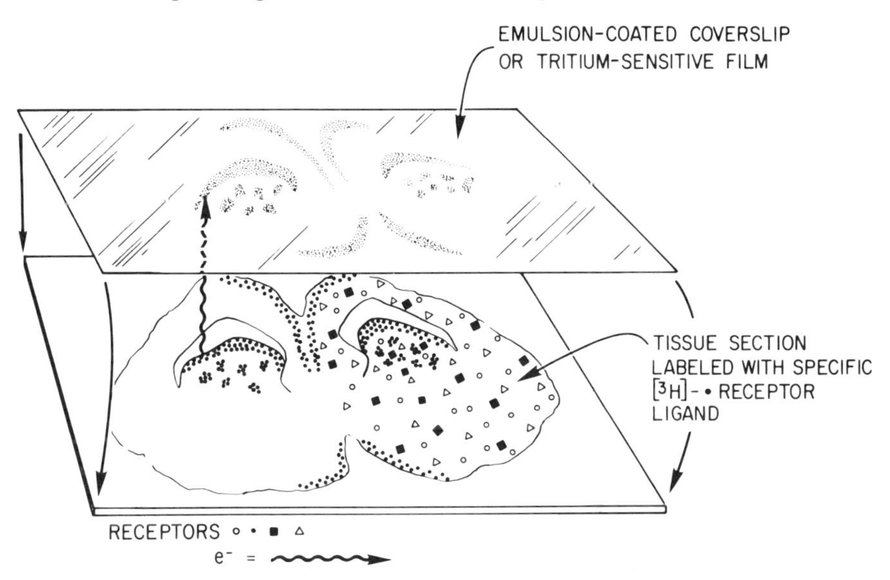

Figure 1. Schematic demonstrating the autoradiographic image formation in the emulsion overlying a tissue section with a particular type of receptor labeled with a specific tritiated ligand.

result in a rinsing away of the emulsion. Thus, for *in vitro* labeling it is necessary to cover the labeled tissue section with a dry-emulsion-coated coverslip or with a sheet of tritium-sensitive film (Figure 1).

3 *IN VITRO* LABELING OF DRUG AND NEUROTRANSMITTER RECEPTORS

To autoradiographically localize a subpopulation of specific receptors it is first necessary to selectively attach a radioactive compound to that particular group of receptors. Some definitions of terminology may be helpful at this point. *A receptor, in this context, refers to a membrane site that will interact with a certain group of compounds known to be specific for a certain type of receptor, but will not interact with reasonable concentrations of unrelated compounds. A ligand is a pharmacological agent that will attach to a receptor. The process of interacting or attaching a ligand to a receptor is referred to as binding or labeling the receptor.* It is absolutely imperative that steps be taken to show that under the conditions used, the ligand will bind to the particular receptor of interest, more so than to any other binding site present in the tissue.

3.1 Preparation of Tissues for Labeling

Tissues to be used for receptor autoradiography need not be fixed or perfused. It has been demonstrated, in the case of opiate receptor binding, that concentrations of fixative (paraformaldehyde or glutaraldehyde) significantly interfere with receptor binding (17). The deterioration rate of the tissue, however, is probably slowed by perfusing the animal with an ice-cold isotonic solution. A very small amount of fixative can be included in this solution to at least attempt to maintain tissue morphology without sacrificing too much receptor binding. Ideally, the exact concentration of fixative that can be used with each ligand should be determined prior to the labeling procedure. To date, most of the receptor localizations using apposition techniques of autoradiography have employed tissues taken from animals perfused intracardially with ice-cold 0.01% formalin (37% formaldehyde solution) in an isotonic buffered solution (0.05 *M* phosphate buffered saline, pH 7.4). The animal is anesthetized with an intraperitoneally administered injection of pentobarbital (65 mg/kg). A midline incision is made along the ventral thoracic and abdominal wall and the skin is dissected away. The abdominal wall is severed and the thorax penetrated by cutting through the rib cage in the cephalic direction along the midaxillary line on both sides. The diaphragm is then detached from the ventral part of the thoracic wall along with adherent portions of the pericardial and pleural membranes, and the resulting flap of ventral thoracic wall is lifted up, exposing the underlying heart. The sternal flap can be clamped with a hemostat at its cephalic end. This procedure holds the ventral thoracic wall up away from the heart and eliminates loss of blood and perfusate from the internal thoracic vessels. A small cannula (19-gauge needle) is inserted into the left ventricle and the flow of perfusate is started into the systemic circulation under the force of gravity. (This can be preceded 30 sec by an injection of 0.05 cc sodium heparin). The right atrium is then severed to provide an exit for the blood and perfusate. Depending on the location of the tissue of interest and the amount of perfusate available, the descending aorta and inferior vena cava may be clamped off so that only the head, neck, and upper extremity are perfused. A whole-body perfusion is accomplished over a 10 min period utilizing approximately 200 to 250 mL perfusate. The brain, spinal cord, or other tissue is then dissected from the surrounding tissues and frozen onto a microtome chuck (object disc) coated with a plastic embedding medium (O.C.T. Compound, Lab-Tek Products, Naperville, IL) or with brain paste (homogenized beef or pork brains). The plastic embedding compounds are satisfactory for sectioning large pieces of tissue. But since these media are water soluble they rinse away during the incubation procedure, often taking the tissue sections along. Thus, for small tissues, especially peripheral tissues that contain large amounts of connective tissue, it is advisable to embed in brain paste and include the surrounding paste in the sectioning procedure. Embedded tissues are stored in liquid nitrogen or wrapped in aluminum foil and placed in a −70°C freezer.

Prior to sectioning, the embedded tissues are allowed to equilibrate with the temperature of a cryostat set to maintain a temperature of approximately −15° to −20°C. Six to ten micron sections (about the limit of tissue thickness penetrable by tritium) are then cut and picked up on cold, acid-cleaned, subbed microscope slides. The acid cleaning procedure involves placing the slides in glass-slide racks and immersing them for several hours in a chromic and sulfuric acid solution (Chromerge, Fisher Scientific Co., Pittsburgh, PA). The slide-containing racks are then rinsed by placing them in a tray continuously overflowing with freely running water. Finally, the slides are dipped into several changes of distilled water and allowed to dry. Subbing of the acid-cleaned slides is accomplished by immersing them in a solution containing 0.5% gelatin (0.5 g/ 100 mL) and 0.05% chromium potassium sulfate (0.05 g/100 mL) that has been boiled, filtered, and cooled. The slides are then covered and allowed to dry. This procedure is facilitated by placing the slides in a slide holder (e.g., Peel-A-Way slide grips, VWR Scientific, San Fancisco, CA) and dipping several slides at a time.

Once the tissue section is picked up on the microscope slide, the sectioner should place his or her finger against the undersurface of the slide at one edge of the tissue section and slowly draw the resulting warmth under the tissue to the opposite border of the section, "melting" the tissue onto the slide. The slide containing the mounted tissue section should then be numbered, labeled (slides frosted on one end should be used for this purpose), and stored in the cryostat or in a frost-free freezer so that the tissue can slowly desiccate and more strongly adhere to the slide.

Since these tissue sections will be labeled by noninvasive means, it is possible to study the receptor array of human tissues as well. Thus, unfixed human post-mortem tissues can be frozen and sectioned in the same manner as that described for animal tissues. When working with diseased human tissue it is advisable to protect oneself against possible undiagnosed viral contamination. Cleaning contacted apparatus and utensils with a 0.05% solution of sodium hypochlorite (bleach) is indicated for protection against some particularly communicable slow virus disorders that may be resistant to other forms of decontamination (56).

Attempts should be made to determine the effects of prolonged post-mortem time on receptor density. The influence of these effects on the data, however, can be minimized by using post-mortem time-matched controls.

3.2 Determination of Binding Parameters

Initial experiments utilize binding parameters established in membrane homogenate studies. Boxes of slides containing sections of brain from similar regions (preferably regions known to contain high concentrations of the receptor of interest) are placed on dry ice immediately before beginning the incubation. These slides are removed one at a time at 30 sec intervals and placed on a slide warmer until the moisture dissipates from the slide surface (40°C slide warmer, 3–5 sec). The tissue-containing end of the slide is then immersed in a Coplin jar containing the radioactive ligand diluted to a predetermined concentration with an appropriate buffer. Usually triplicate sections are used for each variable, so slides 1–3 are placed in the incubation medium. Slides 4–6 serve as their controls and are incubated in a separate jar containing the same buffer and concentration of radioactive ligand but in the added presence of a displacer. *A displacer is a compound known to be specific for the receptor of interest and it will thus compete with the radioactive ligand for binding of the receptor site.* Since the displacer is present in excess, it will effectively occupy the receptor sites and prevent the radioactive ligand from binding. Radioactivity bound in these tissues will indicate nonreceptor or unrelated receptor binding sites. The concentration of radioactive ligand and displacer, the incubation time and temperature, the buffer used, its concentration and pH are all initially patterned after the results found to be effective in homogenate binding assay.

After the incubation period, each slide is transferred sequentially to a Coplin jar containing buffer alone for a rinsing period. The first experiment should involve varying the amount of time the sections are exposed to the rinse. Since the rinsing time can be precisely controlled, it is possible to accurately determine the amount of rinse time that will produce high specific-to-nonspecific (signal-to-noise) ratios.

Total binding (total in the sense that all the binding sites for that particular concentration of ligand are occupied) is represented by the amount of radioactivity present in the tissues which are incubated in the presence of the radioactive ligand alone. Nonspecific binding (nonspecific in the sense that the binding of the radioactive ligand is not displaced by a drug known to be specific for the receptor of interest) is established by determining the radioactivity remaining in sections incubated in the presence of the radioactive ligand and excess displacer. By subtracting nonspecific binding from total binding the amount of radioactivity attributable to specific receptor binding can be determined.

The amount of radioactivity bound in each tissue section is determined by wiping the tissue from the slide and counting it in a liquid scintillation system. This is accomplished by removing the slide-mounted tissue sections from the rinse and dipping them quickly into fresh buffer so that only radioactivity in the rinse solution adhering to the slide is washed away. The excess moisture is quickly dabbed dry and the tissue sections are wiped from the slide with a microfiber glass filter disc (Whatman GF/B, Whatman, U.K.) held with a hemostat. The disc is then placed in a scintillation vial containing scintillation cocktail (Formula 947, New England Nuclear, Boston, MA). The vials are shaken for 1 hr or are left in a refrigerator overnight before being counted in a liquid scintillation counter. The counts per minute (cpm) from each triplicate set of sections are averaged and total counts are compared with nonspecific counts for each different rinse period employed. By plotting cpm vs. rinse time, a so-called dissociation curve is generated. It is desirable to determine the point on this curve where the nonspecific binding is low and, at the same time where the specific binding has not been seriously perturbed. This rinse time provides the highest specific-to-nonspecific ratio and will be used in all subsequent experiments.

The next experiment should establish the appropriate incubation time. This is accomplished by exposing tissue sections to the incubation media for various periods of time while holding the other variables fixed, determining the bound radioactivity, and then plotting cpm vs. incubation time. Triplicate sets of sections should be incubated in the presence or absence of displacer, and all slides should be rinsed the optimal rinse time established in the previous experiment. By comparing total binding to nonspecific binding, specific to nonspecific ratios can be determined. The optimal incubation time is represented by the time point on the curves where the specific binding is high and the nonspecific binding is low.

Now that the optimal incubation and rinse times are established, the optimal concentration of radioactive ligand to have present in the incubation media can be determined. Again triplicate sets of tissue sections are used; one set each for several different concentrations of radioactive ligand. A corresponding set of sections is incubated in the same concentration of radioactive ligand plus the displacer. Plotting cpm vs. incubation concentration gives rise to a saturation curve. To select the optimal incubation concentration it is necessary to examine the specific-to-nonspecific ratios at each concentration point. It is advisable to select a concentration near the point at which half of the receptors are occupied, that is, halfway between the origin and the point where the curve plateaus. This point represents the K_D of the ligand. *The K_D is the dissociation constant and is equal to the concentration of the ligand required to occupy half of the total number of receptors available.* The K_D inversely reflects the affinity of the ligand for the receptor. Thus, the lower the K_D the higher the affinity and vice versa.

The K_D can be calculated by performing a so-called Scatchard analysis of the saturation data. This involves plotting the concentration of bound ligand (ordinate) vs. the concentration of the bound ligand divided by the concentration of free ligand (abscissa). The slope of the generated line is the K_D. The y-intercept of this line is the B_{max}. *The B_{max} is the number of binding sites present in the tissue sample.* If the Scatchard plot is not linear it may indicate that binding is taking place to more than one class of receptors. Hill plots can be performed to determine if cooperativity is taking place between the receptor and the ligand, for example, if the binding of one molecule facilitates or suppresses the binding of another molecule. (For more information on examination of the binding kinetics and calculations involved, see ref. 57.)

Finally, one of the most important aspects to demonstrate is the specificity of the ligand for the receptor of interest. This experiment involves the demonstration of the pharmacology of the receptor site. It is accomplished by generating a displacement curve represented by plotting cpm vs. concentration of displacer. The previously determined optimal binding conditions are employed for this experiment. The variable in the experiment is the amount of displacer found in the incubation medium. Triplicate samples are incubated in one of several solutions. The first solution has no displacer and represents total binding. The next group of solutions contains increasing quantities of a displacer. This dissociation analysis should be accomplished for several compounds known to be specific for the receptor of interest and a few closely related compounds that do not bind to the receptor. One or two reasonably high concentrations of a few unrelated compounds should also be tried to show that they have no effect on the total binding.

In all cases, the kinetics and the pharmacology of the receptor binding should appear in accordance with those previously determined by binding assay in homogenates. These data will show that the binding of the ligand is taking place to the pharmacologically relevant receptor. Once these studies have been accomplished, it is advisable to routinely repeat one or two points on the curves to demonstrate that in the investigator's hands the binding is taking place in a fashion similar to that previously described.

3.3 Binding Parameters

Research communications that examine the binding parameters, kinetics, and pharmacology of specific ligands are listed in Table 1. The optimal conditions for the labeling of many different types of receptors for apposition techniques of receptor autoradiography, and appropriate references where examples can be found, are given below.

3.3.1 GABA Receptors

High-affinity GABA receptors can be labeled by incubating slide-mounted tissue sections for 40 minutes at 4°C in the presence of 5 n*M* {^{3}H}-muscimol (New England Nuclear Co., Boston, MA) in 0.31 *M* Tris-citrate buffer. The tissues are rinsed for one minute in buffer alone to remove unbound radioactive ligand. Blanks are generated by incubating adjacent slides in a separate incubation medium with the addition of 200 μ*M* unlabeled GABA (36, 37).

Table 1. Published Receptor Binding Studies in which Data Was Obtained with Slide-Mounted Tissue Sections

Receptor	Ligand	Reference
GABA	{^{3}H}-Muscimol	36
Benzodiazepine	{^{3}H}-Flunitrazepam	39
BZ-1 vs. BZ-2		40, 58
Muscarinic cholinergic	{^{3}H}-Quinuclidinyl benzilate	42
High vs. low affinity	{^{3}H}-*N*-Methyl scopolamine	41
Dopamine	{^{3}H}-Spiperone	44
Serotonin 2		
Serotonin 1 and 2	{^{3}H}-LSD	46
Opioid	{^{3}H}-Diprenorphine	17
Beta adrenergic	{^{3}H}-Dihydroalprenolol	26
Neurotensin	{^{3}H}-Neurotensin	30
Alpha adrenergic, α_1 and α_2	{^{3}H}-Paraaminoclonidine	28
	{^{3}H}-WB-4101	
Histamine H_1	{^{3}H} Mepyramine	47
Glycine	{^{3}H}-Strychnine	25

3.3.2 *Benzodiazepine Receptors*

Benzodiazepine receptors can be labeled by incubating sections in 0.17 *M* Tris-HCl buffer containing 1 n*M* {^{3}H}-flunitrazepam (NEN) for 40 min at 4°C followed by a 2-min cold rinse. To determine regions where GABA enhances benzodiazepine binding, 10^{-4} *M* GABA can be added to the incubation medium of adjacent sections. Clonazepam or diazepam (1 μ*M*) can be used to generate controls (39). Different types of benzodiazepine receptors have been localized autoradiographically (40, 58). The {^{3}H}-flunitrazepam binding to the type I-BZ receptor can be displaced with appropriate concentrations of triazolopyridazine compounds, which have a higher affinity for the type I than type II receptor. Thus, sections can be incubated in the presence of {^{3}H}-flunitrazepam to label all benzodiazepine receptors, and other sections can be incubated in {^{3}H}-flunitrazepam plus 200 n*M* CL 218-872 to displace the type I receptors. Differences in grain densities resulting from these two incubation conditions indicate regions containing type I receptors. Beta-carboline-carboxylate esters have also been used to autoradiographically distinguish between the two benzodiazepine receptor subtypes (58).

3.3.3 *Muscarinic Cholinergic Receptors*

Muscarinic cholinergic receptors can be labeled for apposition techniques of autoradiography using 1 n*M* concentrations of {^{3}H}-quinuclidinyl benzilate (Amersham Corp., Arlington Heights, IL) in phosphate-buffered saline, using a 2-hr incubation at room temperature followed by two 5-min rinses (42). If differentiation of high- and low-affinity muscarinic receptors is desired, {^{3}H} *N*-methyl scopolamine (NMS obtainable from New England Nuclear Corporation, NEN) can be used to label the muscarinic cholinergic receptors, using the same parameters as that presented above except with a 1 hr incubation at room temperature. An adjacent set of slides is then incubated in a similar solution but with the addition of 10^{-4} *M* carbachol to displace the high-affinity muscarinic cholinergic binding sites. A further set of adjacent sections should be incubated in the presence of {^{3}H}-NMS with 1 μ*M* atropine added to the incubation medium to eliminate specific binding and thus generate blanks (41).

3.3.4 *Dopamine and 5HT-2 Receptors*

Dopamine and serotonin (5HT-2) sites can be labeled with 0.4n*M* {^{3}H}-spiperone (NEN) in a buffering medium consisting of 0.17 *M* Tris-HCl containing 120 m*M* NaCl, 5 m*M* KCl, 2 m*M* $CaCl_2$, 1 m*M* $MgCl_2$ and 0.001% ascorbic acid. A 60-min incubation at room temperature is followed by two 5-min rinses. This set of slides represents total binding. An adjacent set of slides is incubated in the same fashion but in the added presence of 0.3 μ*M* cinanserin to displace the {^{3}H}-spiperone from the 5HT-2 site. Thus, areas where there are differences in grain density (radioactivity) indicate sites of localization of the 5HT-2 receptor. Another set of tissue is incubated in {^{3}H}-spiperone plus 1.0 μ*M* ADTN to displace {^{3}H}-spiperone from binding to dopamine receptors. Blanks are generated by incubating in the presence of 0.4 μ*M* haloperidol or 0.1 μ*M* unlabeled spiperone (44).

3.3.5 *Serotonin Receptors*

Both serotonin sites, 5HT-1 and 5HT-2, are labeled with {^{3}H}-lysergic acid diethylamide (LSD) while only the 5HT-1 site is labeled with serotonin itself (59). Thus, tritiated serotonin can be used to differentially label the 5HT-1 site. The 5HT-1 receptor is labeled by preincubating 5 min in 0.17 *M* Tris-HCl buffer (pH 7.6) containing 0.001% ascorbic acid, 10 μ*M* pargyline, and 4 m*M* $CaCl_2$ followed by incubation in this buffer containing 1.0 n*M* {^{3}H}-5HT (NEN). Both serotonin receptors can be labeled by incubating sections using conditions similar to those described above but substituting 3.1 n*M* {^{3}H}-LSD (Amersham) for the radioactive ligand. Blanks can be generated in either experiment by adding 1 μ*M* unlabeled LSD to the incubation medium (45, 46).

3.3.6 *Opioid Receptors*

Opioid receptors can be labeled with a variety of compounds, including opioid peptides (60). Many studies have employed slide-mounted tissue sections preincubated 5 min in 0.17 *M* Tris-HCl buffer containing 50 μ*M* GTP and 100 m*M* NaCl (to dissociate and rinse away endogenous ligands) followed by a 15-min rinse in buffer alone. This is followed by a 60-min incubation period at room temperature in buffer containing {^{3}H}-dihydromorphine (Amersham) under subdued light, and then succeeded by two 5-min rinses in buffer at 4°C. Blanks are generated with the addition of 1 μ*M* naloxone or levallorphan to the incubation medium (17). Opioid receptor subtypes have also been localized using apposition techniques of autoradiography. This was accomplished by labeling "delta" receptors with 0.1–0.2 n*M* {^{125}I}-D-Ala^2-D-Leu^5-enkephalin plus 1 n*M* FK 33-824. "Mu" receptors were labeled with 0.1–0.2 n*M* {^{125}I}-FK 33-824 (23).

3.3.7 *Beta-Adrenergic Receptors*

Beta receptors are labeled by incubation of sections in 0.17 *M* Tris-HCl buffer containing 2 n*M* {^{3}H}-dihydroalprenolol (NEN) for 30 min at room temperature followed by two 10-min rinses. An adjacent set of sections can be incubated in the same manner but in the added presence of 10^{-7} *M* zinterol (a predominantly β_2 agonist) to displace the binding of the radioactivity to the β_2 receptor. Comparison of the latter slides with those representing total binding allow the differentiation of β_1 and β_2 adrenergic receptor sites. Blanks are produced with excess (10^{-4} *M*) zinterol or 10^{-5} *M* DL-propranolol (26).

3.3.8 *Neurotensin Receptors*

Neurotensin receptors have been successfully labeled for autoradiography by incubating sections in 0.17 *M* Tris-HCl, pH 7.6, containing 4.0 n*M* {^{3}H}-neurotensin (NEN) for 60 min at 4°C followed by two 5-min rinses. Blanks can be generated on adjacent slides by displacing with 5 μ*M* unlabeled neurotensin (29, 30).

3.3.9 *Alpha-1 and Alpha-2 Adrenergic Receptors*

Alpha-1 and alpha-2 receptors can be selectively labeled for autoradiography using WB-4101 and paraaminoclonidine respectively. Alpha-2 receptors are labeled by incubating sections in 0.17 *M* Tris-HCl buffer containing 2.5 n*M* {^{3}H}-paraaminoclonidine (NEN) for 60 min followed by a 5-min rinse and a 10 min rinse. Alpha-1 receptors can be labeled by incubating sections in 0.17 *M* Tris-HCl, pH 7.6, containing 0.001% ascorbic acid and 1.0 n*M* {^{3}H}-WB-4101 (NEN) for 70 min at 4°C followed by a 5-min then a 20 min rinse in buffer alone (27, 28). More recently, α_1 receptors have been successfully labeled with {^{3}H}-prazosin (NEN) with a 60 min rinse at room temperature followed by a 1-min rinse (Unnerstall et al., in preparation).

3.3.10 *Insulin Receptors*

Insulin receptors can be labeled by incubating slide-mounted tissue sections for 60 min at room temperature in 0.17 M Tris-HCl, pH 7.4, containing 0.17 n*M* {^{125}I}-insulin. The incubation medium should also contain 10^{-4} *M* bacitracin and 1.3 mg/mL bovine serum albumen. The sections are then rinsed for 5 min in buffer alone followed by another 15 min rinse in fresh buffer. Blanks are produced by displacing with 3.3×10^{-6} *M* unlabeled insulin (31).

3.3.11 *Cholecystokinin Receptors*

Cholecystokinin (CCK) receptors are labeled by incubating sections in 50 m*M* Tris-HCl, pH 7.7, containing 0.2% bovine serum albumin, 5 m*M* $MgCl_2$, 1 m*M* dithiothreitol, 0.1 m*M* bacitracin and 50 p*M* {^{125}I}-CCK-33 (Bolton Hunter). This is accomplished with a 45-min incubation at room temperature followed by a 30-min wash in buffer at 4°C. Nonspecific binding is determined by using 10^{-7} *M* unlabeled sulfated CCK octapeptide as a displacer (32, 33).

3.3.12 *Histamine Receptors*

Histamine-H_1 receptors can be labeled using a 40-min incubation at 0°C with 5 n*M* {^{3}H}-mepyramine (NEN) in 0.3 *M* sodium potassium phosphate buffer. The sections are then given two 5-min rinses in buffer alone. Nonspecific binding is determined by incubating sections in the incubation medium plus 2 μ*M* triprolidine (47, 48).

3.3.13 *Glycine Receptors*

Binding glycine receptors for apposition techniques of autoradiography is accomplished by incubating sections for 20 min at 4°C in 0.05 *M* sodium potassium phosphate buffer, pH 7.4, containing 4 n*M* {^{3}H}-strychnine. The incubation is followed by a 5-min buffer rinse. Control sections involve incubation under the same conditions presented above but in the added presence of 10^{-2} *M* glycine as a displacer (25).

3.3.14 *Kainic Acid Binding Sites*

Binding sites of the glutamate analog kainate have been localized by autoradiographic techniques (34, 35), but the binding has, to date, not been well characterized on slide-mounted tissue sections although the binding has been studied in detail by homogenate binding assay (61). For autoradiography, incubation conditions have varied from 100 n*M* {^{3}H}-kainic acid (NEN) using an ice-cold 30 min incubation (rat) to 1.5 n*M* (pigeon) or 60 n*M* (rat, pigeon) {^{3}H}-kainic acid with a 60-min room temperature incubation in 50 m*M* Tris-HCl buffer at pH 7.4. In the latter study, the high affinity binding site was displaced using 500 n*M* unlabeled kainic acid in the incubation medium with 45 nM {^{3}H}-kainic acid followed by a few-second rinse. Blanks were generated with 0.1 mM kainic acid.

4 APPOSITION OF EMULSION FOR AUTORADIOGRAPHY

After using the optimal conditions to bind a high concentration of specific receptor sites with an appropriate ligand, the slide mounted tissue sections are quickly dipped into distilled water (to rinse away the buffer salts) and placed on a cold plate (a metal pan over ice). The sections are then quickly blown dry by using cool, dry, and filtered air. This procedure involves running a forced air stream into a trap immersed in a dry-ice acetone mixture. The cooled air is then routed into another trap filled with desiccant (Drierite, W.A. Hammond Drierite, Xenia, OH), and finally filtered through a hose packed at one end with angel's hair, before being passed across the tissue surface. Using this procedure, the tissue sections appear dry after about 30 sec. Thus, the incubation of each slide should be staggered about a minute to allow ample drying time and for manipulation of each slide. The slides are then placed in plastic slide boxes under the cool forced-air stream of a hair dryer until the box is filled with slides. Desiccant is then placed in the slide boxes and they are taped shut and stored in the refrigerator overnight.

4.1 Coverslip Method

Thin, long, flexible, acid-cleaned coverslips (25 × 77 mm, Corning No. 0, Corning Glass Works, Corning, NY) are dipped (under safelight conditions) in Kodak (Rochester, NY) NTB-3 liquid emulsion diluted 1:1 with H_2O. The emulsion must be placed in a warm-water bath before being diluted. The mixture is then stirred with a microscope slide and must be allowed to stand a couple of hours so that the bubbles created in the solution will rise to the surface. The coverslips are dipped into this solution and are suspended from wooden clothespins to dry for 2 hr. The emulsion will run to the end of the slides, leaving behind a uniformly thin layer except at the lower edge. Previous tests have demonstrated no significant differences in grain density resulting from placing the section at different locations under the emulsion-coated coverslip. The coverslip edge extends slightly over the length of the slide, and thus the slight build-up of emulsion at the end of the slide will not cause problems. Dry emulsion-coated coverslips are sealed in slide boxes containing desiccant and stored in the refrigerator for one to seven days.

When the coverslips and slide-mounted tissue sections have been prepared as described above, both will be allowed to come to room temperature before the boxes are opened (to prevent condensation of moisture on the slides or coverslips). Then, under safelight conditions (Wratten No. 2), a drop of glue is placed on the frosted end of the slide and the coverslip is glued to it such that the emulsion coating is positioned over the labeled tissue section. The long coverslip should extend slightly over the tissue-containing end of the slide. A 1-in square of Teflon is then placed on the coverslip over the tissue section and the layers are clamped tight with a No. 20 binder clip. These assemblies are placed in light-tight wooden slide boxes containing desiccant. The slide boxes are placed in sealable plastic bread boxes (Tupperware) containing desiccant, and these are in turn stored in lead containing plexiglass storage boxes kept in a cold room at 4°C.

After allowing an appropriate exposure period (which depends on the concentration of radioactive ligand bound and its specific activity) the slide boxes are allowed to equilibrate with the room temperature before the boxes are opened (under safelight conditions). The binder clip and Teflon are removed and the coverslip is bent (slightly) back away from the tissue-containing end of the slide. The coverslip is maintained in this position by inserting a spacer (a small piece of plastic or a fragment of toothpick) between the coverslip and slide. This usually results in formation of stress or tension lines in the emulsion where the bend is taking place, usually about halfway along the length of the coverslip. Thus, it is important at the time of sectioning to position the tissue sections towards the end of the slide rather than in the middle.

The autoradiographic image on the coverslip is developed by immersing the assembly for two min in Kodak developer (Dektol diluted 1:1 with H_2O) at 17°C, then 15 sec in liquid hardener (Kodak Liquid Hardener diluted 1:13 with H_2O) and then in Kodak Rapid Fix for 3 min. The slide assemblies are next rinsed several minutes in running water, and the sections are fixed 10–15 min in Carnoy's solution (6 parts ethanol to 3 parts chloroform to 1 part glacial acetic acid). The slides are again rinsed with water, and the sections are then stained in pyronin-Y by immersing them 1 min in 0.2 *M* solution of Na_2HPO_4 containing 0.1 *M* citric acid (pH 5.3), followed by 1 min in the same buffer solution containing 1% pyronin-Y. The slides are rinsed free of the stain with water and they are placed on a 40°C slide warmer to dry for 2 hr or more. The assemblies are then quickly dipped in xylene, and several drops of permanent mounting fluid (Permount, Fisher Scientific Co., Pittsburgh, PA) are placed on the slide. The spacer is then removed and the coverslip is pressed back into place over the tissue section.

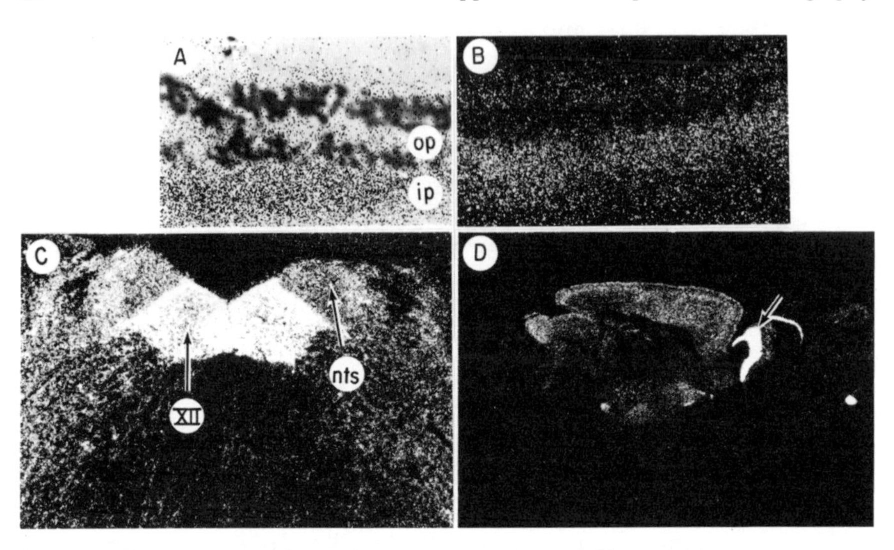

Figure 2. Receptor autoradiography using coverslips and film. (A) Photomicrograph of the autoradiographic grains on an emulsion-coated coverslip associated with a section of human retina incubated with {^{3}H}-QNB to label the muscarinic cholinergic receptors. The grains, which appear as small black dots when using brightfield illumination, are in focus. Slightly below the plane of focus the heavily stained nuclear regions of the retina can be seen. *op*, outer plexiform layer; *ip*, inner plexiform layer. (B) Section is similar to that shown in (A) but is photographed using darkfield illumination. Now the grains appear as white dots and the underlying tissue cannot be seen. Muscarinic cholinergic receptors are plentiful in the inner plexiform layer. A slight increase (over background) in grain density is also seen in the photoreceptor cell layer. (C) Darkfield photomicrograph of the grain density appearing over a section of rat brain stem, with muscarinic cholinergic receptors labeled with {^{3}H}-QNB. Very high grain densities appear in the nucleus of the hypoglossal nerve (*XII*) with somewhat fewer grains over the nucleus of the tractus solitarius (*nts*). (D) Low-magnification print made directly from the autoradiographic image as it appeared on a sheet of tritium-sensitive film placed over a saggital section of rat brain incubated in {^{3}H}-dihydromorphine to label opiate receptors. Arrow indicates the high concentration of these receptors found in the olfactory bulb.

Examination of the autoradiographic image on the coverslip is facilitated by using darkfield (incident light) illumination. The grains appear white against a dark background (Figure 2). Low-power examination of the tissue sections and the autoradiograms can be accomplished with a low-cost Olympus JM microscope equipped with a light and darkfield base. Wild makes a much more expensive model (although much more versatile), called Photomakroskop, which can be outfitted with a brightfield-darkfield base. Low-power observation of slides using only darkfield illumination may result in some erroneous conclusions since areas of heavily stained nuclei can give rise to reflectance of light by the tissue. It is necessary, then, to observe areas of interest under high-power magnification and brightfield illumination to assure that the reflected light is actually due to grain densities. An all-purpose microscope for autoradiography is the Leitz Orthoplan microscope (West Germany) equipped with a Hinsch-Goldman box (Bunton Instruments Co., Rockville, MD). Grain counts can be performed using a microscope with a 40× or 100× oil-immersion objective and a grid-containing eyepiece.

4.2 Tritium Film Method

Slide-mounted tissue sections are prepared in the same manner described above, but instead of placing coverslips on the slides, a sheet of LKB (Rockville, MD) Ultrofilm is placed against them. This is accomplished by taping the back of the slide (double-stick tape) onto a 10 × 12 in. piece of photographic mounting board. The mounted slides are then placed in an X-ray cassette (Wolf Cassette) and, in the dark, a sheet of tritium sensitive film is placed on top of the slides (emulsion-side down). Another piece of mounting board is placed on top of the film, and the X-ray cassette is clamped shut and stored in a cold room for an exposure period. The films are then removed, developed 5 min (20°C) in D-19 (full strength), rinsed in water, fixed 5 min (20°C) in Rapid Fix, rinsed again in running water (20 min), then rinsed in Photo-Flo solution, and finally allowed to dry. The tissue sections can be stained in any suitable histological stains for comparison with the autoradiographic images.

Photographic prints can be made directly from the film by placing the film autoradiogram into an enlarger and exposing photographic paper under it in much the same fashion as one would a 35 mm negative (Figure 1). The grain densities on the film can also be examined by cutting the autoradiographic image out of the film and placing it under a microscope.

4.3 Comparison of the Two Apposition Techniques

The coverslip method affords the highest level of resolution currently available for the localization of many different types of receptors labeled with diffusible ligands. Grain densities in minute regions of the brain or in small tissues can be differentiated with the coverslip method. Anatomically distinct areas containing significant elevations of grains (over background) can be defined by focusing on the grain density and then switching to brightfield illumination to examine the stained tissue section. Quantification of the grain densities requires counting individual grains under the microscope since photometry readings will be adversely affected by the regions of tissue reflectance.

The disadvantages of the film method are a lower resolution and a separation of the autoradiographic image from the stained tissue sections, thus making identification of anatomical regions of grain densities highly dependent on the neuroanatomical expertise of the examiner. The main advantage of the film technique is the ability to quantify regions containing various grain densities using microdensitometry. The autoradiographic images on the tritium-sensitive film resemble those produced on X-ray films when measuring glucose utilization with the {^{14}C}-2-deoxyglucose (2-DG) method (62). The same sort of quantitation methods used with 2-DG utilization can thus be applied to autoradiograms of receptor density. The simplest method of densitometry involves the use of photometry systems that measure transmitted light. Small models of these photometers can be purchased that read the optical density of a tiny beam of light, which can be focused on various parts of the autoradiographic image on the film. Photometry systems can also be attached to microscopes for analysis of the grain density in microscopically identified regions. Computerized image-analysis systems have also been developed for autoradiographic localization of areas with increased glucose utilization (63). These computerized methods can also be applied to the analysis of receptor densities after autoradiographic localization on the tritium-sensitive film (19). The analysis of the image is performed by scanning the autoradiogram with a microdensitometer that feeds the grain densities into a computer system. The computer then reconstructs the image and encodes the corresponding densities on a scale made up of shades of gray or can actually color code the image so the different densities appear as different shades of color on the computer screen (Figure 3).

Intricate methods for analyzing and quantitating the autoradiographic images are still not sufficient for measuring receptor density without a proper study of the response of the film to tritium. A method for the production of standards to include when using the film has been described (58). This method involves mixing various concentrations of radioactivity in brain paste and then freezing and sectioning these samples so that sections containing several different levels of radioactivity can be picked up on a single slide to include in the film exposure. The optical density measurements are optimal at certain exposure periods, i.e., at certain grain densities. The film saturates at a certain point and does not respond to further increases in bound radioactivity. At the other end of the scale, very low levels of radioactivity do not provide a linear relationship as exposure periods increase. Thus, certain exposure periods may not provide optimal grain densities in all regions of the brain. It may be necessary to use shorter exposure periods to analyze grain densities in regions of high receptor concentrations and longer exposure periods to analyze

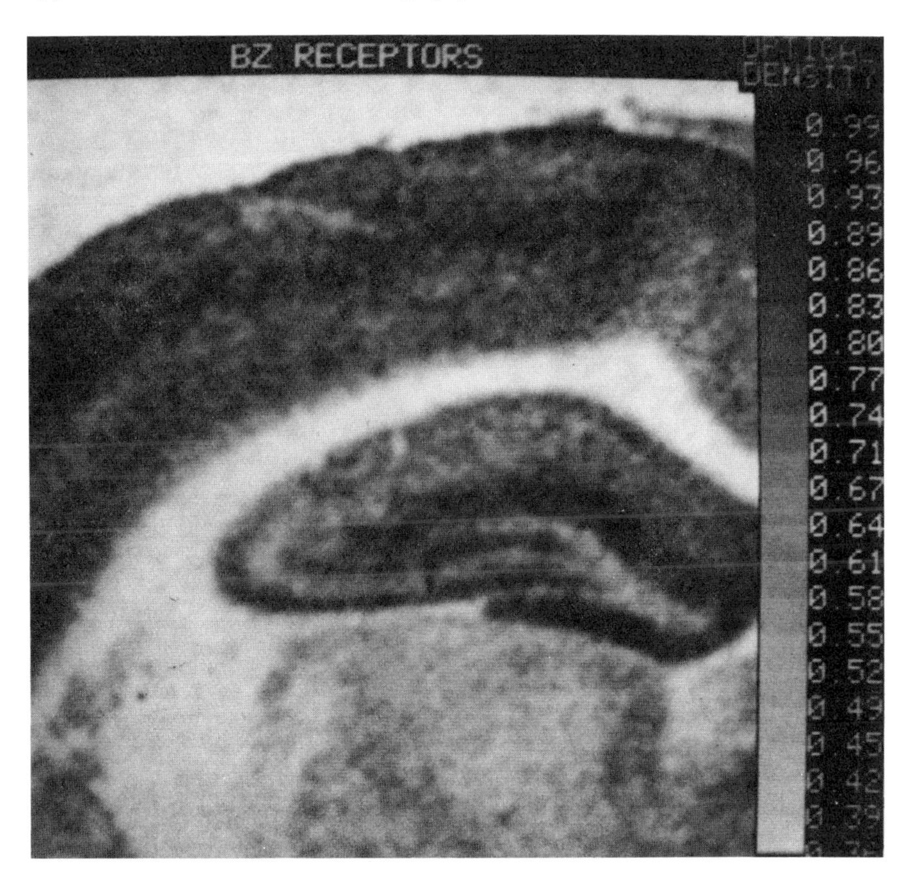

Figure 3. Photograph of a computer-reconstructed image of the grain density found in rat brain after incubation with {^{3}H}-flunitrazepam to label benzodiazepine receptors. The computer image shades the regions of the brain according to grain density. Scale on the right depicts the different shades according to optical density. The image analysis was performed in the Laboratory of Cerebral Metabolism, NIMH, using its computerized microdensitometric equipment.

grain densities in regions of low receptor concentration. By determining the response of the film to certain concentrations of radioactivity in predetermined tissue concentrations, it is possible to determine the concentration of radioactive ligand bound to the slide-mounted tissue sections.

As with the coverslip technique, it is possible to incubate sections in a series of different concentrations of the radioactive ligand. By plotting grain density vs. concentration of radioactivity, it is possible to generate saturation curves of bound radioactivity from the autoradiograms. This makes possible Scatchard analysis and thus determination of the K_D and B_{max} in discrete regions of the nervous system. Since the coverslip method provides higher anatomical resolution, this method is indicated when examining receptor densities in small tissues or in tiny regions of the brain or spinal cord. Receptors in easily definable neuroanatomical areas can best be analyzed with the tritium-sensitive film, especially when one is interested in changes in receptor density following experimental manipulations. When both maximal anatomical resolution and repetitive quantitation are desired, it is advisable to employ both techniques simultaneously.

Interestingly, it has been established that as long as the labeled tissue sections are kept dry, the sections can be exposed to the tritium-sensitive film to generate film autoradiograms and then removed from the X-ray cassettes and used with the coverslip technique of receptor autoradiography (Palacios, Wamsley, and Kuhar, unpublished observations). Both the film autoradiogram and the coverslip autoradiogram can be compared with the stained tissue section, allowing complete utilization of the advantages of both techniques.

Both techniques should include appropriate controls for positive and negative chemography. Positive chemography is spurious grain formation simply resulting from the presence of tissue under the emulsion. Thus, a few unlabeled sections should be exposed to the emulsion to determine if and where grains are produced. Negative chemography is the prevention of grain formation by the presence of the tissue. Thus, a slide-mounted tissue section covered with an emulsion-coated coverslip or with a piece of tritium-sensitive film should be exposed to light before being developed to see if the tissue interfered in any way with the exposure of the emulsion and the formation of grains.

5 APPLICATIONS OF RECEPTOR AUTORADIOGRAPHY

The most productive first step in receptor autoradiography would be to map the normal receptor distribution in serial sections throughout the brain. This would provide a wealth of information on where the neurotransmitter specific for that particular receptor may be having its effects. These studies would also suggest regions where drug effects or unwanted side effects may be mediated. Receptor distributions should be valuable to the physiologist, indicating regions of importance for further studies on recording the effects of administration of receptor-specific compounds. The behavioral scientist should also find receptor-mapping studies helpful in estimating the location mediating behavioral changes found after drug

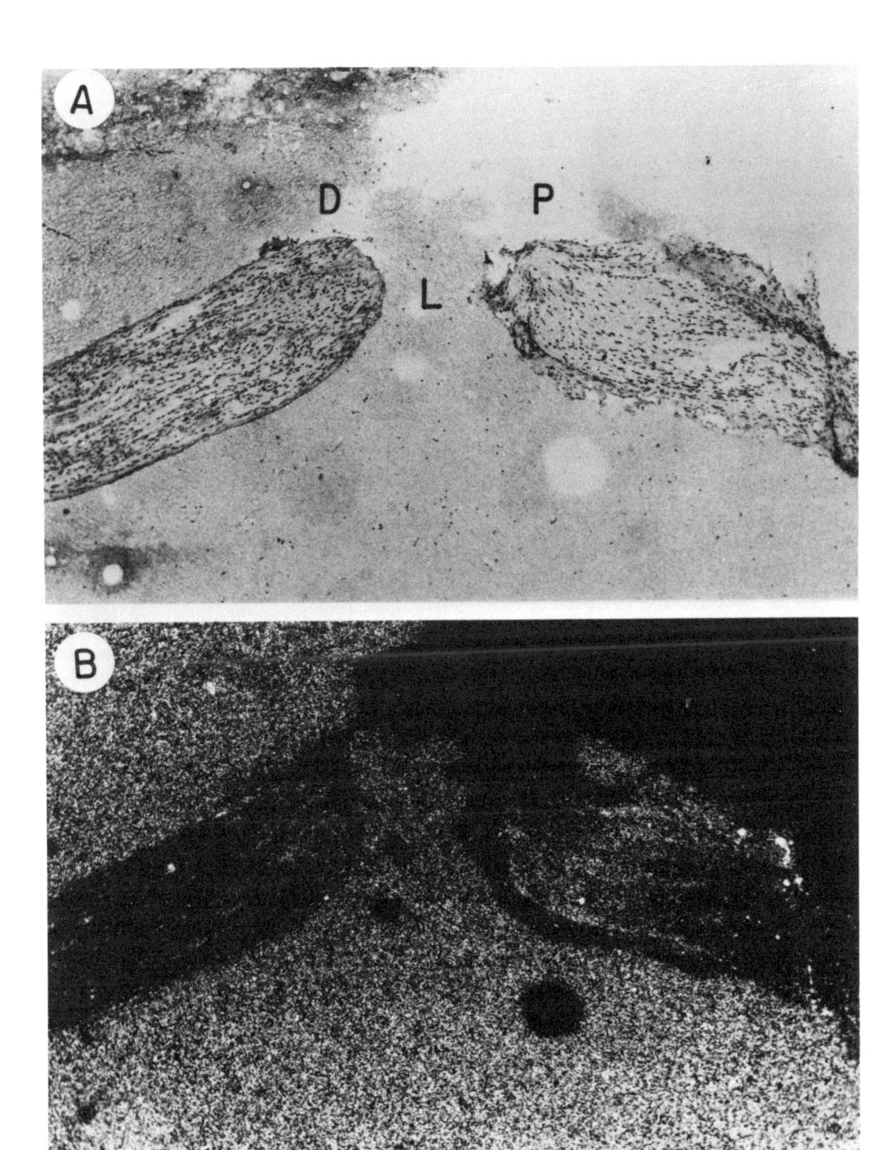

Figure 4. Muscarinic cholinergic receptors flow in the vagus nerve. (A) Photomicrograph shows a stained section of vagus nerve which was ligated 24 hr before being removed from the animal, sectioned longitudinally, and incubated in {^{3}H}-NMS to label the muscarinic cholinergic receptors. The autoradiographic grains are not in the plane of focus and cannot be seen. Depicted are the regions proximal (*P*, toward the nodose ganglion cell bodies near the base of the brain) and distal (*D*) to the ligature (*L*). (B) Without moving the microscope stage, the illuminator was switched to darkfield and this photomicrograph was taken to demonstrate the autoradiographic grains; the areas are in register with those seen in (A). Autoradiographic grains build up proximal to the ligature, indicating an increased density of muscarinic cholinergic receptors. A slight build-up of grains distal to the ligature can also be seen; they might indicate retrograde receptor flow. The grain densities in regions surrounding the nerve are randomly distributed receptors in the homogenized brain that was used to embed the nerve for sectioning.

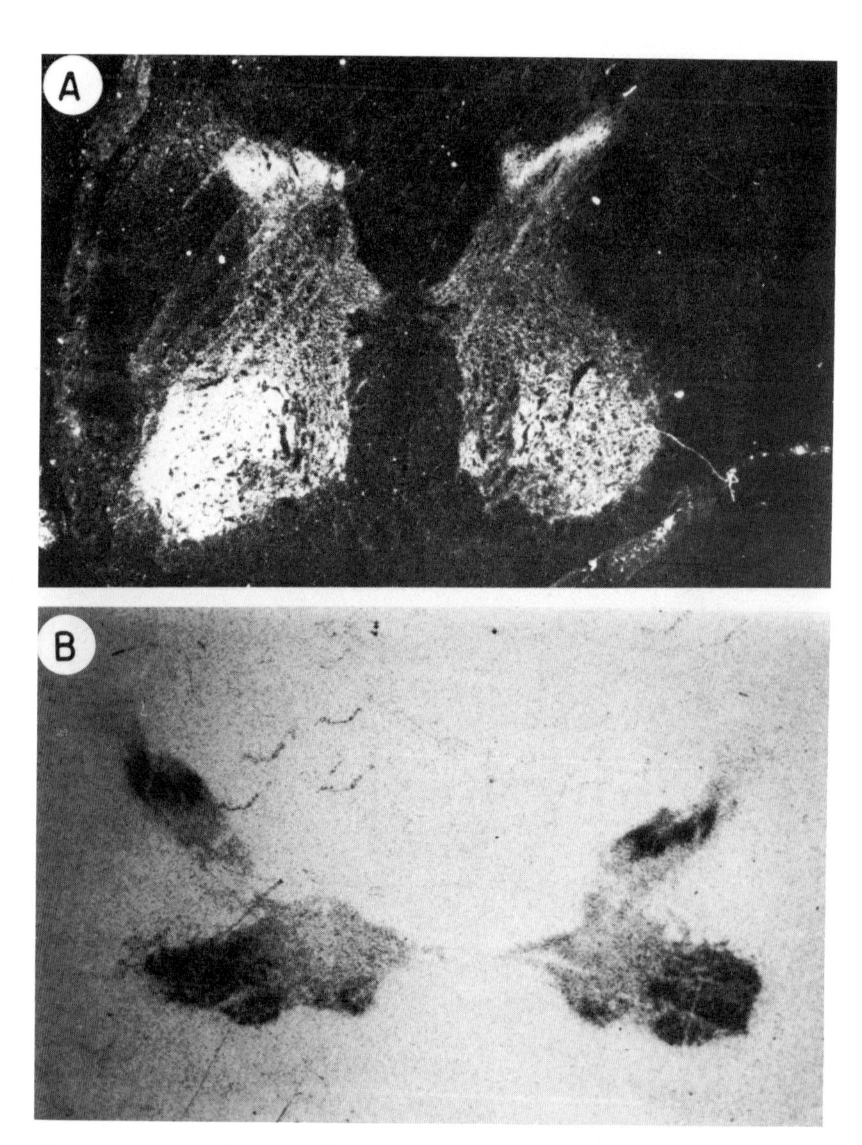

Figure 5. Autoradiograms of human spinal cord. (A) Brightfield photomicrograph of the autoradiogram produced on a coverslip over a section of human lumbar spinal cord with the muscarinic receptors labeled by {^{3}H}-QNB. Grain densities elevated over densities found in white matter regions can be seen throughout the gray matter. The highest concentrations of muscarinic receptors are thus found in the lateral aspects of the ventral horn and in the substantia gelatinosa of the dorsal horn. (B) This is a photographic print made directly from the autoradiographic image seen on the LKB Ultrofilm. The section used to generate the autoradiogram was taken from a cervical level of human spinal cord and was labeled with {^{3}H}-NMS to indicate the position of muscarinic cholinergic receptors.

administration. Ablation of the indicated regions followed by demonstration of receptor removal, followed in turn by the disappearance of the drug effect, would help verify the structure–function relationship.

Once the normal receptor distribution is known, experimental manipulations can be performed which alter the receptor densities or distributions. Kainic acid (a neurotoxin) injection in certain brain regions has been shown to cause the degeneration of neurons with their cell bodies in the region of the injection while leaving the majority of the neurons in passage undamaged (64). Injection of kainic acid then could help identify receptors associated with certain neuronal populations by comparing receptor density before and after the lesion. This approach has been used to supply further evidence for the association of histamine-H_1, GABA, and benzodiazepine receptors with specific cell populations in the cerebellum (65). Chemical or surgical lesions have also been used to autoradiographically identify a population of opiate receptors (9) and GABA receptors (20) on intrinsic neurons in the striatum. Removal of the catecholaminergic neurons from the substantia nigra with 6-hydroxydopamine lesions has associated the neurotensin receptors in the substantia nigra with these neurons (Palacios et al., in preparation). The utility of apposition techniques of receptor autoradiography for determining the differential distribution of receptor subtypes has also been well established (23, 26–28, 40, 41, 44, 45, 47, 58).

Apposition techniques of receptor autoradiography have made possible the demonstration of axonal transport of neurotransmitter receptors (22, 33, 43). This has been accomplished by ligating peripheral nerves and demonstrating a time-dependent build-up of receptors on one side of the ligature (Figure 4). Opioid, cholecystokinin, insulin, and muscarinic cholinergic receptors have been shown to be flowing in the vagus nerve axons while muscarinic cholinergic, β-adrenergic, and histamine-H_1 receptors flow in the sciatic nerve axons.

In vitro incubation of tissues for receptor autoradiography makes possible the localization of receptors in human tissues (Figure 5). The distribution of receptors in some post-mortem human brain regions have been examined (24, 38, 47, 66).

6 FURTHER USES OF RECEPTOR AUTORADIOGRAPHY

Some studies that are already under way seek to "visualize" the effect of chronic drug treatment on certain receptor populations. These studies may be enlightening as to the role certain receptors play in drug tolerance, withdrawal, and addiction. Using autoradiographic receptor techniques, it should also be possible to study such receptor responses as supersensitivity and subsensitivity, or to help define a role for receptors in mediating such phenomena as tardive dyskinesia.

Receptor binding assays have demonstrated changes in receptor densities in certain nuclei of the human brain in persons who suffer from various psychiatric and neurological disorders (67). Many autoradiographic studies are under way to determine exactly where, within the various nuclei, the neurotransmitter receptor deficits are taking place. These studies will be valuable when the use of receptor localization by PET (position emission tomography) scanning for disease diagnosis in living humans "comes of age." These studies may also suggest possible drug treatments to alleviate some of disease symptoms.

Attempts should still be made to improve the methods of receptor autoradiography, especially the resolution. Resolution problems in receptor autoradiography may be encountered even with the magnification capabilities of the electron microscope (68). Still, apposition techniques of receptor autoradiography will undoubtedly prove a useful adjunct to investigations that attempt to localize sites of drug or neurotransmitter action.

ACKNOWLEDGMENTS

We wish to express our appreciation to Leslie Bangerter for her phenomenal secretarial assistance. We are also grateful to Dr. Michael J. Kuhar and Drs. W. Scott Young III, Marco A. Zarbin, James R. Unnerstall, Debra L. Niehoff, and all others who have contributed to the development and use of these receptor autoradiographic techniques.

REFERENCES

1. S. H. Snyder, "Overview of Neurotransmitter Receptor Binding," in H. I. Yamamura, S. J. Enna, and M. J. Kuhar, Eds., *Neurotransmitter Receptor Binding*, Raven Press, New York, 1978, pp. 1–11.
2. M. J. Kuhar and H. I. Yamamura, *Nature* **253**, 560 (1975).
3. M. J. Kuhar and H. I. Yamamura, *Brain Res.* **110,** 229 (1976).
4. C. B. Pert, M. J. Kuhar, and S. H. Snyder, *Life Sci.* **16,** 1849 (1975).
5. S. F. Atweh and M. J. Kuhar, *Brain Res.* **124,** 53 (1977a).
6. S. F. Atweh and M. J. Kuhar, *Brain Res.* **129,** 1 (1977b).
7. S. F. Atweh and M. J. Kuhar, *Brain Res.* **134,** 393 (1977c).
8. C. B. Pert, M. J. Kuhar, and S. H. Snyder, *Proc. Natl. Acad. Sci. USA* **73,** 3729 (1976).

9. L. C. Murrin, J. T. Coyle, and M. J. Kuhar, *Life Sci.* **27,** 1175 (1980).
10. J. Pearson, L. Brandeis, E. Simon, and J. Hiller, *Life Sci.* **26,** 1047 (1980).
11. J. K. Wamsley, M. A. Zarbin, W. S. Young III, and M. J. Kuhar, *Neuroscience* **7**, 595 (1982).
12. P. Schubert, V. Hollt, and A. Herz, *Life Sci.* **16,** 1855 (1975).
13. M. J. Kuhar, L. C. Murrin, A. Malouf, and N. Klemm, *Life Sci.* **22,** 203 (1978).
14. L. C. Murrin, K. Gale, and M. J. Kuhar, *Eur. J. Pharmacol.* **60,** 229 (1979).
15. L. C. Murrin and M. J. Kuhar, *Brain Res.* **177,** 279 (1979).
16. N. Klemm, L. C. Murrin, and M. J. Kuhar, *Brain Res.* **169,** 1 (1979).
17. W. S. Young III and M. J. Kuhar, *Brain Res.* **179,** 255 (1979).
18. L. J. Roth, I. M. Diab, M. Watanabe, and R. Dinerstein, *Mol. Pharmacol.* **10,** 986 (1974).
19. J. M. Palacios, D. L. Niehoff, and M. J. Kuhar, *Neurosci. Lett.* **25**, 101 (1981).
20. J. B. Penney, K. Frey, and A. B. Young, *Eur. J. Pharmacol.* **72,** 421 (1981).
21. T. C. Rainbow, W. Bleisch, A. Biegon, and B. S. McEwen, *J. Neurosci. Methods* **5**, 127 (1982).
22. W. S. Young III, J. K. Wamsley, M. A. Zarbin, and M. J. Kuhar, *Science* **210,** 76 (1980).
23. R. R. Goodman, S. H. Snyder, M. J. Kuhar, and W. S. Young III, *Proc. Natl. Acad. Sci. USA* **77,** 6239 (1980).
24. J. K. Wamsley, J. M. Palacios, and M. J. Kuhar, *Neurosci. Lett.* **27**, 19 (1981).
25. M. A. Zarbin, J. K. Wamsley, and M. J. Kuhar, *J. Neurosci.* **1,** 532 (1981).
26. J. M. Palacios and M. J. Kuhar, *Science* **208,** 1378 (1980).
27. W. S. Young III and M. J. Kuhar, *Eur. J. Pharmacol.* **59,** 317 (1979).
28. W. S. Young III and M. J. Kuhar, *Proc. Natl. Acad. Sci. USA* **77,** 1696 (1980).
29. W. S. Young III and M. J. Kuhar, *Eur. J. Pharmacol.* **59,** 161 (1979).
30. W. S. Young III and M. J. Kuhar, *Brain Res.* **206,** 273 (1981).
31. W. S. Young III, M. J. Kuhar, J. Roth, and M. J. Brownstein, *Neuropeptides* **1,** 15 (1980).
32. M. A. Zarbin, R. B. Innis, J. K. Wamsley, S. H. Snyder, and M. J. Kuhar, *Eur. J. Pharmacol.* **71,** 349 (1981).
33. M. A. Zarbin, J. K. Wamsley, R. B. Innis, and M. J. Kuhar, *Life Sci.* **29,** 697 (1981).
34. A. C. Foster, E. E. Mena, D. T. Monaghan, and C. W. Cotman, *Nature* **289,** 73 (1981).
35. H. Henke, A. Beaudet, and M. Cuenod, *Brain Res.* **219,** 95 (1981).
36. J. M. Palacios, W. S. Young III, and M. J. Kuhar, *Proc. Natl. Acad. Sci. USA* **77,** 670 (1980).
37. J. M. Palacios, J. K. Wamsley, and M. J. Kuhar, *Brain Res.* **222**, 285 (1981).
38. W. S. Young III and M. J. Kuhar, *Nature* **280,** 393 (1979).
39. W. S. Young III and M. J. Kuhar, *J. Pharmacol. Exp. Ther.* **212,** 337 (1980).
40. W. S. Young III, D. Niehoff, M. J. Kuhar, B. Beer, and A. S. Lippa, *J. Pharmacol. Exp. Ther.* **216,** 425 (1981).
41. J. K. Wamsley, M. A. Zarbin, N. J. M. Birdsall, and M. J. Kuhar, *Brain Res.* **200,** 1 (1980).
42. J. K. Wamsley, M. Lewis, W. S. Young III, and M. J. Kuhar, *J. Neurosci.* **1,** 176 (1981).
43. J. K. Wamsley, M. A. Zarbin, and M. J. Kuhar, *Brain Res.* **217**, 155 (1981).
44. J. M. Palacios, D. L. Niehoff, and M. J. Kuhar, *Brain Res.* **213,** 277 (1981).

45. W. S. Young III and M. J. Kuhar, *Eur. J. Pharmacol.* **62,** 237 (1980).

46. R. C. Meibach, S. Maayani, and J. P Green, *Eur. J. Pharmacol.* **67,** 371 (1980).

47. J. M. Palacios, W. S. Young III, and M. J. Kuhar, *Eur. J. Pharmacol.* **58,** 295 (1979).

48. J. M. Palacios, J. K. Wamsley, and M. J. Kuhar, *Neurosci.* **6,** 15 (1981).

49. A. W. Rogers, *Techniques of Autoradiography*, Elsevier/North-Holland, Amsterdam, 1979.

50. G. Polz-Tejera, J. Schmidt, and H. J. Karten, *Nature* **258,** 349 (1975).

51. S. Hunt and J. Schmidt, *Brain Res.* **157,** 213 (1978).

52. M. Segal, Y. Dudai, and A. Amsterdam, *Brain Res.* **148,** 105 (1978).

53. Y. Arimatsu, A. Seto, and T. Amano, *Brain Res.* **147,** 165 (1978).

54. A. Rotter, N. J. M. Birdsall, A. S. V. Burgen, P. M. Field, E. C. Hulme, and G. Raisman, *Brain Res. Rev.* **1,** 141 (1979).

55. L. J. Roth and W. E. Stumpf, *Autoradiography of Diffusible Substances*, Academic Press, New York, 1969.

56. D. C. Gajdusek, C. J. Gibbs, D. M. Asher, P. Brown, A. Diwan, P. Hoffman, G. Nemo, R. Rohwer, and L. White, *New Engl. J. Med.* **297,** 1253 (1977).

57. J. P. Bennett Jr., "Methods in Binding Studies," in H. I. Yamamura, S. J. Enna, and M. J. Kuhar, Eds., *Neurotransmitter Receptor Binding*, Raven Press, New York, 1978, pp. 57–90.

58. J. R. Unnerstall, D. L. Niehoff, M. J. Kuhar, and J. M. Palacios, in press.

59. S. J. Peroutka and S. H. Snyder, *Mol. Pharmacol.* **16,** 687 (1979).

60. J. K. Wamsley, "Opioid Receptors: Autoradiography," *Pharmacological Reviews*, in press.

61. E. D. London and J. T. Coyle, *Mol. Pharmacol.* **15,** 492 (1979).

62. L. Sokoloff, *J. Neurochem.* **29,** 13 (1978).

63. C. Goochee, W. Rasband, and L. Sokoloff, *Ann. Neurol.* **7,** 359 (1980).

64. J. T. Coyle, M. E. Molliver, and M. J. Kuhar, *J. Comp. Neurol.* **180,** 301 (1978).

65. J. M. Palacios, J. K. Wamsley, and M. J. Kuhar, *Brain Res.* **214,** 155 (1981).

66. W. S. Young III and M. J. Kuhar, "Opioid Receptor Autoradiography in Brains of Humans and Animals," in E. L. Way, Ed., *Endogenous and Exogenous Opiate Agonists and Antagonists*, Pergamon Press, New York, 1980, pp. 131-134.

67. R. W. Olsen, T. D. Reisine, and H. I. Yamamura, *Life Sci.* **27,** 801 (1980).

68. M. J. Kuhar, N. Taylor, J. K. Wamsley, E. C. Hulme, and N. J. M. Birdsall, *Brain Res.* **216,** 1 (1981).

Chapter 9

Localization of Activity-Associated Changes in Metabolism of the Central Nervous System with the Deoxyglucose Method: Prospects for Cellular Resolution

Carolyn B. Smith

Laboratory of Cerebral Metabolism
National Institute of Mental Health
Bethesda, Maryland

The deoxyglucose method was developed by Sokoloff et al. (1) in order to determine quantitatively the rates of glucose utilization simultaneously in all structural and functional components of the central nervous system. Numerous studies with the method have established that a close relationship exists between functional activity and energy metabolism in the central nervous system (2). As originally described, the method employs quantitative autoradiography in which 2-[^{14}C]deoxy-D-glucose (DG)* is the radiolabeled tracer. The autoradiographic images produced by the method resemble histological sections of nervous tissue, and the method is often taken for a neuroanatomical technique. This classification obscures a unique and most significant feature of the method. The images are not of structure; they are images of a dynamic biochemical process, glucose utilization, which may be as much a function of the physiological state of nerve cells as the biophysical properties of their membranes. The quantitative resolution of the [^{14}C]DG autoradiographs is at the very best about 200 μm. Even with this limited resolving power the method has made it possible to map changes in metabolic and functional activity in regions of the brain as small as cortical layers, whisker barrels, and parts of brain nuclei. The possibility of visualizing and measuring the metabolic rate in single cells and even single synapses remains an important goal.

There are a number of reports in the literature of cellular resolution achieved with the deoxyglucose method. In this chapter the basic principles of the deoxyglucose method and the techniques necessary to achieve its present limits of resolution are described, followed by a consideration of the factors in general that affect autoradiographic resolution and those factors in particular which are especially relevant to the deoxyglucose method. Previously proposed methods to achieve finer resolution with the deoxyglucose method are critically examined, and guidelines and criteria to be met are presented. Finally, those problems associated with quantitative autoradiography at the cellular level which must be resolved in order to measure actual rates of glucose metabolism in single cells will be analyzed.

1 THE DEOXYGLUCOSE METHOD

1.1 Theoretical Basis for the Deoxyglucose Method

The deoxyglucose method is an *in vivo* biochemical method. It is based on several fundamental principles. One is tracer theory, which states that the tracer, 2-deoxy-D-glucose, when introduced into a system in a steady state, behaves in a manner similar to the natural molecules for which it is a tracer, in this case glucose, and also that the concentration of tracer is negligible and insufficient to alter the kinetic parameters of

* The abbreviations used in this chapter are DG, 2-deoxy-D-glucose; DG-6-P, 2-deoxy-D-glucose-6-phosphate; G-6-P, glucose-6-phosphate; ATP, adenosine triphosphate.

the process being traced. Another principle is kinetic compartmental analysis, which makes it possible to define concentrations of the tracer in compartments that cannot be measured directly. The third principle is that of Michaelis–Menten enzyme kinetics, modified for the kinetic behavior of competitive substrates. The kinetic model (Figure 1) from which the deoxyglucose method was derived is based on the known biochemical properties of 2-deoxyglucose and glucose in brain. 2-Deoxyglucose is transported bidirectionally between blood and brain by the same carrier that transports glucose across the blood–brain barrier (3). In brain it is phosphorylated to 2-deoxyglucose-6-phosphate (DG-6-P) by hexokinase (4). 2-Deoxyglucose and glucose are, therefore, competitive substrates for both blood–brain transport and hexokinase-catalyzed phosphorylation. Unlike glucose-6-phosphate (G-6-P), however, which is metabolized further, eventually to CO_2 and water and to a lesser degree via the hexosemonophosphate shunt, DG-6-P cannot be converted to

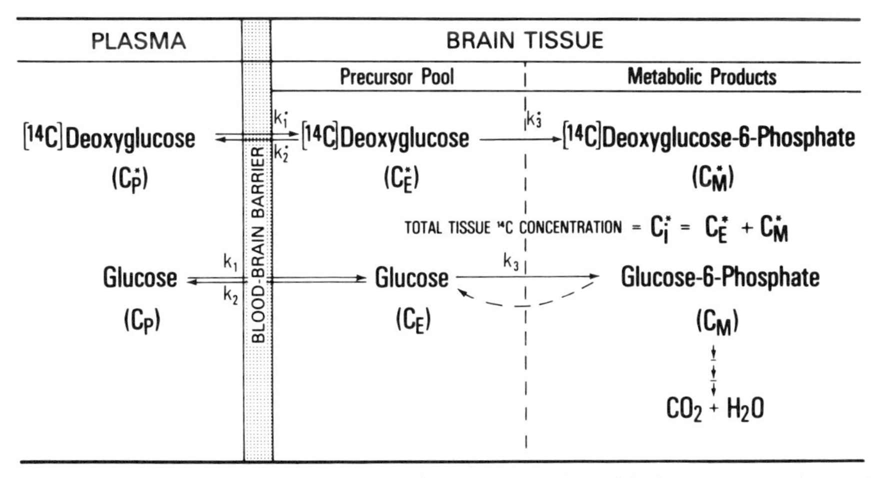

Figure 1. Diagrammatic representation of the theoretical model. C_i^* represents the total ^{14}C concentration in a single homogeneous tissue of the brain. C_p^* and C_p represent the concentrations of [^{14}C]deoxyglucose and glucose, respectively, in the arterial plasma; C_E^* and C_E represent their respective concentrations in the tissue pools that serve as substrates for hexokinase. C_M^* represents the concentration of [^{14}C]deoxyglucose-6-phosphate in the tissue. The constants k_1^*, k_2^*, and k_3^* represent the rate constants for carrier-mediated transport of [^{14}C]deoxyglucose from plasma to tissue, for carrier-mediated transport back from tissue to plasma, and for phosphorylation by hexokinase, respectively. The constants k_1, k_2, and k_3 are the equivalent rate constants for glucose. [^{14}C]Deoxyglucose and glucose share and compete for the carrier that transports both between plasma and tissue and for hexokinase, which phosphorylates them to their respective hexose-6-phosphates. The dashed arrow represents the possibility of glucose-6-phosphate hydrolysis by glucose-6-phosphatase activity, if any. From Sokoloff et al. (1).

fructose-6-phosphate and is not a substrate for G-6-P dehydrogenase (4). There is very little glucose-6-phosphatase activity in brain (5) and even less deoxyglucose-6-phosphatase activity (1). DG-6-P, therefore, once formed, is essentially trapped in the cerebral tissues, at least long enough for the duration of the measurement (1).

If the interval of time is kept short enough to allow the assumption of negligible loss of [^{14}C]DG-6-P from the tissues—(less than an hour)—then the quantity of [^{14}C]DG-6-P accumulated in any cerebral tissue at any given time following the introduction of [^{14}C]DG into the circulation is equal to the integral of the rate of [^{14}C]DG phosphorylation by hexokinase in that tissue during that interval of time. This integral is in turn related to the amount of glucose that has been phosphorylated over the same interval, depending on the time courses of the relative concentrations of [^{14}C]DG and glucose in the precursor pools and the Michaelis–Menten kinetic constants for hexokinase with respect to both [^{14}C]DG and glucose. With cerebral glucose consumption in a steady state, the amount of

General Equation for Measurement of Reaction Rates with Tracers:

$$\text{Rate of Reaction} = \frac{\text{Labeled Product Formed in Interval of Time, 0 to T}}{\left[\begin{array}{c}\text{Isotope Effect}\\ \text{Correction Factor}\end{array}\right]\left[\begin{array}{c}\text{Integrated Specific Activity}\\ \text{of Precursor}\end{array}\right]}$$

Operational Equation of [^{14}C]Deoxyglucose Method:

$$R_i = \frac{\overbrace{\underbrace{C_i^*(T)}_{\text{Total }^{14}\text{C in Tissue at Time, T}} - \underbrace{k_1^* e^{-(k_2^*+k_3^*)T} \int_0^T C_p^* e^{(k_2^*+k_3^*)t}\,dt}_{^{14}\text{C in Precursor Remaining in Tissue at Time, T}}}^{\text{Labeled Product Formed in Interval of Time, 0 to T}}}{\underbrace{\left[\frac{\lambda \cdot V_m^* \cdot K_m}{\Phi \cdot V_m \cdot K_m^*}\right]}_{\text{"Isotope Effect" Correction Factor}} \underbrace{\left[\underbrace{\int_0^T \left(\frac{C_p^*}{C_p}\right) dt}_{\text{Integrated Plasma-Specific Activity}} - \underbrace{e^{-(k_2^*+k_3^*)T} \int_0^T \left(\frac{C_p^*}{C_p}\right) e^{(k_2^*+k_3^*)t}\,dt}_{\text{Correction for Lag in Tissue Equilibration with Plasma}}\right]}_{\text{Integrated Precursor-Specific Activity in Tissue}}} \quad \text{[Equation 1]}$$

Figure 2. Operational equation of radioactive deoxyglucose method and its functional anatomy. T represents the time at the termination of the experimental period; λ equals the ratio of the distribution space of deoxyglucose in the tissue to that of glucose; Φ equals the fraction of glucose that, once phosphorylated, continues down the glycolytic pathway; K_m^* and V_m^* represent the familiar Michaelis–Menten kinetic constants of hexokinase for deoxyglucose, and K_m and V_m for glucose. The other symbols are the same as those defined in Figure 1. From L. Sokoloff (6).

Table 1. Values of Rate Constants and Lumped Constant in Normal Conscious Albino Rats

	Rate Constants (min^{-1})		
	k_1^*	k_2^*	k_3^*
Mean gray matter structures[a]	0.189	0.245	0.647
± SEM	±0.012	±0.040	±0.073
Mean white matter structures[b]	0.079	0.133	0.020
± SEM	±0.008	±0.046	±0.020
	Lumped Constant		
$\lambda V_m^* K_m / \Phi V_m K_m^*$ [c]	0.481 ± 0.023		

[a] Values given are the means ± SEM for 15 structures in 75 rats.
[b] Values given are the means ± SEM for 3 structures in 15 rats.
[c] Value given is the mean ± SEM for 26 rats: 15 conscious, 9 anesthetized, 2 conscious with 5% CO_2. There were no significant differences among these three groups.

glucose phosphorylated during the interval of time equals the steady state flux of glucose through the hexokinase-catalyzed step times the duration of the interval, and the net rate of flux of glucose through this step equals the rate of glucose utilization.

These relationships can be mathematically defined and an operational equation derived if the following assumptions are made: (i) a steady state for glucose (that is, constant plasma glucose concentration and constant rate of glucose consumption) throughout the period of the procedure; (ii) homogeneous tissue compartment within which the concentrations of [^{14}C]DG and glucose are uniform and exchange directly with the plasma; and (iii) tracer concentrations of [^{14}C]DG, that is, molecular concentrations of free [^{14}C]DG essentially equal to zero. The operational equation that defines R_i, the rate of glucose consumption per unit mass of tissue, i, in terms of measurable variables is presented in Figure 2.

The rate constants, k_1^*, k_2^*, and k_3^*, have been determined (Table 1) in the normal, conscious rat by a nonlinear, iterative process. This process provides the least-squares best fit of an equation that defines the time course of total tissue ^{14}C concentration in terms of the time, the history of the plasma concentration, and the rate constants to the experimentally determined time courses of tissue and plasma concentrations of ^{14}C.

The λ, ϕ, and the enzyme kinetic constants are grouped together to constitute a single, lumped constant (Figure 2) (6). It can be shown mathematically that this lumped constant is equal to the asymptotic value of the product of the ratio of the cerebral extraction ratios of [^{14}C]DG and glucose and the ratio of the arterial blood to plasma-specific activities when the arterial plasma [^{14}C]DG concentration is maintained constant (1). The lumped constant (Table 1) is also determined in a separate group of animals from arterial and cerebral venous blood samples drawn during a programmed intravenous infusion that produces and maintains a constant arterial plasma [^{14}C]DG concentration (1). The lumped constant has been determined in the albino rat, monkey, cat, and dog. It appears to be characteristic of the species and does not appear to change significantly in a wide range of physiological conditions (1), except in severe hyper- and hypoglycemic states, in which it has been shown to vary with the plasma glucose concentration (7, 8).

1.2 Experimental Procedure

The operational equation of the method (Figure 2) defines the variables to be measured in each determination: (i) the entire history of the arterial plasma [^{14}C]DG concentration, C_p^*, from zero time to the time of killing, T; (ii) the steady-state arterial plasma glucose level, C_p, over the same interval; and (3) the local concentration of ^{14}C in the tissue at the time of killing, $C_i^*(T)$. The values for the rate constants and the lumped constant that are used are those determined in a separate group of animals (Table 1). The values for the rate constants will vary with plasma glucose concentration, blood flow, blood-brain barrier transport, and with the experimental conditions. The procedure has been designed to minimize the importance of their values. Thus, if the [^{14}C]DG is administered as an intravenous pulse and allowed to clear from the plasma over 45 minutes, the value of C_p^* approaches zero, and consequently the exponential terms in both the numerator and denominator which contain the rate constants become so small as to have very little influence on the final result.

1.2.1 Protocol

The animals are prepared for the experiment by the insertion of plastic catheters in a femoral or tail artery and vein while under light halothane anesthesia. Their hindlimbs are restrained by encasing them in a plaster cast taped to a lead brick. At least two hours are allowed for recovery from the surgery and anesthesia.

The procedure is initiated with an intravenous pulse injection of 125 μCi of [^{14}C]DG per kg body weight. Arterial sampling is initiated with the onset of the pulse, and timed 50 to 100 μl samples of arterial blood are collected consecutively as rapidly as possible during the early period so as not to miss the peak of the arterial curve. Arterial sampling is continued at sufficient frequency to fully define the arterial curve. The arterial blood samples are immediately centrifuged to separate the plasma, which is stored on ice until assayed for [^{14}C]DG by liquid scintillation counting and glucose concentrations by standard enzymatic methods. At about 45 min the animal is killed and the brain is removed and frozen in isopentane or Freon XII at −40° to −60°C. The brain is then coated with embedding medium and stored in plastic bags at −70°C until sectioned and autoradiographed.

1.2.2 *Quantitative Autoradiography*

The ^{14}C concentrations in localized regions of the brain are measured by a modification of the quantitative autoradiographic technique previously described (9). Brain sections precisely 20 μm thick are prepared in a cryostat maintained at −18° to −22°C. The brain sections are picked up on glass slides, dried on a hot plate at 60°C for at least 5 min, and placed sequentially in an X-ray cassette. A set of [^{14}C]methylmethacrylate standards (Amersham Corp., Arlington Heights, IL), which include a blank and a series of progressively increasing ^{14}C concentrations, is also placed in the cassette. These standards must previously have been calibrated for their autoradiographic equivalence to the ^{14}C concentrations in 20 μm thick brain sections prepared as described above. The method of calibration has been described previously (9).

Autoradiographs are prepared from these sections directly in the X-ray cassette with Kodak single-coated, blue-sensitive Medical X-ray Film, Type SB-5 (Eastman Kodak Co., Rochester, NY). The exposure time is generally 5 to 6 days with the doses used as described above, and the exposed films are developed according to the instructions supplied with the film. The SB-5 X-ray film is rapid but coarse-grained. For finer grained autoradiographs, and therefore sharper images, it is possible to use mammographic films such as DuPont LoDose or Kodak MR-1 films, or fine-grain panchromatic film such as Kodak Plus-X, but the exposure times are two to three times longer. The autoradiographs provide a pictorial representation of the relative ^{14}C concentrations in the various cerebral structures and the plastic standards (Figure 3). A calibration curve of the relationship between optical density and tissue ^{14}C concentrations for each film is obtained by densitometric measurements of the portions of the film representing the various standards. The local tissue concentrations are then determined from the calibration curve and the optical densities of the film in the regions representing the cerebral structures of interest. Local cerebral glucose utilization is calculated from

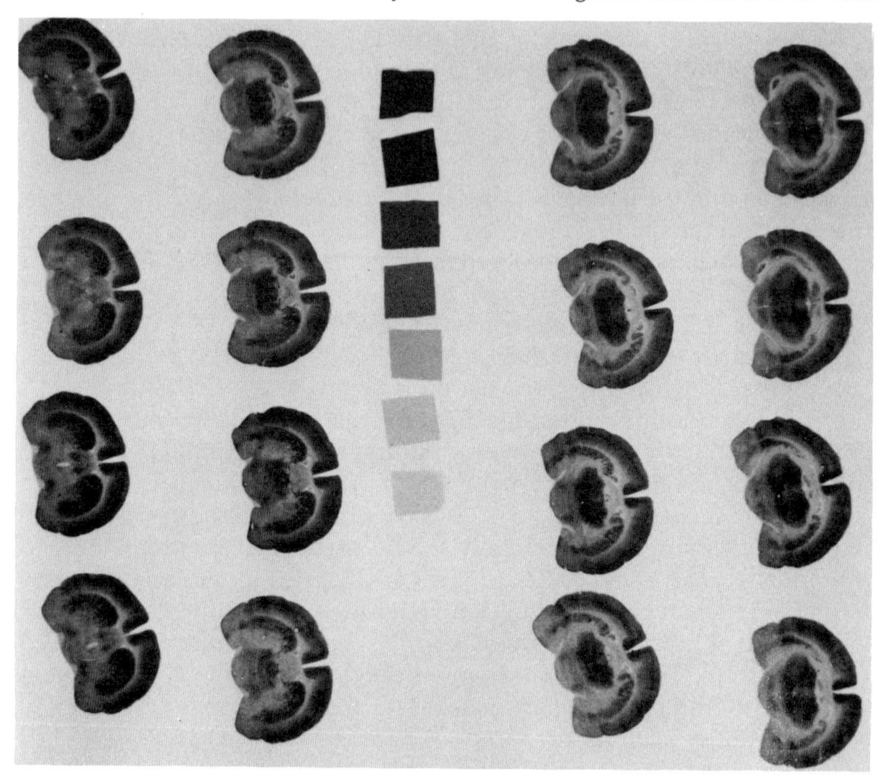

Figure 3. Typical [^{14}C]deoxyglucose autoradiograph. Brain sections 20 μm in thickness were cut at −18°C, thaw-mounted on glass coverslips, and placed in a cassette with calibrated [^{14}C]methylmethacrylate standards and Kodak SB-5 X-ray film. After an exposure of 5 days the film was developed. The resulting image, shown here, illustrates the heterogeneity of brain tissue with regard to the rate of glucose utilization. The darker areas in the autoradiograph (e.g., Layer IV in the cortex, the lateral habenula) represent regions with greater amounts of product formed and therefore higher levels of glucose utilization. The squares in the center are the autoradiographic images of the standards with a 50-fold range of [^{14}C] concentration.

the local tissue concentrations of ^{14}C and the plasma [^{14}C]DG and glucose concentrations according to the operational equation (Figure 2).

1.2.3 *Image Processing*

The [^{14}C]DG autoradiographs are pictorial representations of relative rates of glucose utilization in the various structures of the brain. These images can be digitized and analyzed through the use of a computer-based scanning technique adapted (10) from the techniques used by the intelligence community for the analysis of reconnaissance photographs. With this technique autoradiographs are scanned automatically by a computer-controlled microdensitometer with optional aperture sizes ranging from 12.5 to 400 μm. The optical density of each spot is stored in a computer, converted to ^{14}C concentration on the basis of the optical densities of the calibrated ^{14}C plastic standards, and then converted to rates of glucose utilization by solution of the operational equation of the method. Colors are assigned to narrow ranges of glucose utilization, and the autoradiographs are reconstructed and displayed in color on a color monitor, along with a color calibration scale for identifying from its color the rate of glucose utilization in each spot on the autoradiograph. This technique provides an excellent means of presenting the data obtained with the deoxyglucose method. The use of color coding enhances the resolution of differences in rates of glucose utilization because of the greater ability of the human visual system to discriminate colors than shades of gray. Reconstructed color-coded autoradiographs greatly increase perception of marked heterogeneity of metabolic activities throughout the brain.

Although image-processing techniques cannot improve upon the actual resolution of the autoradiographs, they do make it possible to make use of the resolving power to its limits.

1.3 The Role of the Deoxyglucose Method in Neurobiology

The deoxyglucose method has enabled neuroscientists to map and quantify changes in glucose utilization simultaneously in all structural and functional components of the nervous system in conscious animals in a relatively noninvasive manner. It has been established in numerous studies (2) that there is a close relationship between the rate of glucose utilization and the functional state of nervous tissue. Regions of the nervous system activated by some physiological or pharmacological manipulation of the animal exhibit an increase in the rate of glucose utilization. Conversely, regions which are "quieted" by some manipulation exhibit a decrease in the rate of glucose utilization. The terms "activated" and "quieted" are used in the most general sense because the relationship between the rate of energy metabolism and the precise neurophysiological state of a neuron is not understood.

A direct relationship has been shown (11) in the rat superior cervical ganglion between the frequency of action potentials and the rate of glucose utilization. This relationship is presumably the result of the coupling between energy consumption and the maintenance and restoration of the Na+ and K+ gradients across neuronal membranes.

Following depolarization and the ensuing flux of Na^+ and K^+ ions across the membranes, the ionic gradients are reestablished by the action of the sodium pump. The energy required to transport the ions back across the cell membrane is presumably derived from the splitting of ATP by Na^+,K^+-ATPase (12, 13). Once ATP is split, there are adequate biochemical mechanisms to explain the increased glucose utilization. *In vitro* experiments with the rat posterior pituitary strongly support this explanation of the coupling between neuronal activation and glucose utilization (14). Depolarization of the pituitary preparation either electrically or by veratridine, an alkaloid that opens Na^+ channels, resulted in increased glucose utilization, which could be blocked by either ouabain or tetrodotoxin.

The relationship between the rate of glucose utilization and sodium pump activity is reflected in the variability in the metabolic responsiveness among brain regions. The DG autoradiographs obtained from the brain of a resting conscious rat reveal that the rate of glucose consumption of gray matter is higher than that of white matter. Among gray matter regions there is a fourfold range in values. Regions that are particularly prominent in the autoradiographs are areas rich in neuropil, e.g., cortical Layer IV and the molecular layer of the hippocampus. Similarly, it has been shown in the hypothalamo-neurohypophysial system of the rat (15) that changes in glucose utilization which occur as a result of a physiological manipulation are found primarily in the neuropil, while changes in perikarya are more difficult to detect. Physiological stimulation of this pathway by salt-loading is known to increase the firing rate of the magnocellular neurons in the supraoptic and paraventricular nuclei as well as the secretion of vasopressin at the terminals of these neurons in the posterior pituitary. The changes in glucose utilization under these conditions were limited to the posterior pituitary, in which there was a profound increase. The effect on the nuclei, however, was not detectable. These results suggest a greater sensitivity of neuropil than perikarya to metabolic stimulation by functional activation. This is most likely a reflection of the greater surface-area-to-volume ratio in dendrites and terminals compared with cell bodies. The greater the surface-area-to-volume ratio the greater is the flux of ions per unit volume per action potential and the greater is the activity of the Na^+,K^+-ATPase per unit volume. In general, changes in glucose utilization in response to neuronal activation are most evident in regions rich in terminals and dendrites.

The deoxyglucose method has been applied to a wide range of problems in neurobiology. As these have been reviewed extensively elsewhere (16), only a few examples are presented here. These examples illustrate the sort of resolution attainable as well as the wide range of problems to which the method lends itself. The visual system has been studied

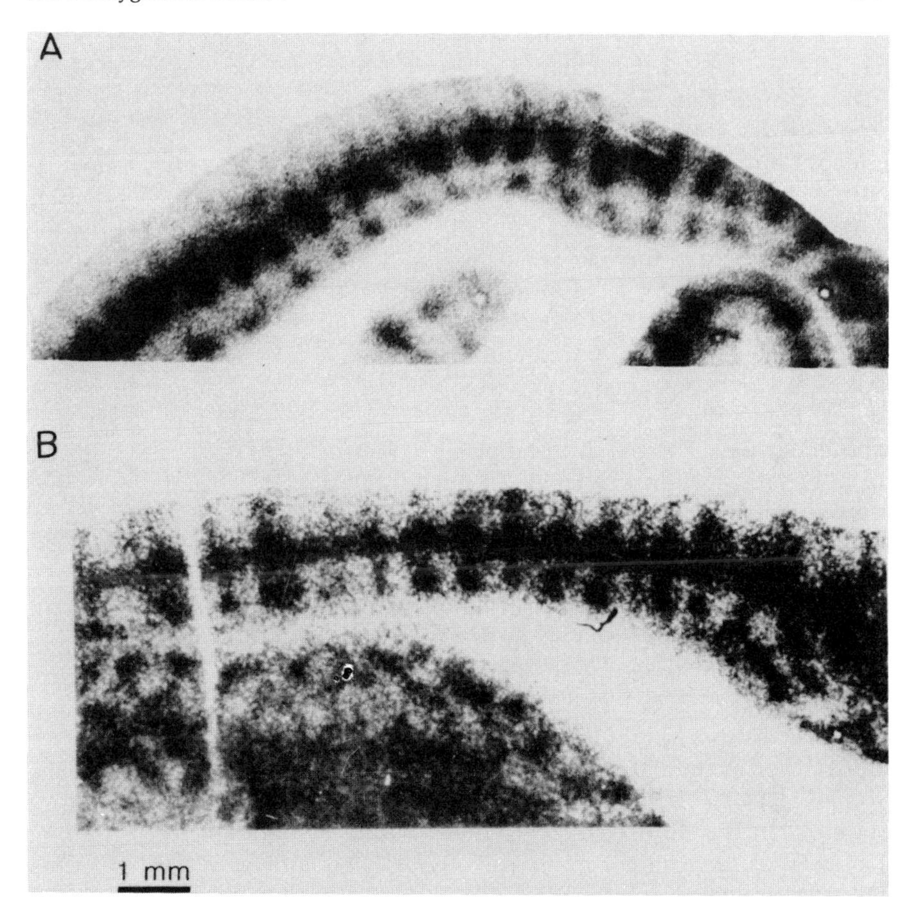

Figure 4. [^{14}C]Deoxyglucose autoradiographs of monkey striate cortex. (A) Ocular dominance columns. The right eye of an adult monkey was occluded during the study. The alternate dark and light striations, each approximately 300–400 μm in width and most prominent in Layer IV, were visible in both hemispheres. They represent the ocular dominance columns (20). (B) Orientation columns. The entire visual field of both eyes of the monkey was stimulated with moving vertical white stripes, irregularly spaced. Labeled regions are columns perpendicular to the cortical surface approximately 600 μm apart. Layer IVC is uniformly labeled, as expected from the absence of orientation-specific cells in that layer. From Hubel et al. (21).

extensively with the method. In the rhesus monkey, it has been shown with electrophysiological (17) and neuroanatomical (18, 19) methods that the striate cortex is composed of a mosaic of columns about 400 μm in width alternately representing the monocular inputs of the two eyes. The cortex is organized in such a manner that the information for a given point in the visual field converges to two adjacent columns, one column for each of the two eyes. The projections to the cortex originate in the cells of the lateral geniculate bodies and terminate in Layer IV of the striate cortex. The cells responding to the input of each monocular terminal zone are distributed through the thickness of the cortical column. Deoxyglucose autoradiographs obtained from monkeys with only monocular visual input (Figure 4A) exhibit a regularly alternating pattern of intensely labeled and lightly labeled cortex (20). This pattern of labeling is oriented perpendicularly to the cortical surface, and the dimensions, arrangement, and distribution of the columns are identical to those of the ocular dominance columns (17–19). Orientation columns in the monkey striate cortex have also been demonstrated with the deoxyglucose method (21). These columns can be visualized with the DG method by stimulating the entire visual field of both eyes with moving, specifically oriented, irregularly spaced stripes. The DG autoradiograph (Figure 4B) of the cortex consists of alternating intensely labeled and lightly labeled columns arranged perpendicularly to the cortical surface. The columns differ in several respects from the ocular dominance columns. The periodicity of the orientation columns is 570 μm as opposed to 770 μm for the ocular dominance columns. In Layer IVC in which the ocular dominance columns are most prominent, the DG autoradiograph shows uniform labeling in the orientation experiment. This agrees with the neurophysiological findings that cells in Layer IVC show no orientation specificity (22) and are strictly monocular (17).

Another system that has been mapped with the DG method is the somatosensory cortical representation of the whiskers or vibrissae of rodents. In the mouse, discrete multicellular cytoarchitectonic units in Layer IV of the SmI cortex (barrels) have been shown to receive a one-to-one projection from the vibrissae on the contralateral face. The pattern of cortical barrels in the posteromedial barrel subfield is identical to that of the large mystacial vibrissae (23). Removal of a single row of whiskers from the left side of a mouse 24 hr before the injection of [^{14}C]DG resulted in an autoradiograph (Figure 5) that exhibited decreased label in the corresponding barrel in the right SmI cortex (24). These studies show that the barrels are morphological correlates in Layer IV of the functional cortical columns found in the monkey striate cortex.

Figure 5. The right SI cortex of a mouse following the removal of row C vibrissae from the left face. The uppermost figure is a tracing of the posteromedial barrel subfield reconstructed from photographs of the thionin-stained section (middle) and [^{14}C]DG autoradiograph (bottom). In the tracing the stippling indicates major section artifacts. The bar is 1 mm. In the autoradiograph a strip of ligher density corresponding to row C can be seen flanked by two wider dark areas indicating increased activity in rows A, B, and D, for which the associated vibrissae remained intact. From Durham and Woolsey (24).

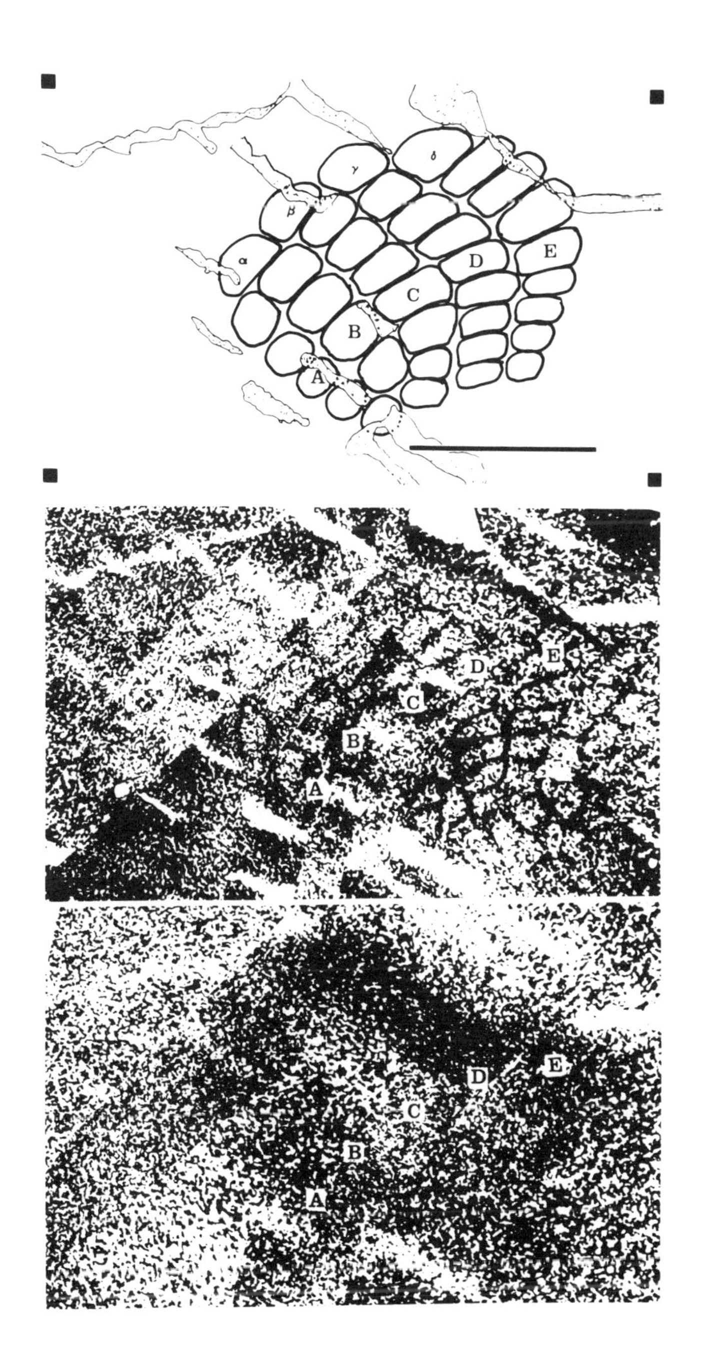
α
β
γ
δ
A
B
C
D
E
A
B
C
D
E
A
B
C
D
E

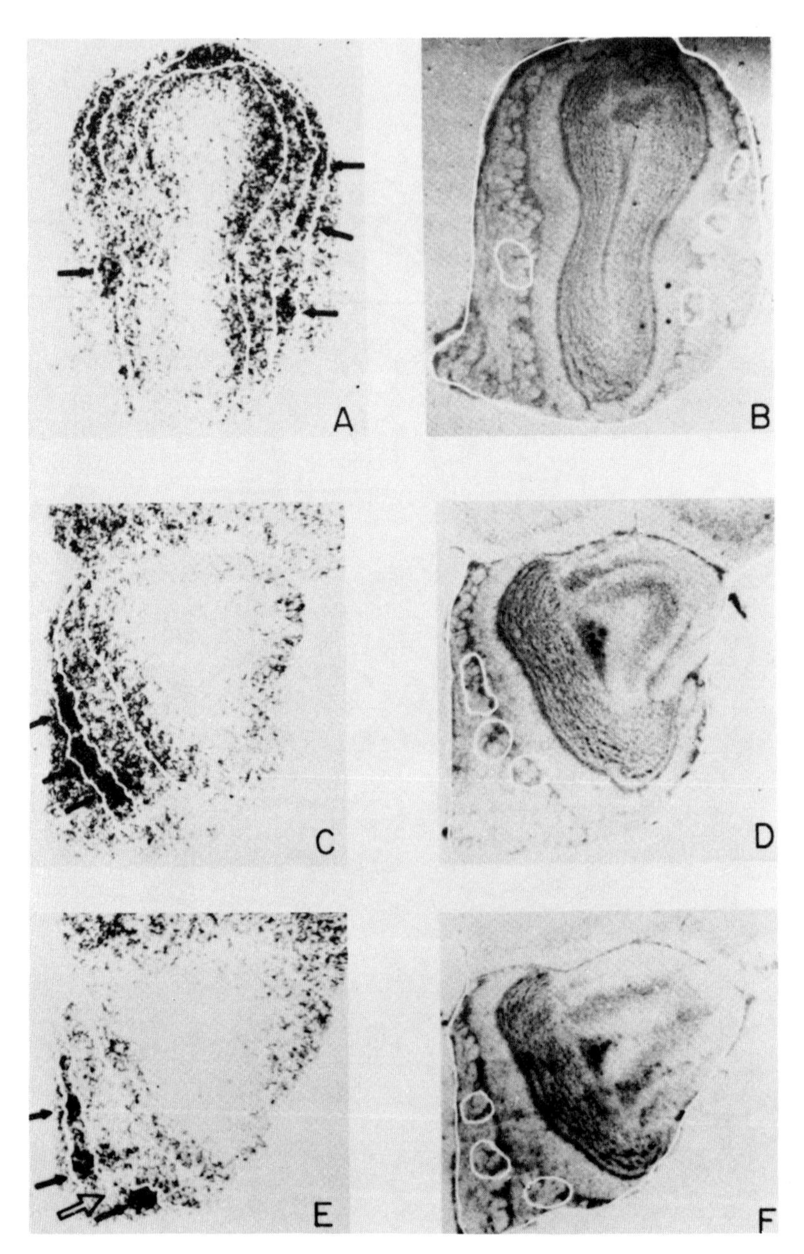

Figure 6. The right olfactory bulb of a rat with amylacetate olfactory stimulation. The [^{14}C]DG autoradiograph is on the left and the Nissl-stained section is on the right. The positions of the histological laminae are superimposed on the autoradiograph (white lines). The closed arrows indicate regions of focally increased activity. The open arrow indicates a region of low activity. The regions exhibiting high activity lie over the glomerular layer. From Stewart et al. (25).

The functional organization of the rat olfactory bulb has also been extensively mapped with the deoxyglucose method. In DG experiments (25) in which rats were exposed to controlled odor conditions (e.g., amylacetate, camphor, cage air, dimethyldisulfide, pure air) at various concentrations, the glomerular layer of the olfactory bulb exhibited some specificity for both odor modality and intensity. Exposure to single odors such as amylacetate (Figure 6) resulted in punctate regions of intense labeling in the glomerular layer. At high concentrations the labeled areas were expanded but remained in the same spatial pattern. The pattern of labeling varied with different odors, with some overlap. These results may be correlated with anatomical studies of the projections from olfactory receptors in the nose to the olfactory bulb, in which a regional relationship was established between parts of the nasal sheet and punctate parts of the olfactory bulb (26, 27). The implication is that specific regions of labeling with DG induced by a specific odor could be due to input from receptors in certain regions of the receptor sheet in the nasal cavity.

These studies on the monkey visual system, the mouse somatosensory system, and the rat olfactory system all illustrate the feasibility of mapping nervous system function in discrete brain regions. The purpose of these studies was to a large extent to test the applicability of the DG method to systems already mapped by neurophysiological and neuroanatomical techniques. Currently several laboratories are investigating the functional plasticity in these three systems (28–30). The studies illustrated here (Figures 4–6) show that the [^{14}C]DG autoradiographs can clearly delineate brain regions ranging in size from 150 to 400 μm. These dimensions are very close to the theoretical limit of resolution of the method with [^{14}C]DG (see Section 2.2.1).

2 THE RESOLUTION OF THE DEOXYGLUCOSE METHOD

2.1 Factors that Affect the Resolution of Autoradiographs

Resolution in photography is usually defined as the distance that must separate two objects before they can be distinguished from each other. More precisely, resolution can be defined as the distance that must separate two sources of equal strength such that the grain density between them falls to half that seen over each source (31).

The grain density D at a distance d from the source can be described by the equation

$$D = D_0 e^{-\mu d} \qquad \text{[Equation 2]}$$

where D_0 is the grain density over the source and μ is the linear absorption coefficient for the particular conditions used in the autoradiographic preparation. The distance from the source at which $D = D_0/2$ is the half-distance ($d\frac{1}{2}$). The grain density for two equal point sources at various distances from each other is illustrated in Figure 7. When the sources are separated by one $d\frac{1}{2}$, a single broad peak of grain density will result (Figure 7A). Increasing the number of half distances (Figure 7B, C, and D) between them lowers the trough between them until at four half-distances the trough is 50% of the peaks (Figure 7D). It is at this point that one could with some certainty decide that there are two sources. It is also at this point that the grain density peaks over the sources are within 2% of what they would be were the other source not present. For the quantitative autoradiographic measurements made in the deoxyglucose method, this is an important consideration.

Theoretically, the resolution of autoradiographs depends on: (i) the geometrical relationship of the radioactive source and the emulsion; (ii) the energy and intensity of the radiation; (iii) the characteristics of the photographic emulsion. The following discussion is intended to examine these factors briefly in an attempt to define the conditions for obtaining the best resolution of autoradiographs. More detailed discussions can be found in Gross et al. (32) and A. W. Rogers (33).

2.1.1 *Geometry*

An autoradiographic preparation consists of the section, the overlying emulsion, and the space between the two, that is, the interspace. Any variation in the geometry of these components, such as variations in their respective thicknesses, will affect the resolution of the autoradiographic image. Theoretically, reduction of the section thickness and interspace to a minimum will give the finest resolution of a point source, whereas variation in emulsion thickness has little effect. In biological systems the source is more likely to be irregular and three-dimensional in shape. Analysis of the theoretical autoradiographic effect of linear

Figure 7. The theoretical resolution of the autoradiographic images of two point-sources of equal concentration. In A–D the sources are separated from each other by 1, 2, 3, and 4 half-distances ($d_{1/2}$) respectively. In each graph, A–D, the upper and lower abscissas refer to the sources on the right and left, respectively. The sources are located at the 0 on their respective abscissas. The pattern of distribution of silver grains that would be produced by each source if autoradiographed separately is shown by the dashed lines. The decay of the grain density from each point is defined by a single exponential equation (2) where μ is the linear absorption coefficient and the half-distance $d_{1/2}$ is ln $2/\mu$. At one $d\frac{1}{2}$ from the source the grain density is half of that over the source. The solid line in each graph is the distribution of silver grains that would result from the autoradiography of the two sources together.

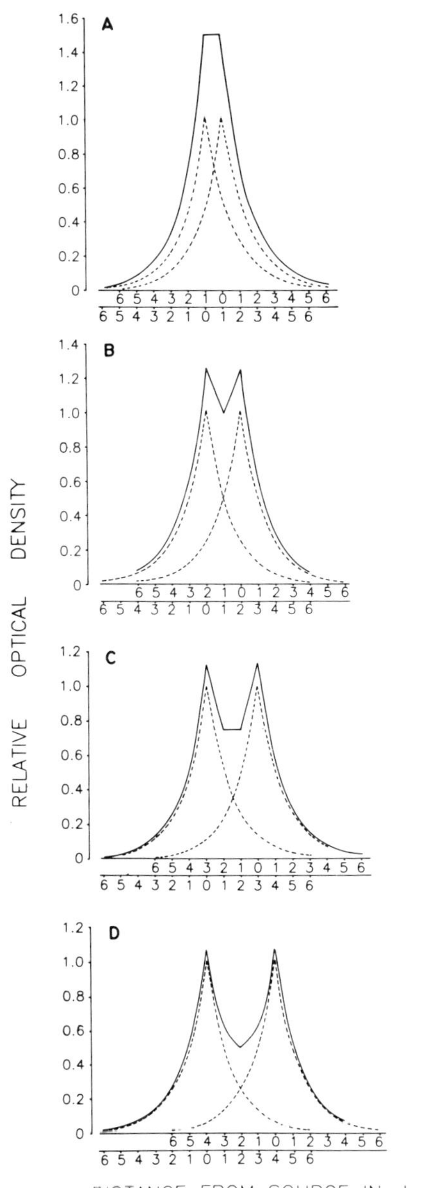
A
B
C
D
RELATIVE OPTICAL DENSITY
DISTANCE FROM SOURCE IN $d_{1/2}$ UNITS

Table 2. Theoretical Effect of Variations in the Geometry of an Autoradiographic Preparation on the Resolution[a]

Source	Interspace (μm)	Section Thickness (μm)	Emulsion Thickness (μm)	Resolution[b] (μm)
Linear	0.5	5	5	11.4
	0.1	5	5	4.6
	0.1	10	5	5.6
	0.1	5	10	5.4
Cylindrical	0.5	5	5	12.8
	0.1	5	5	12.0
	0.1	2	5	10.0
	0.1	5	10	16.0

[a] Data from Gross et al. (32), Tables I and II.
[b] Resolution is defined as 4 $d½$. The $d½$ of the cylindrical source defined as the distance from the edge of the source at which the density is 1/2 the density at the edge.

and cylindrical sources (Table 2) shows that decreasing the interspace produces a significant improvement in the resolution of the line source, whereas the cylindrical source is relatively insensitive to changes in the geometry of the preparation.

2.1.2 *Energy of the Radiation*

There is an inverse relationship between the energy of the source β-particles and the autoradiographic resolution. Higher energy particles will have greater range in the emulsion. In addition, as they travel in the emulsion they lose energy and consequently have increasing photographic effectiveness. Thus, toward the end of the range of a β-particle the grain density will increase. This factor will have an adverse effect on the resolution. The two isotopes of particular interest in this discussion, ^{3}H and ^{14}C, emit β-particles with maximum energies of 18.5 and 155 kev, respectively. As a result their track lengths in nuclear track emulsions are very different. A tritium-labeled specimen would produce developed grains in a nuclear track emulsion at a maximum distance of 3 μm from the source; the majority of the grains would be within 1 μm. Carbon-14 will give rise to tracks up to 40 μm from the source, most of them lying within 10 μm (33).

Exposure is another factor that has an effect on the resolving power of an autoradiograph. The density curves shown in Figure 7 were constructed on the basis of the density being directly proportional to the amount of radiation. This relationship is true for moderate densities. As fewer halide crystals become available for development, the probability of β-particles striking unexposed grains decreases. Accordingly, with increasing exposure the incremental increase in the density will decrease. In the autoradiograph this effect will be greatest in regions of maximum density and will result in a flattening of the density curves with a considerable loss in resolution. This is a particularly serious problem with tritium because in this case the source can affect only the limited number of crystals that lie within a 1 μm radius.

2.1.3 Properties of the Emulsion

The same influences of the characteristics of the emulsion on the resultant photographic image apply equally to autoradiography. The properties of the silver halide crystals—size, density, and uniformity—all affect the resolving power and the sensitivity of the emulsion. In general, the larger the crystal diameter, the higher is the probability of its being traversed by a β-particle, the greater is the energy absorbed, and the greater is the quantity of silver atoms reduced for the same amount of incident radiation. Thus, the larger the crystal size the greater the sensitivity. The increase in sensitivity, however, is accompanied by a decrease in resolving power. Crystal diameters in X-ray films range from 0.2 to 3.0 μm, whereas in nuclear track emulsions they range from 0.02 to 0.5 μm diameter. Although the smaller crystals record with greater precision events occurring within the emulsion, there is a loss of sensitivity. The small size of the crystal limits the portion of the trajectory of a charged particle that can traverse it, so that the total energy liberated within the crystal is relatively small.

A corollary to the effect of crystal size on resolution and efficiency is that the uniformity of crystal size affects the contrast. Crystals of a similar size tend to respond similarly. Therefore, emulsions with uniform grain size have a high contrast, and emulsions with a wide range of grain sizes have a relatively low contrast.

The density of the silver halide crystals is another property of the emulsion that affects both resolution and sensitivity. Increasing the density of the crystals increases the probability of a β-particle hit and thus increases the sensitivity of the emulsion. At the same time the "stopping power" of the emulsion is increased with a resultant increase in the resolution. Probably the main difference between photographic and nuclear track emulsions is the much higher density of silver halide crystals in the latter.

The conditions of development can also have an effect on autoradiographic results, that is, on both the intensity of the image produced and on the size of the developed grains. Detailed explanations of the photographic process can be found in Rogers (33) and Mees and James (34). The definition of a "developed" silver grain depends on how it is viewed (by electron microscopy, light microscope, darkfield illumination, etc.). Conditions of development optimal for the particular conditions of viewing one's experimental result should be determined. The variables that are most important are the dilution of the developer, the amount of agitation, the temperature, and the time of development.

2.2 Special Problems Encountered in the Autoradiography of Water-Soluble Substances

A number of methods are available for preparing tissue sections for autoradiography at the light or electron microscopic level that involve aldehyde fixation and dehydration followed by embedding in paraffin, celloidin, or an araldite or epoxy resin. These techniques are quite appropriate in experiments designed to look at the incorporation of radioactive label into a macromolecule that has been "fixed" in the tissue by these procedures. When the radioactive label in the tissue is contained in a water-soluble molecule, however, heroic measures may have to be taken to avoid the loss or diffusion of the labeled molecules during the preparation of the tissue. The level of resolution required determines the degree of technical difficulty involved in preparing the tissue.

There are two major areas of difficulty, the preparation of the tissue sections and the autoradiographic process itself. In both processes labeled water-soluble molecules can be moved or removed.

2.2.1 The Limiting Factors in the Resolution of 2-DG Autoradiography

2.2.1a The Autoradiographic Process. The procedures used in the deoxyglucose method as originally described (1) are to thaw-mount and heat-dry 20 μm cryostat-cut sections followed by exposure of X-ray film to the sections. The limiting factors in the resolution of these autoradiographs have been systematically analyzed by C. R. Gallistel and S. Nichols (unpublished). With microcircuit lithography they prepared radioactive swept-frequency gratings with alternating equal-width radioactive and nonradioactive bars in groups of bar widths ranging from 500 to 10 μm. A density scan of a contact print of the mask was made with an Optronics 1000 HS (12.5 μm aperture, 12.5 μm raster) (Figure 8A).

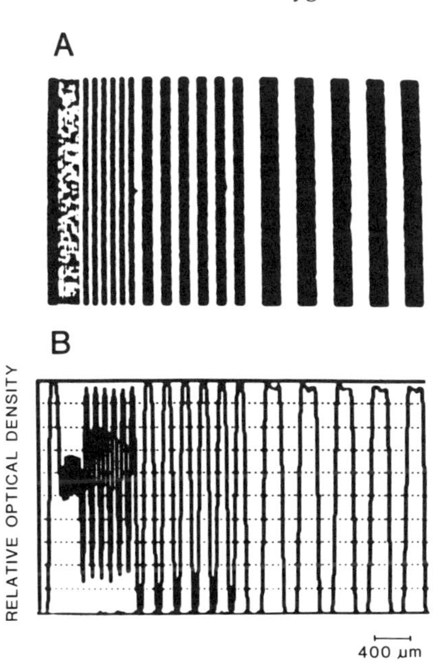

Figure 8. The resolution of the Optronics P-1000 HS scanning microdensitometer (12.5 μm apertures, 12.5 μm raster). (A) Density scan of a copy of a contact print of the mask used in studies of autoradiographic resolution (provided by Gallistel and Nichols). The mask is the image from which frequency gratings with alternating equal-width radioactive and nonradioactive bars were made. The bar widths shown here in groups are, from right to left: 200, 100, 50, 20, and 10 μm. (B) A plot of the relative optical density along the mask. With the use of image processing, the optical density was averaged at each point along the abscissa over the entire width of the image.

A plot of relative optical density along the mask (Figure 8B) shows that the densitometer itself can limit the resolution. Of the four densitometers examined, including the Optronics instrument with three different apertures, only the Joyce Loebl single-line scanning densitometer with a 20× objective could resolve the 20 μm bars at 25 bars/mm. This instrument was used to scan the autoradiographic images produced by exposure of ^{3}H and ^{14}C gratings at several thicknesses (2.25–20 μm) on several different types of film. The results of these studies show that grating (section) thickness and film type made little difference to the resolution (Table 3). The energy of the isotope was the most serious limitation. It was possible to resolve 25 lines/mm (20 μm bar width) with ^{3}H swept gratings, whereas only 10 lines/mm (50 μm bar widths) could be resolved on ^{14}C swept gratings.

2.2.1b Preparation of the Tissue Sections. The results obtained with these gratings establish that cellular or near-cellular levels of resolution are, in principle, obtainable in conventional 2-DG autoradiography with [^{3}H]DG and LKB tritium film *provided* that there is no diffusion of the isotope during the preparation of the tissue sections. The limitation on resolution imposed by diffusion of the label in a conventional [^{14}C]DG study has been assessed by the following experiment. A rat brain was

Table 3. Resolution of Autoradiographic Images of Radiolabeled Frequency Gratings as a Function of Isotope, Film, and Section Thickness[a]

Conditions			Resolution[b]	
Isotope	Film	Section Thickness (μm)	Relative Contrast[c] (50 μm)	Bar Widths 20 μm
^{3}H	LKB Ultrofilm	2	1.0	0.76
^{14}C	LKB Ultrofilm	2	0.56	0.27
^{14}C	Kodak SB-5	2	0.62	0.31
^{14}C	Kodak SB-5	20	0.51	0.24
^{14}C	XTL	20	0.56	0.12

[a] Data from C. R. Gallistel and S. Nichols (unpublished).
[b] The bar width at which the relative contrast is 0.5 is equivalent to 4 $d½$ as shown in Figure 7, the level of resolution of the autoradiograph.
[c] Relative contrast is defined as the ratio of the contrast at the specified bar width to the contrast at a bar width of 500 μm. The contrast is the maximum optical density over the bar minus the minimum optical density between the bars.

prepared that contained a plug of uniformly distributed [^{14}C]DG-labeled brain tissue. The plug was inserted into a hole drilled in a frozen and unlabeled rat brain. Cryostat sections 20 μm thick were thaw-mounted, dried at 60°C, and placed in a cassette along with [^{14}C]methylmethacrylate calibrated standards. A sheet of Kodak MR-1 film was apposed to the sections and the standards and exposed for 9 days. The resultant autoradiographs (Figure 9, inset) were analyzed with an Optronics scanning densitometer (apertures at 25 μm, raster at 25 μm). The optical densities produced by the standards and the optical densities within and at various distances from the plug were determined. It can be shown that the decay of the measured ^{14}C concentration from the edge of the plug is described by a single exponential (Equation 2) and the slope of the line formed by plotting the natural logarithm of the concentration of ^{14}C against distance from the edge is equal to the linear absorption coefficient, μ. The half distance ($d½$), the distance from the edge at which the determined concentration is one-half the actual concentration in the plug, is then equal to $\ln 2/\mu$.

A semilogarithmic plot of the data (Figure 9) is linear in the region past the edge of the plug, and the mean ± SD of the half-distance obtained from 22 determinations was 47.3 ± 8.0 μm. Accordingly, in a conventional [^{14}C]DG autoradiograph two point-sources with equal concentrations of [^{14}C]DG can be resolved if they are separated by a distance of 200 μm (4 $d½$).

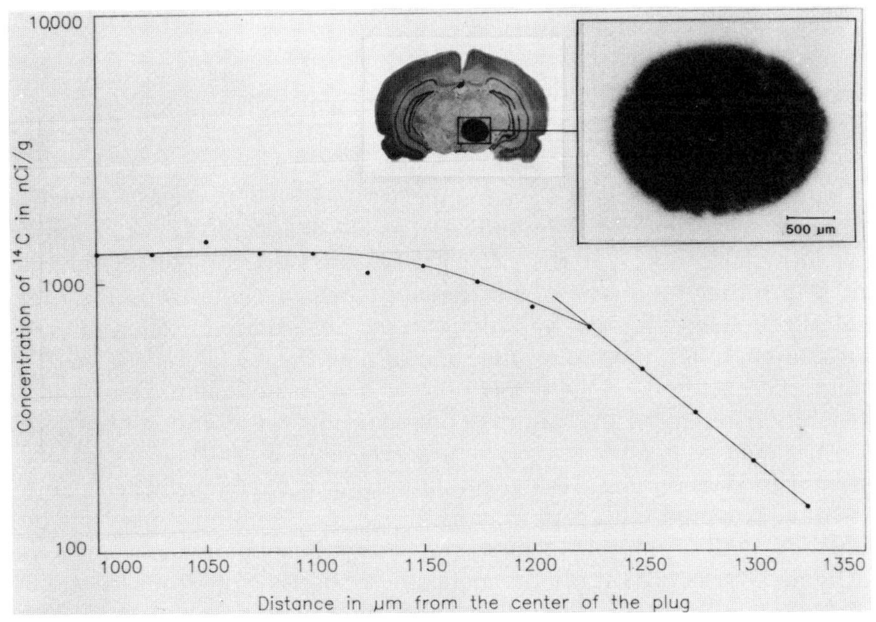

Figure 9. The resolution of [^{14}C]deoxyglucose autoradiographs. Autoradiographs on Kodak MR-1 film were made from 20 µm frozen and thaw-mounted sections of an unlabeled brain into which a plug of uniformly distributed [^{14}C]DG had been inserted. A typical autoradiograph is shown in the inset adjacent to a reconstruction of the autoradiographic image superimposed on a stained section of unlabeled brain. The autoradiographs were analyzed with an Optronics P-1000 scanning microdensitometer with 25 µm apertures and raster. Readings were made every 25 µm from the center of the plug and across the "edge" of the autoradiographic image. The semilog plot illustrated here was obtained from one such analysis. The slope of the linear portion of the plot is (−0.0156), the linear absorption coefficient, μ, for this particular autoradiograph. The horizontal line is composed of the points taken within the autoradiograph of the plug itself. The sigmoidicity of the curve is due to the source being cylindrical rather than linear, and there is some contribution close to the edge of the source due to radiation from the area adjacent to the edge.

The results of these two experiments show that diffusion of the water-soluble [^{14}C]DG-6-P and/or [^{14}C]DG imposes the greater limit on resolution. The results of the experiments with the gratings (Table 3) showed that the resolution (4 $d\frac{1}{2}$) of the X-ray film ^{14}C autoradiographs was about 50 µm (the bar width at which there was a 50% loss of contrast). In the [^{14}C]DG brain-plug experiment, in which diffusion of the label took place during freezing and thaw-mounting, the resolution was 200 µm. It would appear that cellular or near-cellular resolution could be achieved with the implementation of measures to prevent diffusion and the use of [^{3}H]DG and LKB tritium film. With the development of such techniques it should be possible to attain a resolution of 20 μm with the deoxyglucose method.

2.3 Special Techniques for the Improvement of Deoxyglucose Autoradiographs

A number of investigators have found that the development of techniques to prevent loss and diffusion of the labeled 2-deoxyglucose-6-phosphate is a formidable task. With the use of [^{3}H]DG and LKB Ultrofilm a resolution of 20 μm may be possible, but this approach presents its own special set of problems, those of variations in self-absorption in different brain structures. In addition, consideration of how the deoxyglucose method might be useful at the light-microscopic level leads one to think that a resolution better than 20 μm may be required. A resolution of only 1 or 2 μm would be needed to study energy metabolism at the level of nerve terminals. With this high level of resolution, one could map which synapses become activated with certain types of stimulation, and if used in conjunction with immunocytochemistry, which types of synapses become activated. One could study the metabolic response of neurons to various ion currents and their rectification, and one could examine the question how glial cells respond metabolically to neuronal activation.

The deoxyglucose method was originally described for [^{14}C]DG (1) because of the efficacy of ^{14}C in the autoradiographic process. Theoretically the method applies to deoxyglucose, however labeled. [^{3}H]Deoxyglucose offers several advantages: (i) specific activities higher than that of [^{14}C]DG are available with the advantage that smaller molar amounts of DG can be introduced into the system; (ii) ^{3}H-labeled brain sections are infinitely thick at 5 μm (35), thus eliminating errors due to imprecision in the microtome; (iii) ^{3}H, with its low energy β particle, should yield images with higher resolution.

2.3.1 *The Use of [^{3}H]Deoxyglucose and LKB Ultrofilm*

The use of ^{3}H with X-ray film has been limited because the low-energy β particles will not penetrate the protective coating on standard X-ray

Figure 10. Self-absorption of ^{14}C and ^{3}H in 20 μm rat brain sections. (A) Rat was infused with [^{14}C]methylglucose, a sugar that is not metabolized and that distributes uniformly throughout the brain. That some structures can be discerned, particularly the corpus callosum, indicates some difference in the self-absorption characteristics of gray and white matter. (B) Rat was infused with [^{3}H]methylglucose. Corpus callosum is even more visible in this ^{3}H labeled autoradiograph because of the very significant differences in self-absorption characteristics of gray and white matter.

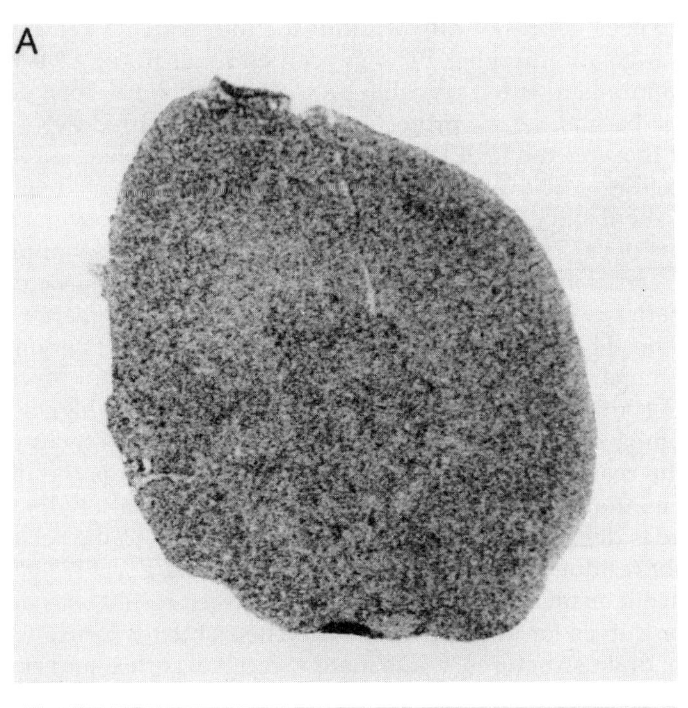
A

B

film. However, an X-ray film without the antiscratch layer and with a high silver/gelatin ratio has been developed by LKB, the LKB Ultrofilm (36). A tritium film with a very thin protective coating has been developed by Kodak; because of the protective coating the required exposure time is longer than for the LKB Ultrofilm, but the images obtained appear to be satisfactory. The availability of ^{3}H-sensitive film has made it possible to adapt the deoxyglucose method for use with [^{3}H]deoxyglucose.

The use of [^{3}H]DG and LKB Ultrofilm presents a new set of problems, those of variations in self-absorption in different brain structures (35). White matter exhibits greater self-absorption than gray matter for both ^{14}C (37) and ^{3}H (Figure 10). However, the difference in attenuation in a 20 μm section between gray and white matter is only 13% for ^{14}C compared with about 50% for ^{3}H. Self-absorption is quantitatively a bigger problem for ^{3}H compared with ^{14}C because of the lower energy of the β-radiation of ^{3}H. The reason for the difference in self-absorption is the difference in water content: 80% in gray matter, 70% in white matter (38). Because the tissue is dehydrated before exposure of the film to the sections, the amount of residue remaining in each structure (mg/g wet weight) varies with its white matter content. F. Orzi (unpublished results) has developed calibration curves for [^{3}H]methylmethacrylate standards to use with three groups of brain structures: white matter, cerebral cortex, and subcortical structures. At a particular optical density the equivalent concentrations of ^{3}H in subcortical and white matter structures are about 25 and 50% higher, respectively, than in cerebral cortical structures. Thus, corrections can be made for variations in seslf-absorption among these three groups. Both Orzi (unpublished data) and Alexander et al. (35) have found that the values obtained for local rates of glucose utilization with the deoxyglucose method are the same with both ^{14}C- and ^{3}H-labeled DG, provided that corrections are made for variations in self-absorption.

A comparison of ^{14}C- and ^{3}H-labeled DG autoradiographs of cat spinal cord (Figure 11) shows that what appear to be single motoneurons can readily be discerned in the [^{3}H]DG autoradiograph but are not visible in the [^{14}C]DG autoradiograph. One must be cautious about the appearance of the autoradiographs. The fact that they look "sharper," as in the studies of Alexander et al. (35), probably reflects the increased contrast between gray and white matter structures, again as a result of the differences in self-absorption. "Structures" not visible heretofore in [^{14}C]DG autoradiographs may be visible because of the increased contrast produced by the differences in self-absorption rather than a real improvement in resolution.

In the first report of the problem of differences in self-absorption, the deoxyglucose method was carried out in the conventional manner, except

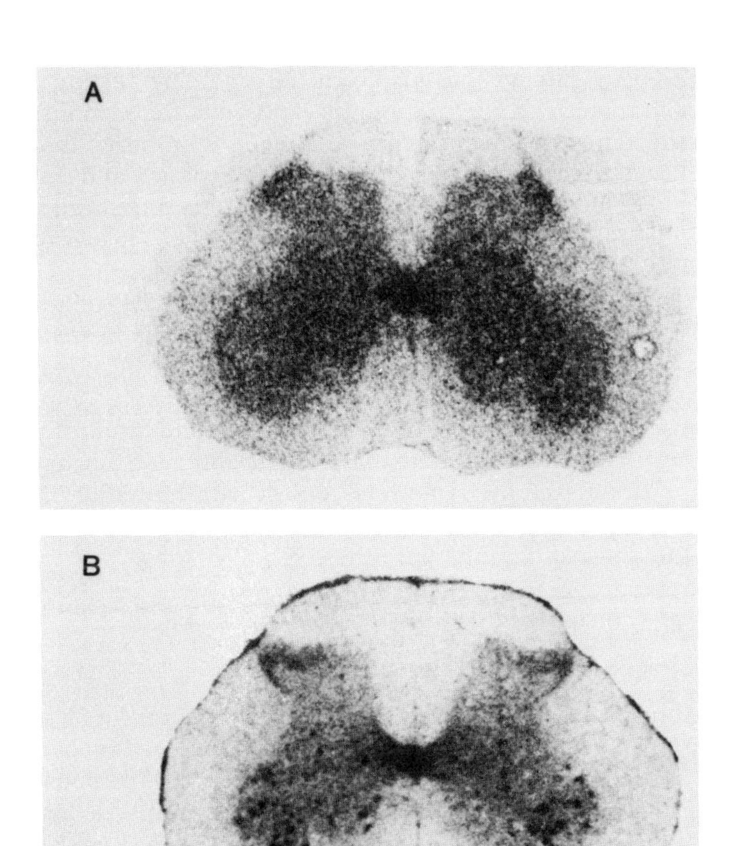

Figure 11. Comparison of ^{14}C and ^{3}H labeled DG autoradiographs of cat lumbar spinal cord. The cats were subjects in a study of the effects of unilateral acupuncturelike stimulation on the metabolic activity in the cervical dorsal horn. The cats were decerebrated at the midcollicular level and spinalized at the midthoracic level. (A) [^{14}C]DG autoradiograph. There is some heterogeneity through the anterior horns but discrete patches of activity cannot be discerned. (B) [^{3}H]DG autoradiograph. Punctate areas of activity can be easily discerned in the anterior horns with diameters of about 100 μm. The areas were probably produced by single motoneurons. (Sjölund and Crosby, unpublished results).

that [^{3}H]DG and LKB Ultrofilm were used (35) and 20 μm frozen sections were cut and thaw-mounted. LKB Ultrofilm was exposed to the sections at room temperature. Thus, no real attempt was made to improve the resolution by the prevention of diffusion of the labeled product, and no attempt was made to determine the resolution with these modifications. Our analysis would suggest that it was not improved merely because of the use of [^{3}H]DG (see Section 2.2.1).

In another report on the use of [^{3}H]deoxyglucose and LKB Ultrofilm (39) the problem of variations in the self-absorption of ^{3}H was not addressed but attempts were made to prevent the diffusion of [^{3}H]DG-6-P. Sections were cut at −18°C and freeze-dried in the cryostat. Subsequently, the Ultrofilm was exposed to the sections at room temperature and humidity for 3 weeks. Undoubtedly there was a potential for some diffusion during the exposure as the dried sections came into contact with humidity in the air (see Section 2.3.3b). Again no attempt was made to determine the resolution of the autoradiographs with these modifications.

2.3.2 *Conventional Fixation and Autoradiography at the Light Microscopic Level*

Conventional methods for preparing a tissue for autoradiography of a nondiffusible substance include glutaraldehyde fixation followed by osmium tetroxide post-fixation, tissue dehydration, and embedding. Sections cut at about 1 μm are then coated with a nuclear track emulsion for autoradiography. One such method, in which the entire fixation, post-fixation, dehydration, and embedding processes are carried out by vascular perfusion, has been devised by Des Rosiers and Descarries (40) for the fixation of labeled deoxyglucose. Work from another laboratory (41), however, has shown that label is lost during these procedures.

Another approach was taken by Basinger et al. (42) in the *in vitro* [^{3}H]DG study of goldfish retina. Retinas were incubated in a Ringer's

Figure 12. Autoradiographs of isolated goldfish retinas incubated for 20 min in teleost Ringer's solution with 1 mCi/m*L* [^{3}H]DG. (*a*) Incubation in the dark. The label appears to be predominantly in photoreceptors, H_1 horizontal cells (open arrow) and presumptive type A (center-hyperpolarizing) bipolar cells (closed arrows). The heavily labeled areas just beneath these cells, in sublamina A of the inner plexiform layer, are probably their axon terminals (open circles). (*b*) Incubation under flashing narrow-band red light (700 nm) at 2×10^{12} quanta S^{-1} cm^{-2}. The label appears to be over H_2 and/or H_3 horizontal cells (open arrow) and type B (center-depolarizing) bipolar cell axon terminals (closed arrows). (*c*) Incubation under flashing broad-band white light at 2×10^{12} quanta S^{-1} cm^{-2}. Very little label is seen in photoreceptors and horizontal cell layers (open arrow); only type B bipolar cell axon terminals are heavily labeled (closed arrows). Scale bars, 20 μm. From Basinger et al. (42).

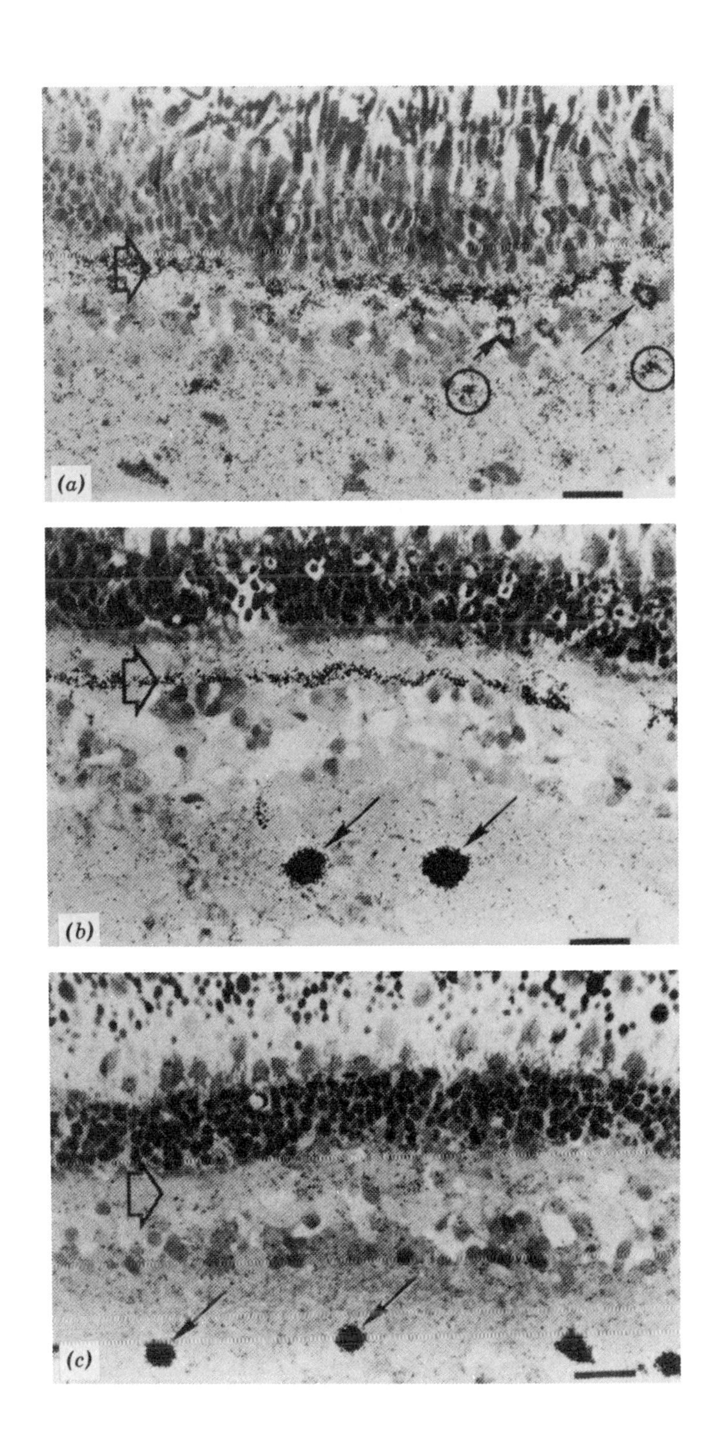
(a)
(b)
(c)

solution containing [^{3}H]DG. During the incubation they were stimulated at a specific wavelength and intensity of light. Following the incubation they were fixed in 6% glutaraldehyde, 3% glutaraldehyde, and 1% osmium tetroxide sequentially. They were dehydrated in graded ethanols and acetone and embedded in epon-araldite. Sections (1 μm) were taken and coated with Kodak NTB-2 emulsion. The results of this study look very impressive (Figure 12), and the results agree to some extent with the known physiology of the retinal cells. Type B bipolar cell terminals were intensely labeled under red and white stimulation, conditions known to depolarize these cells. Two other cell types (H_1 horizontal cells and presumptive type A bipolar cells) were both intensely labeled in the dark, that is, under conditions that are known to depolarize these cells. Furthermore, the H_1 cells accumulated little label under red stimulation, a condition in which they are hyperpolarized. The problem with the interpretation of this and other studies in which conventional fixation methods are used is that more than 90% of the labeled DG-6-P is lost during the fixation procedure (Table 4) (43).

Other approaches that have been taken for the preparation of tissue following an *in vivo* DG experiment include perfusion fixation followed by dehydration and embedding. In one such study (44) rats were perfused with a fixative (0.01 *M* periodate, 0.075 *M* lysine, 2% paraformaldehyde in 0.037 *M* phosphate buffer) designed (45) to stabilize carbohydrates. Following perfusion for 3 min, the brains were removed intact, immersed in 30 ml of 50% ethanol overnight, and subsequently dehydrated in

Table 4. Loss of Labeled Deoxyglucose-6-Phosphate from Cells in Culture Dependent on Method of Fixation[a]

Method	No. of Experiments	[^{14}C]Deoxyglucose Uptake (dpm/mg protein/min)
Control	4	11,100 ± 202
Conventional fixation[b]	4	1,009 ± 202
Freeze substitution	8	12,916 ± 807

[a] From Ornberg et al. (43). *Note*: The values represent the means ± SEM of the results obtained in the number of experiments. Neuroblastoma (NIE-N115) cells grown in culture were incubated for 15 min in a balanced salt solution that contained 1.1 m*M* glucose and 1 μ*M* [^{14}C]deoxyglucose. The incubation was followed by five 2-min washes in balanced salt solution. The washed cells were dissolved in NaOH. Portions were taken for protein determination and liquid scintillation counting. The identity of the product [(^{14}C)DG-6-P] was confirmed in a separate experiment by thin-layer chromatography.

[b] Glutaraldehyde, 3%, followed by 1% osmium tetroxide.

graded alcohols (70%, 4 hr; 95%, 4 hr; butanol, 1 hr; and fresh butanol, overnight). Brains were embedded in Paraplast, sections were cut, and paraffin removed with xylene. The sections were then hydrated through graded alcohols, dried overnight, and covered with nuclear track emulsion. The authors report that most (about 95%) of the labeled DG was lost during the processing of the tissue following the perfusion. The distribution of the remaining 5%, however, was similar to the pattern of metabolic activity found under similar physiological conditions with the original [^{14}C]DG technique. In freely behaving mice activation of the trigeminal pathway differentially by selective patterns of whisker removal resulted in differential labeling within the spinal trigeminal nucleus, pars caudalis, and within Layer IV of the barrel region of the SMI cortex. Examination at higher magnification of the autoradiographs of the sections through the stimulated barrels (Figure 13C) revealed that in the stimulated barrels, in addition to some increased labeling over the neuropil, dense clusters of silver grains could be seen over individual cell somata. In other brain regions also known to be metabolically active in freely moving animals, the autoradiographs obtained in this study (Figure 13) showed intense labeling over cell bodies and in some cases the proximal portions of some processes. The authors noted that although there were substantial losses of isotope in the processing of the tissue, the losses occurred early in the procedure. In addition, analysis of the distribution of grains over a brain section revealed no substantial decline in the radial direction (i.e., from deep to superficial in the brain), as might be expected if the label were not fixed in the tissue. It might be concluded that the label remaining in the tissue after tissue processing in these experiments and in the experiments of Basinger et al. (42) was in fact fixed.

An explanation for these findings can be found in the results of [^{3}H]DG experiments on isolated ganglia of the snail and the leech (46, 47). In these experiments 40–50% of the label was lost during conventional tissue processing for both light and electron microscopy. The electron micrographs showed that 90% of the remaining label was associated with glycogen. Neurons and glia of invertebrates contain substantial amounts of glycogen compared with cells in the mammalian CNS (48). It is possible that the significant retention of label in these invertebrates, compared with the low retention in cultured neuroblastoma cells (43) and mouse brain (44), is related to the differences in their levels of glycogen. That 2-deoxyglucose can be incorporated into glycogen has been clearly shown (49) in an isolated enzyme system. The kinetics of the 2-DG incorporation into glycogen were investigated in this system. The Michaelis constant K_m found for uridine diphosphodeoxyglucose as substrate was 5.6 mM, whereas the K_m for uridine diphosphoglucose was 1.2 mM. Maximal velocity for deoxyglucose incorporation from

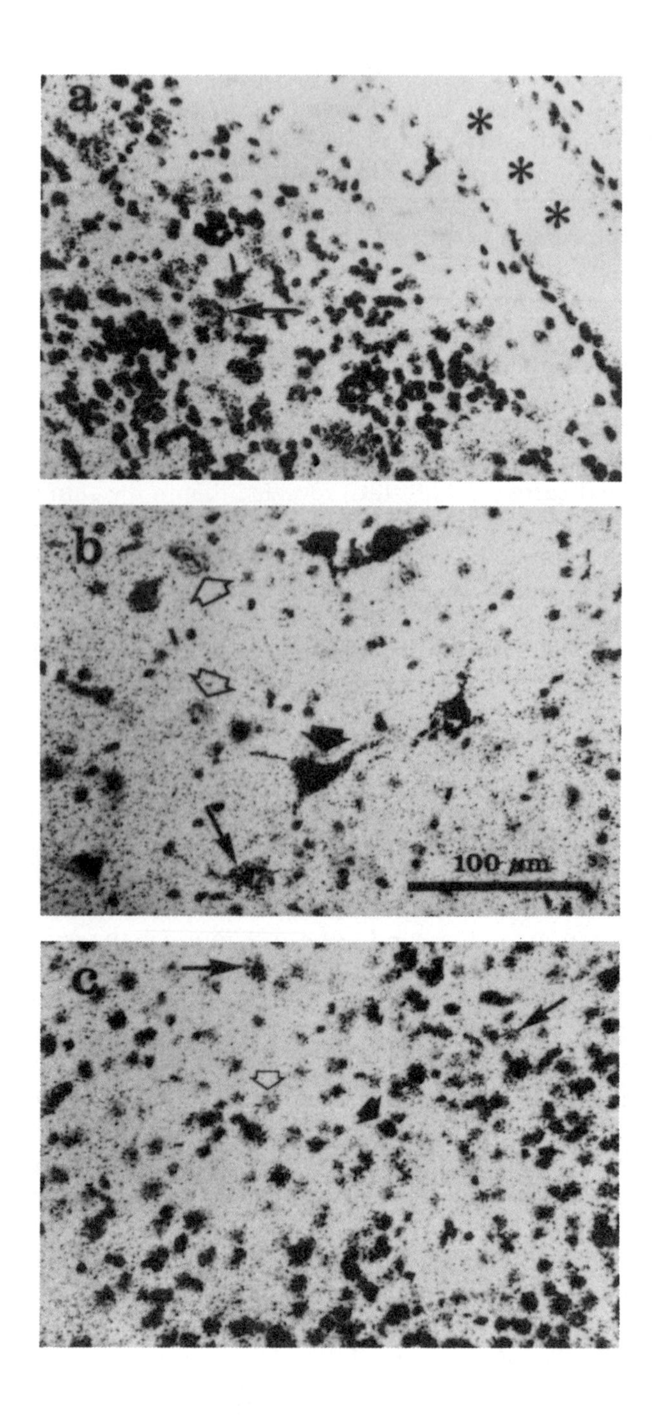
a
b
100 μm
c

uridine diphosphodeoxyglucose into glycogen was 26% of the value for comparable glucose incorporation. Further investigations (47) have shown that 1 hr after intravenous administration of labeled deoxyglucose to mice, 1–2% of the total radioactivity in brain was recovered with high-molecular-weight glycogen. In similar experiments on leech and snail isolated ganglia (46) and on frog isolated retina (50) 1–10% of the label was recovered in the glycogen fraction. Treatment of the extracted labeled glycogen fraction with amyloglucosidase resulted in the release of the label in a manner consistent with the breakdown of labeled glycogen. Similar experiments have been carried out in rats (T. Nelson, unpublished results). In these studies rats were pulse-injected i.v. with a tracer dose of labeled deoxyglucose. After 45 min the rat brains were freeze-blown in order to prevent glycogen breakdown post-mortem. Again 1–2% of the labeled deoxyglucose was recovered in the glycogen fraction. That deoxyglucose can ultimately be incorporated into glycogen does not in any way violate the requirements of the deoxyglucose method for the determination of local rates of glucose utilization. Deoxyglucose must be phosphorylated by hexokinase to form deoxyglucose-6-phosphate, which is the substrate in the initial step in glycogen synthesis. Therefore, the incorporation of deoxyglucose into glycogen merely serves as an additional neurochemical trap for this labeled product of deoxyglucose phosphorylation.

The results obtained with the DG method used in conjunction with conventional fixation procedures, coupled with the probability that a major protion of the labeled product in these experiments is deoxyglycogen, are of particular interest. In the experiments with stimulation of isolated retina (42) and sensory stimulation of the mouse (44) the results are compatible with the known physiology of both systems. Physiological perturbations in isolated ganglia preparations of leech and snail have also been carried out (51). Antidromic stimulation of the nerves entering the ganglia resulted in increased incorporation of labeled DG into the fixation-resistant fraction of the glial cells surrounding the stimulated neurons. The same effect occurred under conditions of high potassium

Figure 13. Autoradiographs of selected brain sections form a freely moving mouse which had had all of the vibrissae except for row C removed from the right side of the face and which had received a pulse injection of 1–2 mCi of [^{3}H]DG. Thick black arrow, heavily labeled somata; fine black arrow, lightly labeled somata; open arrow, unlabeled cells; asterisk, the ventricle. (*a*) Cells in the dorsal cochlear nucleus. Label is absent in the ventricle, and cells are labeled with various intensities. (*b*) Cells in the nucleus gigantocellularis. In heavily labeled cells, proximal processes are also labeled. Several lightly labeled and unlabeled cells are shown. (*c*) Cells in the C_1 barrel in the left hemisphere. Different cells are labeled to different degrees. From Durham et al. (44).

concentration. These results suggest that neuronal activity caused a significant potassium-mediated increase in DG incorporation into glycogen by glial cells. The incorporation of DG into glycogen, therefore, may provide investigators with a convenient neurochemical trap for fixation-resistant labeled DG in cells.

A biochemical rationale for an activity-related increase in deoxyglucose incorporation into glycogen can be constructed. Deoxyglucose-6-phosphate is not a substrate for phosphohexose isomerase and cannot therefore continue down the glycolytic pathway. Thus, with increased metabolic activity, the flux of metabolic substrates through the glycolytic pathway increases and the level of glucose-6-phosphate decreases while the level of deoxyglucose-6-phosphate increases. Since glucose-6-phosphate and deoxyglucose-6-phosphate are the precursors for glycogen synthesis, the specific activity of the precursor pool is increased. Glycogen stores that become depleted during activity may be replenished. On the other hand, an increased labeling of the glycogen may occur without any net glycogen synthesis. Such a sequence of events can explain why there might be a relationship between activity and glycogen-labeling, at least at the sites where glycogen synthesis takes place. This sequence of events coupled with the selective retention of labeled deoxyglycogen after conventional fixation may provide us with a means of mapping activity in cell bodies to which the originally described deoxyglucose method may be relatively insensitive. Such mapping would have to be qualitative, however; it would be difficult if not impossible to overcome the numerous kinetic problems in order to make it quantitative.

2.3.3 *Modified Conventional Procedures*

Localization at a cellular or subcellular level of the primary product of the DG method, deoxyglucose-6-phosphate, remains an unattained but desirable goal. The inherent difficulty is the prevention of either loss or diffusion of the label during tissue processing and autoradiography. A number of procedures (33, 52) have been described for the autoradiographic localization of other water-soluble substances. Some of these have been adapted for microscopic resolution of deoxyglucose autoradiographs.

2.3.3a Dry Mounting. One of the first attempts (53) to improve resolution was through the use of a modified Appleton (54) technique in which the sections are never permitted to thaw. Frozen sections (10 μm thick) were cut at −12°C under safelight illumination. Sections were picked up onto Kodak NTB-2 emulsion-coated slides at −12°C and exposed in light-tight, dessicated boxes at −20°C. The resultant autoradiographs showed rather uniform grain densities over neuronal perikarya and the surrounding neuropil. A slight modification of the Appleton technique includes thaw-mounting by warming the back of the emulsion-coated slide and section by touching it with a finger and then quickly refreezing the section. This method of mounting generally results in better contact between the emulsion and tissue. In studies in which this modification was used (41, 55), the label was found to be particularly localized in cells in the periventricular regions and in certain parts of white matter. In all regions of the diencephalon and telencephalon, however, the label was distributed uniformly. The results obtained in both of these studies with the Appleton technique (41, 53, 55) contrast sharply with the results of the originally described [^{14}C]DG studies by Sokoloff et al. (1) in which there was marked heterogeneity of labeling throughout the brain and spinal cord and in which areas rich in neuropil were most intensely labeled. In view of such differences the reliability of these higher resolution studies is questionable.

This technique applied to muscle tissue cut in cross-section (56) has yielded autoradiograms in which individual muscle fibers could be resolved (Figure 14). In these experiments single motor units in adult cats received repetitive electrical stimulation while [^{14}C]DG was administered intravenously. Frozen 10 μm sections of the muscle were picked up on emulsion-coated slides and dried at 45°C for 1 min. The emulsion–section sandwiches were exposed at −70°C in a dessicator. The results (Figure 14) show that acutely active muscle fibers, independently identified by depletion of intrafiber glycogen, are associated with highly localized accumulations of silver grains over the depleted fibers.

The adequate resolution achieved with this technique when applied to muscle fibers is probably a function of the geometry of the tissue. The label contained in the long cylindrical muscle fibers approximately 50 μm in diameter is less likely to diffuse out of the fibers than if it were contained in nerve processes and terminals 1 or 2 μm in cross-section. For the autoradiographic resolution of these specific structures in nervous tissue with the Appleton technique, dry-mounting and very low (< −20°C) temperatures would be required. This technique has the advantage of simplicity. The frozen section is directly placed in contact with a frozen layer of emulsion. It has been criticized on the ground that the section probably thaws, at the knife edge, like ice under a skate, with consequent redistribution of radioactivity. The zone of thawing, however, is probably very narrow, and if the temperature is kept low enough, refreezing must occur almost instantaneously on the surface of the knife. The other point in the procedure at which transient thawing occurs is at the instant of picking up the section from the knife. With

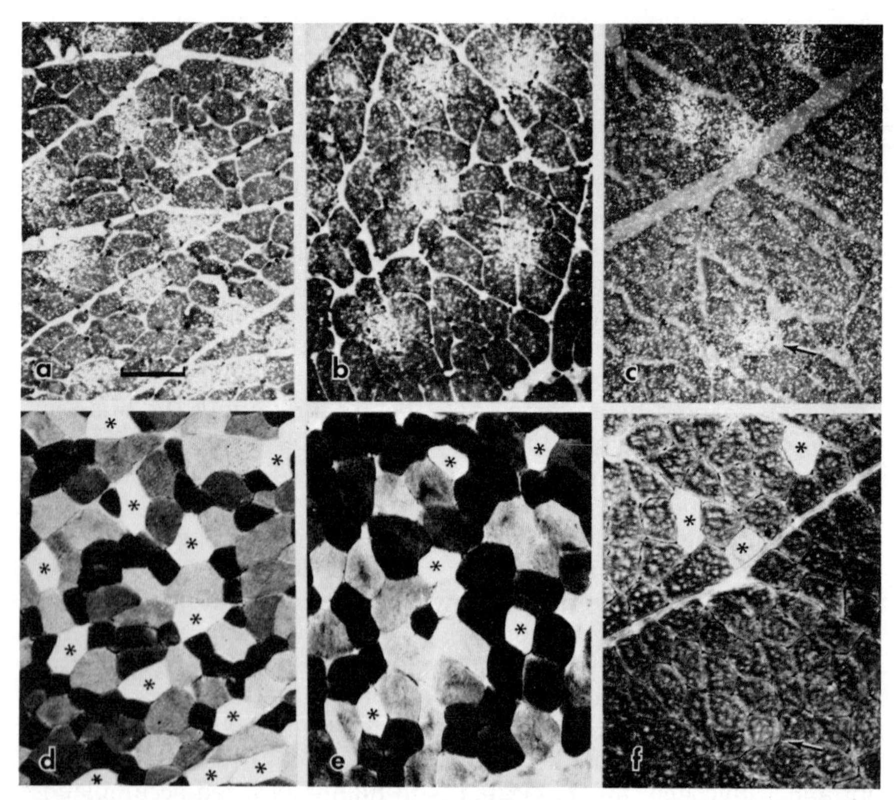

Figure 14. (*a–c*) Photomicrographs of [^{14}C]DG autoradiograms of three cat muscles in which individual motor units were repetitively stimulated. Combination brightfield and darkfield illumination demonstrates the association of increased silver grain densities with individual muscle fibers. Motor unit types were as follows: (*a*) type FF in a normal, fast twitch, fatigable, flexor digitorum longus, FDL muscle; (*b*) type FR, fast twitch, fatigue resistant, also FDL; (*c*) type S, slow twitch, very fatigue-resistant, in normal soleus. (*d–f*) Matching regions in adjacent sections to the above, were stained with Periodic Acid Schiff (PAS) reagent for glycogen. Glycogen-depleted fibers of the motor units are marked with asterisks. In (*a*) and (*d*), and in (*b*) and (*e*), note the correspondence between glycogen-depleted fibers and markedly increased densities of silver grains. In (*c*), three glycogen-depleted fibers apparent in the PAS section are also labeled by silver grains, but there is one fiber (arrow) that exhibits increased silver grains (*c*) without detectable glycogen depletion (*f*). Scale bar, 100 μm. From Toop et al. (56).

slides at the temperature of the cryostat interior and an enclosed cryostat with glove parts, thawing at this stage can be avoided. If used with a great deal of care, particularly with regard to the maintenance of frozen tissue sections, the Appleton technique theoretically could be used to achieve high-resolution autoradiographs.

2.3.3b Freeze Drying and Freeze Substitution. There are a number of reasons why freeze-dried or freeze-substituted and plastic-embedded sections may be preferred over frozen sections. The histology is better, the section thickness is more reproducible, the geometry of section and emulsion is more predictable with plastic embedding, and there is the possibility of extending studies to the electron microscope level.

An approach that has been widely used for the preparation of tissue specimens for the autoradiography of diffusible substances is freeze-drying followed by vapor fixation and plastic embedding (57, 58). As with any approach, the degree of success attainable in the prevention of loss and diffusion of the label depends very much on the chemical nature of the labeled compound. For this reason the work (57) on the high-resolution autoradiographic localization of the sugar [^{3}H]galactose in hamster intestine is particularly encouraging. Although the techniques described to achieve this localization are somewhat tedious and involved, the final results that can be obtained are worthwhile because of their reliability.

The first problem encountered in these methods, and also in methods derived from the Appleton technique, is the formation of ice crystals during the freezing of tissue specimens. In addition to producing distortions in morphology, the growth of an ice crystal results in the redistribution of all the solutes to the rim surrounding the crystal (Figure 15). Ice crystal formation can be minimized by freezing very rapidly at temperatures below −120°C. This can be achieved by freezing at very low temperatures and freezing very small pieces of tissue. Freezing a piece of tissue no larger than 1 mm cube in a large volume of Freon 22 at the temperature of liquid nitrogen has been suggested as the procedure of choice (33). Once frozen, tissue specimens should be kept below −80°C, as it has been shown that the growth of ice crystals occurs at temperatures above −80°C (60). The following description of the processing of the tissue is the method described by Stirling and Kinter (57) for the localization of [^{3}H]galactose.

The frozen tissue fragment is transferred to a freeze-dryer at −184°C and 10^{-5} mm Hg. After 50 hr the temperature is raised to −70°C and then in stages over several days to room temperature. The tissue is then fixed for 12 hr in osmium vapor in a vacuum dessicator containing

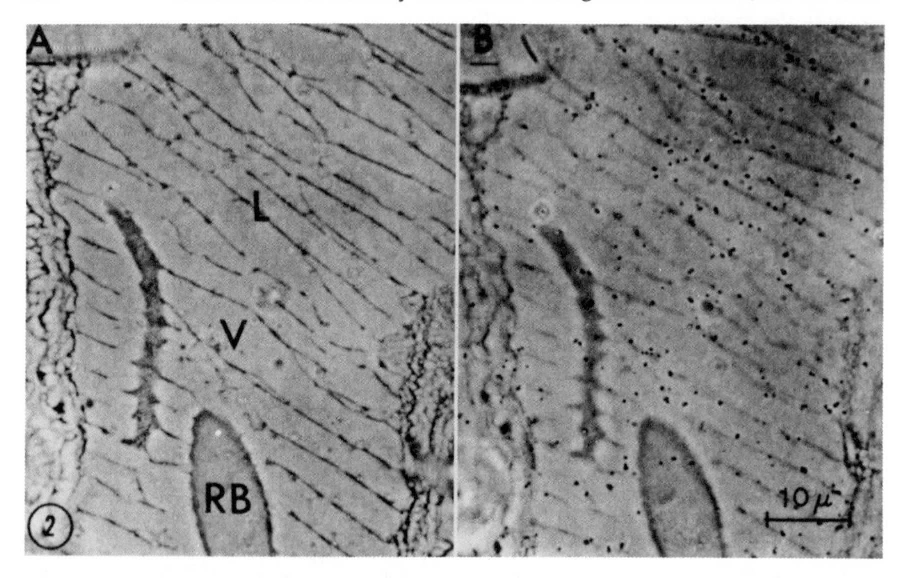

Figure 15. A blood vessel of Necturus that had been incubated in [^{3}H]glucose. It was slowly frozen to produce large ice crystals separated by eutectic lines, which contained all the solutes, including the [^{3}H]glucose. The tissue was prepared for autoradiography by freeze-drying and embedding in Epon. (A) Photomicrograph focused on the section; (B) photomicrograph of the same section focused on the silver grains. Of the silver grains over the plasma, 85% lay within 1 μm of the center of a eutectic line. From Stirling et al. (59). RB: red blood cell; V: vessel; L: lumen of vessel.

phosphorous pentoxide. For embedding, the tissue is placed in the side arm of a Thunberg tube that has about 0.5 ml of embedding medium in the main compartment. After repeated evacuation, the tissue is allowed to fall into the plastic embedding medium, which contains 1% silicone. The tube is warmed to 60°C, evacuated, sealed, and maintained at 60°C for 12 hr. The tissue is transferred to a capsule containing degassed medium with catalyst added and is cured for 2 hr, 0.2 mm Hg at room temperature, followed by 36 hr at 48°C.

Sections (1 μm thick) cut on an ultramicrotome are collected over water, placed on slides dried at 100°C, and dipped in Kodak NTB-2 nuclear track emulsion for autoradiography. At this stage it was noted that in the absence of silicone in the embedding medium, diffusion artifacts could be recognized in the autoradiographs, caused by floating the sections on water as well as dipping in the aqueous emulsion. Improvement was obtained by incorporating the silicone fluid in the embedding medium. This was thought to close off water-permeable channels in the sections. Surface loss to the water on floating out remained high (25%), but the autoradiographs showed no obvious diffusion artifacts (Figure 15). If reliance is to be placed on autoradiographic results in which some losses of radioactivity have occurred, one must assume that the radioactivity was lost randomly and proportionately from all compartments.

The loss observed in Stirling and Kinter's (57) experiments can be prevented with two modifications designed to avoid completely any contact with aqueous media (61). The sections can be cut dry and transferred to a slide with the use of a hair or bristle. They can then be coated with emulsion by the technique of Caro and van Tubingen (62), in which loops of emulsion formed by dipping wire loops into molten emulsion are allowed to dry until gelatinous and are then applied to the section. An alternative method of placing sections on emulsion has been described by Stumpf (63). Emulsion-coated slides are first prepared and dried. Under safelighting, the sections are picked up on a Teflon-coated surface slightly smaller in area than the emulsion-coated slides. An emulsion-coated slide and the Teflon support and section are gently pressed together. The Teflon support is allowed to fall away, and the section remains on the emulsion.

An additional step in the procedure in which the section is coated with a thick (20–30 nm) film of evaporated carbon may also be useful. That this procedure is effective was shown by X-ray microanalysis of freeze-dried sections of mouse kidney (64). The section was examined with and without carbon-coating and before and after a 65 s exposure to a filter paper that had been saturated with water. The results show that the 20–30 nm layer of carbon protected the section from the drastic changes that occur in the peak heights of diffusible elements after exposure to water.

In these procedures there is one other, less obvious source of error due to the introduction of moisture in the section. Stumpf and Roth (65) have shown that freeze-dried sections can easily take up moisture from the ambient air. At room temperature the section weight increases by about 1% of the dry weight for every 10% relative humidity in the air. At high humidities this may well produce movement of labeled material. It is important, therefore, to use dry air to break the vacuum after freeze drying and to keep the tissue dessicated during all procedures.

The technique of freeze-substitution has been employed in the study of extracellular space in nervous tissue at the electron microscopic level (66, 67). With this technique it has been shown that the distribution of water in the tissue can be preserved. With freeze substitution, ice in the tissue is gradually replaced by a solvent, usually acetone. The miscibility of water with acetone diminishes with decreasing temperature such that at −85°C, the temperature of substitution, the water content is about 2 percent. If a frozen-tissue preparation is placed in 100 percent acetone at this temperature, the solvent will proceed to dissolve ice from the tissue. The water released locally diffuses into the surrounding acetone, and in this way the plane of substitution slowly penetrates deeper and deeper into the tissue. At no time, however, will the water concentration rise above 2 percent; the ice in the tissue is in this way replaced directly by 98 percent acetone. The solubility of tissue constituents is likely to be quite different from that in water.

The method of substitution (67, 68) is a relatively simple procedure. The tissue is rapidly frozen in Freon 22 cooled to −184°C with liquid nitrogen, or by bringing it into contact with a copper surface cooled with liquid nitrogen (69) or liquid helium (70), to avoid loss of thermal conductivity associated with direct freezing in this and other coolants at their boiling points. By eliminating the insulating effect of a layer of gas produced by vigorous boiling around the specimen, the rate of freezing can be increased and the growth of ice crystals reduced. An instrument designed for this operation is now commercially available (Polaron Instruments, Doylestown, PA).

The frozen tissue is subjected to substitution fixation at −85°C in 2 percent osmium tetroxide solution in acetone (67) or 10 percent acrolein in acetone (68) for several days. Following substitution the tissue is very gradually brought to room temperature. Tissue blocks are washed in several changes of acetone and then in propylene oxide before embedding. The procedures available for embedding, cutting, and autoradiography are the same as those described for freeze-dried tissue. Because the substituted tissue, like the freeze-dried tissue, is dehydrated, the two methods share the same problems and pitfalls. Once the tissue is rapidly frozen and successfully dehydrated, great care must be taken to avoid any contact of the section with moisture. To repeat, the most likely points in the procedure at which moisture may be reintroduced are: (i) through contact with humid room air; (ii) during sectioning, especially if the sections are floated onto an aqueous medium, and (iii) during the production of contact between the section and a layer of photographic emulsion.

Several laboratories have attempted to apply these approaches to the cellular and subcellular localization of [^{3}H]deoxyglucose-6-phosphate. *In vitro* studies have been carried out on primary cultures of mouse spinal cord and dorsal root ganglia (43). Freeze-substitution of cells in 10 percent acrolein in acetone was followed by araldite embedding and then coating with a dry loop of Kodak NTB-2 emulsion. The resultant autoradiographs revealed that the neuronal/fibroblast grain density was lowest after treatment of the culture with 1 μ*M* tetrodotoxin and highest after 54 m*M* KCl. Results of experiments in which replicate cultures were solubilized and counted with and without freeze substitution showed that no detectable radioactivity was lost during the substitution (Table 4).

The results of other *in vitro* [^{3}H]DG experiments followed by freeze substitution have been reported (71, 72). In these studies the buccal ganglia of the gastropod mollusk *Limax maximus* were incubated in a balanced salt solution with [^{3}H]DG and a high-magnesium low-calcium concentration in order to reduce transmission at chemical synapses. After the incubation and washing, the tissues were rapidly frozen in Freon 22 and substituted in acetone at −70°C. They were then embedded in araldite that contained 1% silicone. The sections were dry-sectioned at 2-4 μm, flattened in anhydrous glycerol, dried, and dipped in Kodak NTB-2 emulsion for autoradiography. Losses of radioactivity were assessed, by liquid scintillation counting of the substitution media, to be less than 0.5% of the total. An autoradiograph of a single cell body obtained with this method is shown in Figure 16. The density of silver grains over the cytoplasm was 40 times the average background density over the tissue. The density of silver grains drops over the cell nucleus and drops abruptly at the cell's edge.

The autoradiographs obtained with this method, at least of cell body regions, are sharp and clearly delineated. The authors report that labeling patterns found for single neurons were the same in both freeze-substituted (−70°C) and acetone-dehydrated (4°C) preparations. This is a puzzling result because the high miscibility of water with acetone at 4°C would have resulted in movement of labeled DG and DG-6-P under these conditions. One possible explanation for the agreement between these two methods is that perhaps a large proportion of the trapped label in the cells of this invertebrate ganglion is in the form of deoxyglycogen.

Some studies on the effects of electrical activity on the degree of cell labeling with [^{3}H]DG were carried out with this preparation (72). In these studies, single neurons were impaled and stimulated at 1–2 pps while intracellular and nerve-root activity were monitored. The results obtained from four such preparations showed some correlation between the number of cells labeled and the number of roots that were firing. In addition, in the three experiments in which a single cell was stimulated, the labeling of a single neuron was 10 to 50 times that of the neuropil, compared with labeling at 3 to 5 times that of the neuropil in an unstimulated but impaled preparation.

Freeze-substitution in acetone has also been used in [^{3}H]DG studies of olfactory stimulation in the rat and in the tiger salamander (73). The spatial distribution of odor-induced activity in the olfactory bulb has

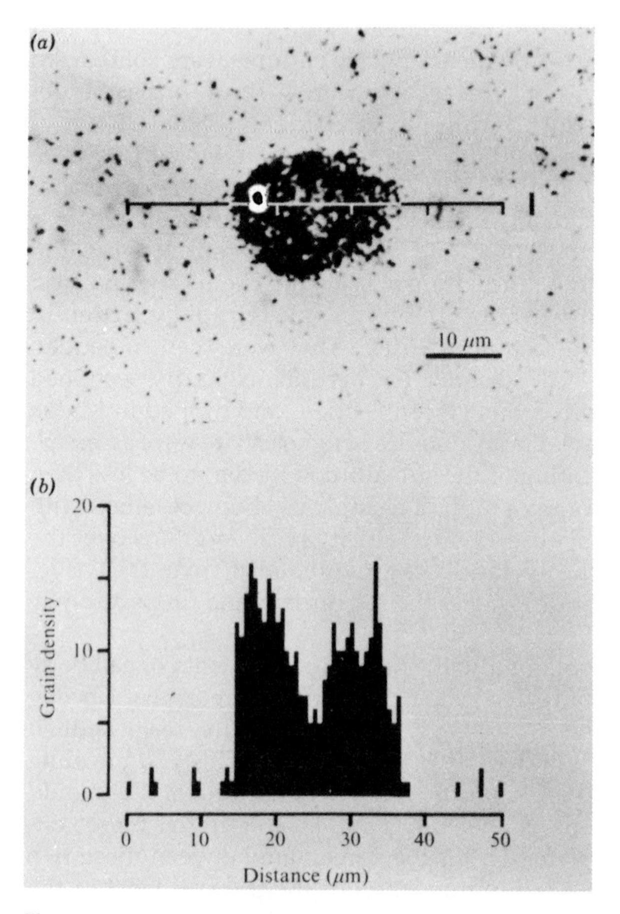

Figure 16. Measurement of silver grain density over a single cell body in a buccal ganglion. (*a*) Light-microscopic autoradiograph from a preparation incubated in [^{3}H]-2-DG for 45 min and freeze-substituted with acetone at −70°C. During the incubation, a neuron in the buccal ganglion was stimulated with an intracellular electrode to elicit action potentials at the rate of 2 pps. Several cells in the buccal ganglion were labeled. Scale bar, 10 μm. (*b*) A reticle with 100 lines was superposed over the cell at high magnification and the silver grains were counted by focusing through the emulsion of the autoradiograph. The number of silver grains between reticle lines [in a rectangle (0.63 × 4.6 μm) as shown to the right of the cell] is plotted as a function of distance along the line drawn through the cell body in (*a*). The drop in density of silver grains over the center of the cell occurs over the nucleus. The density of silver grains over the cytoplasm is 4.1 μm^{-2}, which is 40 times the average background density over the tissue surrounding the cell and 290 times the background density in the emulsion. From Sejnowski et al. (71).

been studied on the macroscopic level with the DG method (25). The results of these studies (Section 1.3 and Figure 6) have shown that there are odor-specific laminar and focal patterns of metabolic activity. These are particularly prominent in the glomerular layer of the olfactory bulb. Similar but high-resolution DG studies show that glomeruli, which are synaptic complexes that contain the first synaptic relay, tend to be uniformly active or inactive during odor exposure. Differential metabolic activity was also shown in the soma of projected neurons (mitral cells) and interneurons (periglomerular and granule cells) (Figure 17).

Some of the most unusual and the most critically evaluated high-resolution [^{3}H]DG studies are those from Buchner's laboratory (74–76) on the visual system of adult *Drosophila*. The DG method was modified somewhat for studies on these 1 mg insects. The labeled DG was administered orally to fasted flies in very high doses (final concentration in the fly was about 0.6 m*M*). The flies were visually stimulated for 8.5 hr. Subsequently they were frozen in melting nitrogen and freeze-dried at −70°C, 10^{-3} mm Hg. They were fixed in osmium vapor, embedded in Epon, and sectioned dry at 2–3 μm. The sections and emulsion-coated slides were sandwiched together for autoradiography at 4°C in a dessicator. These experiments show that it is possible to produce very impressive looking autoradiographs (Figure 18) with this completely nonaqueous method of tissue preparation. The autoradiographic results show that regions that are particularly heavily labeled are tracts (antennal nerve, commissures) composed of numerous thin (2 μm) fibers, while most identifiable large profiles retain very little label. This agrees with the relationship between the rate of glucose utilization and the surface-to-volume ratio of the cell processes in a brain region (see Section 1.3) (15). The effect of visual stimulation on the pattern of labeling in the brain was qualitatively reproducible and highly stimulus-specific. Further experiments at the electron microscopic level (76) have shown that "active" axons 1 μm in diameter can be mapped with this approach. The authors do express some reservations about the technique: (i) a small portion of radioactivity diffuses during embedding in Epon; (ii) some label is displaced from a 3 μm section by applying a drop of water. Therefore, some of the label (probably about 20%) is displaced during the flotation of semithin sections on water.

2.4 Future Directions

The results of the studies in which tissue specimens were either freeze-dried (74–76) or freeze-substituted (43, 71–73) demonstrate that high-resolution autoradiographs can be attained. To achieve this end it is

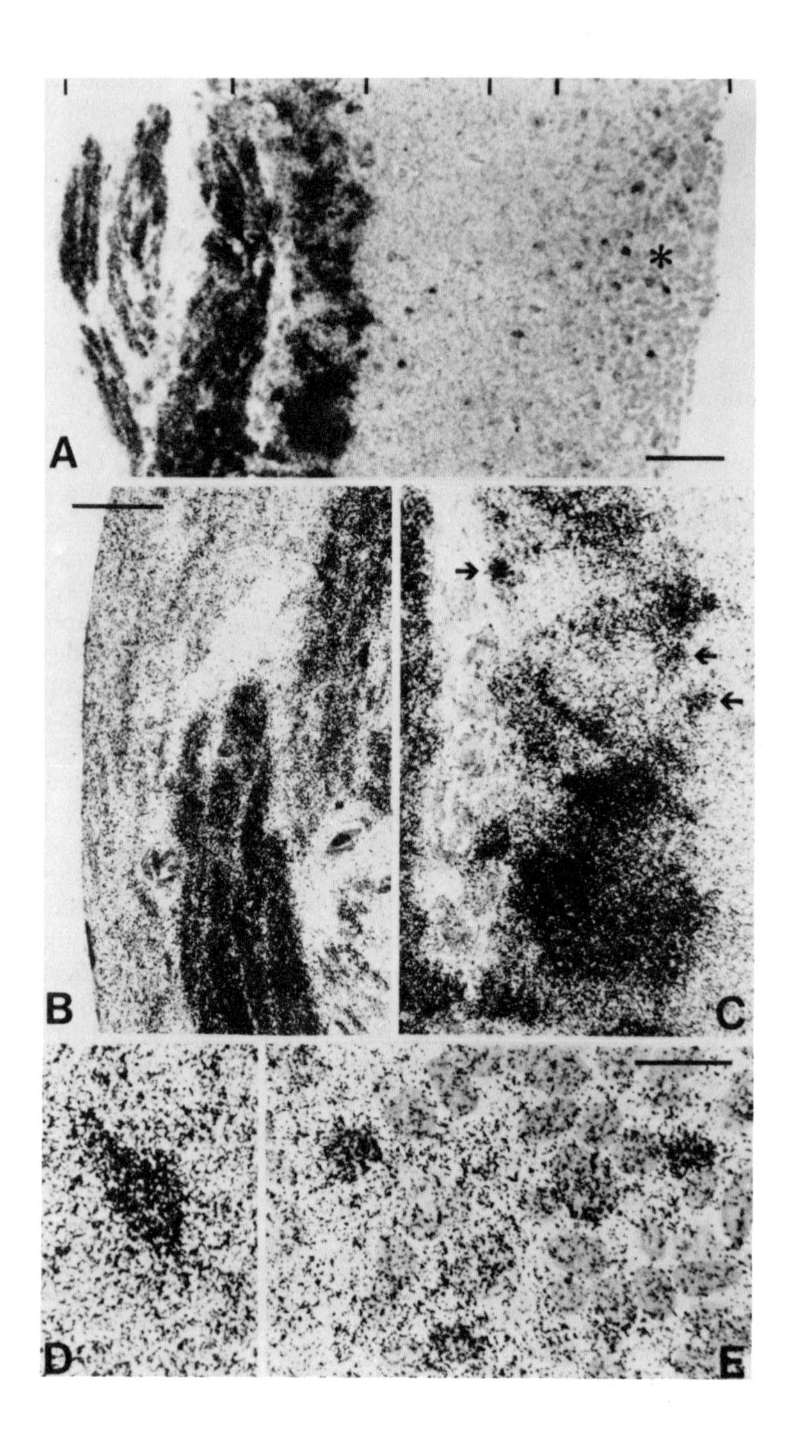
A
B
C
D
E

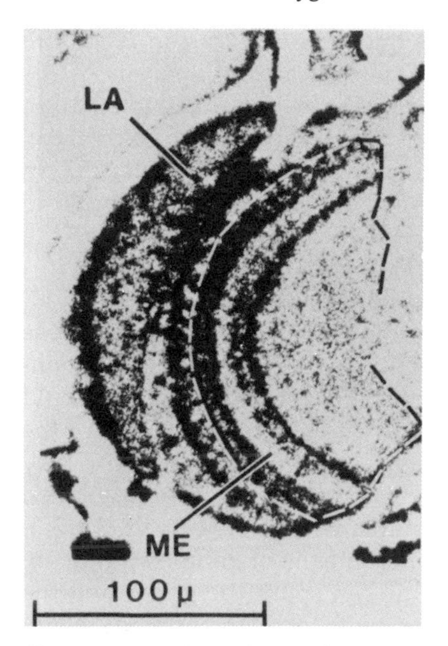

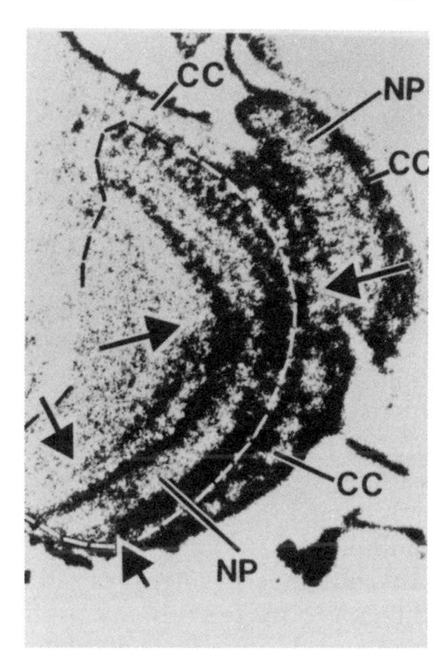

Figure 18. High resolution [^{3}H]DG autoradiograph of right and left optic lobes of a *Drosophila melanogaster* (*LA*, lamina; *ME*, medulla). The drosophila was presented with a flickering visual stimulus to a circular field of the left eye while a moving pattern stimulated the corresponding field of the right eye. The concentric rings of label and their periodic structure reflect the anatomical arrangement of concentric layers and periodic columns of the medulla, respectively. The heavily labeled sector marked by the four arrows in the right medulla corresponds to the visual field that was stimulated by movement. The cellular cortex (CC) of the lamina is easily identified by its enhanced labeling; the boundary between cellular cortex and neuropil (*NP*) of the medulla has been marked by a dashed line. From Buchner and Buchner (75).

necessary to employ procedures that keep ice-crystal formation to a minimum and maintain the specimen in a totally anhydrous state. In each new method described, it is incumbent upon the investigator to show that there are no losses of label during the procedures and that the label has not moved during the procedures. Loss of label can be assessed by counting replicate specimens at different stages of the pro-

Figure 17. High-resolution autoradiograph of [^{3}H]DG in the olfactory bulb of the tiger salamander exposed to isoamyl acetate at 10^{-2} saturation. (A) Overall view (brightfield) Lamina: *N*, nerve; *G*, glomerular; *E*, external plexiform; *M*, mitral cell body; *I*, internal plexiform; *GR*, granule; *Aug.*, accessory olfactory bulb. Bar, 125 µm. (B and C) Regions of high and low uptake in the nerve and glomerular layers (C is from the section shown in A; B is from a different section). Arrows, labeled periglomerular cells. Bar, 50 µm. (D and E) Single cell labeling. (E) is from the section shown in A marked *, (D) is from a different section. The cell in (D) is identified as a mitral cell, those in (E) are granule cells. Bar, 25 µm. From Lancet et al. (73).

cedure or by counting the media. The question of diffusion is a more difficult one. Two approaches to this question are: (i) analysis of the linear absorption coefficient at a sharp edge (Figure 9) (see Section 2.2.1a), and (ii) analysis of an autoradiograph in which the location of the label is known with precision, as in the study of Stirling and Kinter (57) in which tissue containing [^{3}H]glucose was frozen very slowly in order to produce large ice crystals separated by eutectic lines that contain all the solutes including the [^{3}H]glucose (Figure 15).

The deoxyglucose method, as originally described by Sokoloff et al. (1), is a method for the quantitative determination of local rates of glucose utilization. The possibility of employing the deoxyglucose method quantitatively at the cellular and subcellular level is very attractive. It would enable neurochemists to dissect out the metabolic requirements of the various structures in the nervous system and to determine how they are coupled with various cellular and subcellular functions. High-resolution quantitative autoradiography is another area with many pitfalls (33, 77). Absolute measurements of radioactivity by means of grain counting are difficult to make with any precision. One can count the grains and the source and convert this number into the number of disintegrations taking place within the source during the exposure by using one of the figures for grain yield available in the literature (78), but this method may have an error factor of 3 or 4. The soundest approach to this problem is probably the empirical one, that is, the preparation of standard references of known activity and their exposure under conditions identical to those of the unknown sources under study.

ACKNOWLEDGMENTS

I am grateful to Dr. L. Sokoloff for advice and helpful discussions, to Ruth Bower for editorial assistance, and to J. D. Brown for photographic assistance.

REFERENCES

1. L. Sokoloff, M. Reivich, C. Kennedy, M. H. Des Rosiers, C. S. Patlak, K. D. Pettigrew, O. Sakurada, and M. Shinohara, *J. Neurochem.* **28**, 897 (1977).
2. L. Sokoloff, *J. Neurochem.* **29**, 13 (1977).
3. T. G. Bidder, *J. Neurochem.* **15**, 867 (1968).
4. A. Sols and R. K. Crane, *J. Biol. Chem.* **206**, 925 (1954).
5. H. G. Hers, *Le Métabolism du Fructose,* Editions Arscia, Brussels, 1957, p. 102.

6. L. Sokoloff, *Trends Neurosci.* **1**, 75–79 (1978).
7. F. Schuier, F. Orzi, S. Suda, C. Kennedy, and L. Sokoloff, *J. Cereb. Blood Flow Metab.* **1**, S63 (1981).
8. S. Suda, M. Shinohara, M. Miyaoka, C. Kennedy, and L. Sokoloff, *J. Cereb. Blood Flow Metab.* **1**, S62 (1981).
9. M. Reivich, J. Jehle, L. Sokoloff, and S. S. Kety, *J. Appl. Physiol.* **27**, 296 (1969).
10. C. Goochee, W. Rasband, and L. Sokoloff, *Ann. Neurol.* **7**, 359 (1980).
11. P. J. Yarowsky and D. H. Ingvar, *Fed. Proc.* **40**, 2353–2362 (1981).
12. R. W. Albers, *Ann. Rev. Biochem.* **36**, 727–756, (1967).
13. P. C. Caldwell, *Physiol. Rev.* **48**, 1–64 (1968).
14. M. Mata, D. J. Fink, H. Gainer, C. B. Smith, L. Davidsen, H. Savaki, W. J. Schwartz, and L. Sokoloff, *J. Neurochem.* **34**, 213–215 (1980).
15. W. J. Schwartz, C. B. Smith, L. Davidsen, H. Savaki, L. Sokoloff, M. Mata, D. J. Fink, and H. Gainer, *Science* **205**, 723–725 (1979).
16. L. Sokoloff, "The Radioactive Deoxyglucose Method," in B. W. Agranoff and M. H. Aprison, Eds., *Advances in Neurochemistry*, Vol. 4, Plenum, New York, 1982, pp. 1–82.
17. D. H. Hubel and T. N. Wiesel, *J. Physiol.* (London), **195**, 215–243 (1968).
18. D. H. Hubel and T. H. Wiesel, *J. Comp. Neurol.* **146**, 421–450 (1972).
19. T. N. Wiesel, D. H. Hubel, and D. N. K. Lam, *Brain Res.* **79**, 273–279 (1974).
20. C. Kennedy, M. H. Des Rosiers, O. Sakurada, M. Shinohara, M. Reivich, J. W. Jehle, and L. Sokoloff, *Proc. Natl. Acad. Sci. USA* **73**, 4230–4234 (1976).
21. D. H. Hubel, T. N. Wiesel, and M. P. Stryker, *Nature* **269**, 328–330 (1977).
22. D. H. Hubel and T. N. Wiesel, *J. Comp. Neurol.* **158**, 269–294 (1974).
23. T. A. Woolsey and H. Van der Loos, *Brain Res.* **17**, 205–242 (1970).
24. D. Durham and T. A. Woolsey, *J. Comp. Neurol.* **178**, 629–644 (1978).
25. W. B. Stewart, J. S. Kauer, and G. M. Shepherd, *J. Comp. Neurol.* **185**, 715–734 (1979).
26. W. E. LeGros Clark, *J. Neurol. Neurosurg. Psychiat.* **14**, 1–10 (1951).
27. L. J. Land, *Brain Res.* **63**, 153–166 (1973).
28. M. H. Des Rosiers, O. Sakurada, J. Jehle, M. Shinohara, C. Kennedy, and L. Sokoloff, *Science* **200**, 447–449 (1978).
29. P. Hand, "Plasticity of the Rat Cortical Barrel System," in A. R. Morrison and P. L. Strick, Eds., *Changing Concepts of the Nervous System*, Academic Press, New York, 1982, pp. 49–68.
30. M. H. Teicher, W. B. Stewart, J. S. Kauer, and G. M. Shepherd, *Brain Res.* **194**, 530–535 (1980).
31. H. J. Gomberg, University of Michigan Project AT(11-1)-70 No. 3, 1952; through H. W. Rogers, *Techniques of Autoradiography*, Elsevier, Amsterdam, 1979.
32. T. Gross, R. Bogoroch, N. J. Nadler, and C. P. Leblond, *Am. J. Roentgenol. Radium Ther. Nucl. Med.* **65**, 420 (1951).
33. A. W. Rogers, *Techniques of Autoradiography*, Elsevier, Amsterdam, 1979.
34. C. E. K. Mees and T. H. James, Eds., *The Theory of the Photographic Process*, 3rd ed., Macmillan, New York, 1966.
35. G. M. Alexander, R. J. Schwartzman, R. D. Bell, J. Yu, and A. Renthal, *Brain Res.* **223**, 59–67 (1981).

36. E. Ehn and B. Larsson, *Sci. Tools* **26**, 24–29 (1979).
37. R. Hawkins, W. K. Hass, J. Ranschoff, *Stroke* **10**, 690–703 (1979).
38. J. Clausen, "Gray-White Matter Differences," in A. Lajtha, Ed., *Handbook of Neurochemistry*, Vol. 1, Plenum Press, New York, 1969, pp. 273–300.
39. L. Goldberg, J. Courville, J. P. Lund, and J. S. Kauer, *Can. J. Physiol. Pharmacol.* **58**, 1086–1091 (1980).
40. M. H. Des Rosiers and L. Descarries, *C. R. Acad. Sci., Paris, Ser. D* **287**, 153–156 (1978).
41. C. H. Pilgrim and H.-J. Wagner, *J. Histochem. Cytochem.* **29**, 190–194 (1981).
42. S. E. Basinger, W. C. Gordon, and D. M. K. Lam, *Nature* **280**, 682–684 (1979).
43. R. L. Ornberg, E. A. Neale, C. B. Smith, P. Yarowsky, and L. M. Bowers, *J. Cell Biol.* **83**, 142a (1979).
44. D. Durham, T. A. Woolsey, and L. Kruger, *J. Neurosci.* **1**, 519–526 (1981).
45. I. W. McLean and P. K. Nakani, *J. Histochem. Cytochem.* **22**, 1077–1083 (1974).
46. M. A. Kai Kai and V. W. Pentreath, *J. Neurocytology* **10**, 693–708 (1981).
47. V. W. Pentreath, L. H. Seal, and M. A. Kai-Kai, *Neuroscience* **7**, 759–767 (1982).
48. T. Radojcic and V. W. Pentreath, *Prog. Neurobiol.* **12**, 115–179 (1979).
49. J. Zemek, V. Farkas, P. Biely, and S. Bauer, *Biochim. Biophys. Acta* **252**, 432–438 (1971).
50. P. Witkovsky and C.-Y. Yang, *J. Comp. Neurol.* **204**, 105–116 (1982).
51. V. W. Pentreath and M. A. Kai-Kai, *Nature* **295**, 59–61 (1982).
52. L. J. Roth and W. E. Stumpf, *Autoradiography of Diffusible Substances*, Academic Press, New York, 1969.
53. F. R. Sharp, *Brain Res.* **110**, 127–139 (1976).
54. T. C. Appleton, *J. Microscopy* **100**, 49–74 (1974).
55. H.-J. Wagner, C. Pilgrim, and H. Zwuger, *Neurosci. Lett.* **15**, 181–186 (1979).
56. J. Toop, R. E. Burke, R. P. Dum, M. J. O'Donovan, and C. B. Smith, *J. Neurosci. Methods* **5**, 283–289 (1982).
57. C. E. Stirling and W. B. Kinter, *J. Cell. Biol.* **35**, 585–604 (1967).
58. N. J. Nadler, B. Bénard, G. Fitsimons, and C. P. Leblond, "An Autoradiographic Technique to Demonstrate Inorganic Radionuclide in the Thyroid Gland," in L. J. Roth and W. E. Stumpf, Eds., *Autoradiography of Diffusible Substances*, Academic Press, New York, 1969, pp. 121–130.
59. C. E. Stirling, A. J. Schneider, M.-D. Wong, and W. B. Kinter, *J. Clin. Invest.* **51**, 438–451 (1972).
60. T. Nei, *J. Microsc.* **99**, 227–233 (1973).
61. R. W. Baughman and C. R. Bader, *Brain Res.* **138**, 469–485 (1977).
62. L. G. Caro and R. P. van Tubingen, *J. Cell Biol.* **15**, 173–188 (1962).
63. W. E. Stumpf, "Techniques for the Autoradiography of Diffusible Compounds," in D. M. Prescott, Ed., *Methods in Cell Biology*, Vol. 13, Academic Press, New York, 1976.
64. J. R. J. Baker and T. C. Appleton, *J. Microsc.* **108**, 307–315 (1976).
65. W. E. Stumpf and L. J. Roth, *J. Histochem. Cytochem.* **15**, 243 (1967).
66. A. Van Harreveld and J. Crowell, *Anat. Rec.* **119**, 391 (1964).
67. A. Van Harreveld, J. Crowell, and S. K. Malhotea, *J. Cell Biol.* **25**, 117–137 (1965).
68. R. Ornberg and T. Reese, "Quick Freezing and Freeze Substitution for X-ray Mi-

croanalysis of Calcium," in T. E. Hutchinson and A. P. Somiyo, Eds., *Microprobe Analysis of Biological Systems*, Academic Press, New York, 1981, pp. 213–228.

69. A. K. Christensen, *J. Cell Biol.* **51**, 772–804 (1971).
70. J. E. Heuser, T. S. Reese, M. J. Dennis, V. Jan, L. Jan, and L. Evans, *J. Cell Biol.* **81**, 275–300 (1979).
71. T. J. Sejnowski, S. C. Reingold, D. B. Kelley, and A. Gelperin, *Nature* **287**, 449–451 (1980).
72. S. C. Reingold, T. J. Sejnowski, A. Gelperin, and D. B. Kelley, *Brain Res.* **208**, 416–420 (1981).
73. D. Lancet, C. A. Greer, J. S. Kauer, and G. M. Shepherd, *Proc. Natl. Acad. Sci. USA* **79**, 670–674 (1982).
74. E. Buchner, S. Buchner, and R. Hengstenberg, *Science* **205**, 687–688 (1979).
75. E. Buchner and S. Buchner, *Cell Tiss. Res.* **211**, 51–64 (1980).
76. S. Buchner and E. Buchner, *Neurosci. Lett.* **28**, 235–240 (1982).
77. E. A. Barnard, *Int. Rev. Cytol.* **29**, 213–243 (1970).
78. P. Dörmer, in V. Neuhoff, Ed., *Molecular Biology, Biochemistry, and Biophysics*, Vol. 14: *Micromethods in Molecular Biology*, Springer-Verlag, Berlin, 1973.

Index